COSMOS

COSMOS:

A SURVEY

OF THE

GENERAL PHYSICAL HISTORY

OF

THE UNIVERSE.

BY

ALEXANDER VON HUMBOLDT.

ISBN: 978-93-5128-767-4(HB)

First Published in 1845
Indian Reprint in 2017

Published by

Kalpaz Publications
C-30, Satyawati Nagar,
Delhi – 110052
E-mail: kalpaz@hotmail.com
Website: kalpazpublications.com
Ph.: +91-9212142040

Cataloging in Publication Data—DK
Courtesy: D.K. Agencies (P) Ltd. <docinfo@dkagencies.com>

Humboldt, Alexander von, 1769-1859, **author.**
Cosmos : a survey of the general physical history of the universe / by Alexander von Humboldt.
pages cm
"First published in 1845"—Title page verso.
ISBN 9789351287674 (HB)

1. Cosmography. 2. Physical geography. 3. Science—History—19th century. I. Title.

Q158.H86 2017 DDC 500.9 23

TO THE READER.

In presenting the English public with a version, in the vernacular tongue, of the world-renowned Alexander von Humboldt's Cosmos, the Translator begs to say, that he has striven to give a faithful transcript of the original, not less in matter than in manner: he has not taken away from the work, he has not added to it; and he has farther done what in him lay to preserve the lofty tone and imaginative style of the Author.

The Introduction is composed in the manner of an oration or popular discourse, and scarcely admitted of so literal a transfusion into English as the Translator will feel it his duty to secure in the body of the work. The second section, on the Limitation and Scientific Treatment of Physical Cosmography, is extremely abstruse, and cost the Translator no small pains to render it, he trusts intelligibly, into English. With the third section—The Picture of Nature, &c.—the Author enters fairly on his task.

For the use of several compound words, formed after the German originals, the Translator has to apologize to the classical English reader, and for mistakes that may occur in the translation of technical or conventional scientific terms, he must meantime crave the indulgence of the deeply versed in the several disciplines where these occur. He can but refer to the singular difficulties of his task, and solicit indulgence.

June 28th, 1845.

COSMOS:

SKETCH OF A PHYSICAL HISTORY OF THE UNIVERSE.

INTRODUCTION.

THE VARIOUS SOURCES OF OUR ENJOYMENT IN THE CONTEMPLATION OF NATURE.—THE SCIENTIFIC FOUNDATION OF THE LAWS THAT GOVERN THE UNIVERSE.*

In undertaking thus, after so long an absence from my native country, to discourse with you freely on the general physical phenomena of our globe, and to develop the connection of the forces which actuate the universe, I feel myself oppressed with a two-fold difficulty. On one hand, the subject I have to treat is so vast, and the time allowed me is so short, that I am fearful either of appearing superficial, or else, generalizing over much, of proving tiresome to you through aphoristic brevity. On the other, the life of action I have led has prepared me indifferently for the duty of a public teacher; so that, in the embarrassed state of my mind, I fear I may not always succeed in expressing myself with the clearness and precision which the vastness and the variety of my subject require. But the realm of nature is also the realm of freedom; and to exhibit in lively characters the ideas and emotions which a true love of nature inspires, the language must likewise move in harmony with the dignity and freedom of the subject, and this it is only given to high mastery to impart.

He who regards the influences of the study of nature in their relations not to particular grades of civilization or the individual requirements of social life, but in their wider bearings upon mankind at large, promises himself, as the principal fruit of his researches, that the enjoyment of nature will be increased and ennobled through insight into the connection of her phenomena. Such increase, such nobility, however, is the work of observation, of intelligence, and of time, in which all the efforts of the understanding of man are reflected. How the human kind have been striving for thousands of years, amidst eternally recurring changes in the forms of things, to discover that which is stable in the law, and so gradually, by the might of mind, to vanquish all within the wide-spread orbit of the earth, is familiar to him who has traced the trunk of our knowledge through the thick strata of bygone ages to its root. To question these ages is to trace the mysterious course of the idea stamped with the same image as that which, in times of remote antiquity, presented itself to the inward sense in the guise of an harmoniously ordered whole, Cosmos, and which meets us at last as the prize of long and carefully accumulated experience.

In these two epochs in the contemplation of creation—the first dawn of consciousness among men, and the ultimate and simultaneous evolution of every element of human science—two distinct kinds of enjoyment are reflected. The mere presence of unbounded nature, and an obscure feeling of the harmony that reigns amidst the ceaseless changes of her silent workings, are the source of the one. The other belongs to a higher stage of civilization of the species, and the reflection of this upon the individual; it springs from an insight into the order of the universe, and the co-ordination of the physical forces. Even as man now contrives instruments by which he may question nature more closely, and steps beyond the limited circle of his fleeting existence; as he no longer observes only, but has learned to produce phenomena under determinate conditions; as, in fine, the philosophy of nature has doffed her ancient poetical garb, and assumed the earnest character of a thinking impersonation of things observed, positive knowledge and definition have taken the place of obscure imaginings and imperfect inductions. The dogmatical speculations of former ages only exist at present in the prejudices of the vulgar, or in circumstances where, as if conscious of their weakness, they willingly keep themselves in the shade. They also maintain themselves as a heavy inheritance in language, which is disfigured by symbolical words and phrases innumerable. A small number only of the elegant creations of the imagination which have reached us, surrounded as it were with the haze of antiquity, acquire a more definite outline and a renovated shape.

Nature, to the eye of the reflecting observer, is unity in multiplicity; it is combination of the manifold in form and composition; it is the conception of natural things and natural forces as a living whole. The most important consequences of physical researches are therefore these: To acknowledge unity in multiplicity; from the individual to embrace all; amidst the discoveries of later ages to prove and separate the individuals, yet not to be overwhelmed with their mass; to keep the high destinies of man continually in view; and to comprehend the spirit of nature which lies hid beneath the covering of phenomena. In this way our aspirations extend beyond the narrow confines of the world of sense, and we may yet succeed, comprehending nature intimately, in mastering the crude matter of empirical observation through the might of mind.

When, in the first place, we reflect on the different degrees of enjoyment which the contemplation of nature affords, we find that the first or lowest are independent of all insight

* A Discourse delivered on opening the Course of Lectures in the Great Hall of the Singing Academy of Berlin. Many interpolations belong to a later period.

into the operation of her forces, yea, almost of the special character of the objects that are surveyed. When, for instance, the eye rests upon the surface of some mighty plain, covered with a monotonous vegetation, or loses itself in the horizon of a boundless ocean, whose waves are rippling softly to the shore, and strewing the beach with sea-weed, the feeling of free nature penetrates the mind, and an obscure intimation of her "endurance in conformity with inherent everlasting laws," takes possession of the soul. In such emotions there dwells a mysterious power; they are exciting, yet composing; they strengthen and quicken the jaded intellect; they soothe the spirit, painfully commoved by the wild impulses of passion. All of earnest and of solemn that dwells with us, is derived from the almost unconscious sentiment of the exalted order and sublime regularity of nature; from the perception of unity of plan amidst eternally recurring variety of form—for in the most exceptional forms of organization, the General is still faithfully reflected; and from the contrast betwixt the sensuous infinite and the particular finite, from which we seek to escape. In every climate of the globe, wherever the varying forms of animal and vegetable life present themselves, in every grade of intellectual eminence are these beneficent influences vouchsafed to man.

Another kind of enjoyment of nature, which is likewise wholly and solely addressed to the feelings, is that which we experience, not from the simple presence of unbounded nature, but from the individual characters of a country, and for which we have to thank the peculiar physiognomical attributes of the surface of our planet. Impressions of this kind are more lively, more definite, and therefore especially adapted to particular moods of the mind. Here, it is the magnitude of the masses, exposed amidst some wild conflict of the elements, that arrests us; there, it is a picture of the immoveably fixed that meets the eye, as in the waste and stillness of the boundless prairies of the New World and of the steppes of Northern Asia; or it is a softer and more hospitable view that attracts us—a cultivated country, or the first hermitages of man amidst the wilderness, surrounded by craggy peaks, on the margin of the leaping brook. For it is not so much the strength of the emotion that indicates the degree of the particular enjoyment of nature, as the determinate circle of ideas and feelings which induce and give it endurance.

If I might here, for a moment, yield to my own recollections of grand natural scenery, I would revert to the ocean, under the softness of a tropical night, with the vault of heaven pouring down its planetary and steady, not twinkling, starlight upon the heaving surface of the world of waters; or I would call to mind the wooded valleys of the Cordilleras, where, instinct with power, the lofty palm-trees break through the dark canopy of foliage below, and rising like columns, support another wood above the woods(1);" or, I transport myself to the Peake of Teneriffe, and see the cone cut off from the earth beneath by a dense mass of clouds, suddenly becoming visible through an opening pierced by an upward current of air, and the edge of the crater looking down upon the vine-clad hills of Orotava, and the Hesperidian gardens that line the shore. In scenes like these, it is no longer the still creative life of Nature, her peaceful strivings and doings, that address us; it is the individual character of the landscape, a combination of the outlines of cloud and sky, and sea and coast, sleeping in the morning or the evening light; it is the beauty of the forms of the vegetable world, and their groupings, that appeal to us; for the immeasurable, and even the awful in nature—all that surpasses our powers of comprehension—becomes a source of enjoyment in a romantic country. Fancy brings into play her creative powers upon all that cannot be fully attained by the senses, and her workings take a new direction with each varying emotion in the mind of the observer. Deceived, we imagine that we receive from the external world what we ourselves bestow.

When, after a lengthened voyage, and far from home, we for the first time set foot in a tropical land, we are pleased to recognize in the rocks and mountain masses, the same mineral species we have left behind—clay slate, basaltic amygdaloid, and the like, the universal distribution of which seems to assure us, that the old crust of the earth has been formed independently of the external influences of existing climates. But this well-known crust is covered with the forms of a foreign flora. Yet here, surrounded by unwonted vegetable forms, impressed with a sense of the overwhelming amount of the tropical organizing force, in presence of an exotic nature in all things, the native of the northern hemisphere has revealed to him the wonderful power of adaptation inherent in the human mind. We feel ourselves, in fact, akin to all that is organized; and though at first we may fancy that one of our native landscapes, with its appropriate features, like a native dialect, would present itself to us in more attractive colours, and rejoice us more than the foreign scene with its profusion of vegetable life, we nevertheless soon begin to find that we are burghers, even under the shade of the palms of the torrid zone. In virtue of the mysterious connection of all organic forms (and unconsciously the feeling of the necessity of this connection lies within us), these new exotic forms present themselves to our fancy as exalted and ennobled out of those which surrounded our childhood. Blind feeling, therefore, and the enchainment of the phenomena perceived by sense, in the same measure as reason and the combining faculty, lead us to the recognition which now penetrates every grade of humanity, that a common bond, according to determinate laws, and therefore eternal, embraces the whole of animated nature.

It is a bold undertaking to subject the magic of the world of sense to dissection, to a separation of its elements; for the character of grandeur in a landscape is especially determined by this, that the most impressive natural phenomena present themselves at once and together to the mind—that a host of ideas and feelings are simultaneously excited. The extent of mastery over the feelings which is thus gained, is most intimately connected with the unity of the impression. But if we would explain the power of the entire impression by the diversity of the phenomena, we must descend into the realm of determinate natural forms and active forces, and there discriminate and distinguish. The widest and most varied scope for investigations of this kind is afforded by the landscapes of Southern Asia and of the New World; countries where stupendous mountain masses form the bottom and boundary of the atmospheric ocean, and where the same volcanic powers which once forced up the mighty rampart of the Andes, through vast chasms in the earth, sill continue to shake their work to the terror of its inhabitants.

But natural pictures, arranged in succession and in harmony with some leading idea, are not calculated merely to engage the attention agreeably; in their sequence they may farther be made to compose a kind of scale of natural impressions, which, in their gradually increasing intensity, may be followed from the waste without a blade of grass, to the luxuriant vegetation of the torrid zone; from the monotonous level, to the grandest mountain chains. Were we, giving the rein to fancy, to suppose Mount Pilatus piled upon Shreekhorn,(2) or Schneekoppe set upon Mont Blanc, we should still fall short of one of the higher peaks of the Andes, Chimborazo, which has twice the height of Etna; and were we to throne the Rigi, on Mount Athos, on Chimborazo, we should only have an image of the highest summit of the Himalaya, Dhawalagiri. Although the Indian mountains, therefore, far exceed the Andes in colossal massiness, a fact now made certain by repeated measurements, they still present nothing like the variety of feature which characterises the Cordilleras of South America. It is not elevation alone that gives Nature her power of impressing the mind. The Himalaya range lies far beyond the limits of tropical climates; scarcely do we find a palm-tree straying into the beautiful valleys of Nepaul and Kumaon.(3) Between the 28th and 34th parallels of latitude, in the dependencies of the ancient Paropamisus, the vegetable kingdom no longer displays the same luxuriance of arborescent ferns and grasses, or of large-flowered orchideous plants and bananas, as she does within the tropics, even to plateaus some thousands of feet above the level of the sea. Under the shadows of the cedar-like deodwara pines and large-leaved oaks, the vegetable forms of Europe and the north of Asia are found covering the granitic rocks that form the substrata to the soil of the Indian mountains. They are not the same species, indeed, but they are similar forms: junipers, alpine birches, gentians, parnassias, and prickly species of Ribes.(4) The Himalaya, too, is without the varying phenomena of active volcanos, which, among the islands of the Indian Ocean, threateningly remind us of the internal life of the globe. And then, on its southern ridges at least, where the moister air of Hindostan deposits its burthen, the line of eternal snow is mostly met with at an elevation of from eleven to twelve thousand feet, and so sets an earlier limit to the evolution of organic life, than in the equinoctial countries of South America, where organization extends almost two thousand six hundred feet higher.(5)

Mountainous countries near the equator have another peculiarity, not sufficiently regarded: they constitute the portion of the surface of our planet, where, within the narrowest limits, the multiplicity, or variety, of natural impressions attains its maximum. In the deeply-cleft Andes of New Granada and Quito, mankind have the privilege of contemplating all the varieties of vegetable form, and of seeing all the stars in the firmament at once. The same glance rests on heliconias, feathery palms of the loftiest growth, and bambusas; over these characteristic forms of the tropical world, are seen oak forests, mespilus kinds, and umbelliferous tribes, as in our European latitudes; and turning from earth to heaven, the eye takes in the southern cross and Magellanic clouds, and the northern polar star. There, the fruitful bosom of the earth, and both hemispheres of the heavens, display at once the whole stores of their phenomena, their endless variety of forms and features; there are all the climates of the globe, and the vegetable zones they severally determine, superimposed; there are the laws of declining temperature, clearly understood of the careful observer, written in everlasting characters on the precipitous slopes of the mountains. I but lift a corner of the veil from my recollections of tropical landscapes here, that I may not weary this assembly with the repetition of ideas which I have endeavoured to represent in an illustrated work on the "Geographical Distribution of Plants(6)." What to the feelings melts into indefiniteness and indistinctness, like misty mountain air, is only to be comprehended by searching reason, when viewed in its casual connection with general phenomena, resolved into its constituent elements, and as the expression of an individual natural character. But in the circle of science, as in the brilliant circles of descriptive poetry and landscape painting, the representation still gains in clearness and objective animation, as the Individual is more clearly indicated and defined.

If tropical countries be richer in means of impressing the feelings, through the variety and luxuriance of Nature, they are also (and the point of view now taken is the most important in the train of ideas which I am at present pursuing) especially fitted, in the uniform regularity of their meteorological phenomena, in their succession of organic developments, and the sharp separation of forms effected by the perpendicular rise of the surface, to present to the mind the order and harmony of the heavens, mirrored, as it were, in the life of the globe. Let us pause for a moment, and contemplate this picture of harmonious regularity, which is itself connected with numerical relations.

In the burning plains raised but little above the level of the southern ocean, we find, in their greatest luxuriance, Bananas, Cycadeas, and Palms; after them, shaded by the lofty sides of the valleys, arborescent Ferns; next in succession, in full plenitude of growth, and ceaselessly bedewed by cool misty clouds, the Cinchonas, which yield the far-famed and precious febrifuge barks. Where lofty trees no longer grow, we meet with Aralias, Thibaudias, and myrtle-leaved Andromedas, associated and blooming in company. The Alpine rose of the Cordilleras, the Befaria, rich in resinous gum, forms a purple belt about the mountains. In the stormy region of the Paramos, all the more lofty vegetables and large flowering herbs gradually disapper. Glumaceous monocotyledonous tribes now cover the surface without variety, and form unbounded meadows, looking yellow in the distance, where the Llama sheep is seen feeding in solitude, and the cattle introduced by Europeans roam in herds. Upon the naked masses of trachytic rock, which here and there rise above the surface of the turf-clad soil, none but plants of the lowest organization can thrive: the tribe of liverworts, which the atmosphere, now of greatly diminished density, and containing little carbonic acid, supports but sparingly: Parmelias, Lecideas, and Leprarias with their many-coloured sporules, form the flora of this inhospitable zone. Patches or islets of lately fallen snow now begin to cover the last efforts of vegetable life, and then, sharply defined, the line of eternal ice begins. Through the white, and probably hollow, bell-shaped summits of the mountains, the subterranean powers strive, but mostly in vain, to break through. Where they have succeeded in establishing a communication with the atmosphere, through cauldron-shaped fiery throats or far penetrating chasms, they rarely send forth lava, as in the Old World, but carbonic acid, hydrosulphurets, and hot watery vapour in abundance.

So magnificent a spectacle, in its first assault upon the rude natural feelings, could excite nothing but wonder and dull amazement in the mind of natives of the tropical world. The intimate connection of grand periodically recurring phenomena, and the simple laws according to which these phenomena are grouped zone-wise, present themselves there, above all other places, with signal clearness to the senses of mankind; but from causes which, in many portions of this highly favoured quarter of the earth, oppose the local development of high civilization, all the advantages of this more facile study of these laws have remained without effect—so far, at least, as historical data enable us to conclude. The profound researches of recent times have made it more than doubtful that the peculiar seat of the Indian civilization—one of the fairest flowers in the history of humanity, the south-eastern spread of which has been so ably investigated by William von Humboldt(7)—was within the limits of the tropics. Airyana Baedjo, the ancient Zend country, lay to the north-west of the upper Indus; and after the religious disunion or secession of the Iranians from the Brahminical institutes, and their separation from the Hindoos, the original common language acquired its distinguishing features, and the social institutions gained their peculiar characters in Magadha(8), or Madhya Desa, between the little Windhya and the Himalaya chain.

A clear insight into the operations of the physical agencies was first, although, indeed, at a much later period, acquired by the races that people the temperate zone of our northern hemisphere, and this, in spite of all the obstacles which, under higher latitudes, complicate the phenomena of the atmosphere, and render difficult the discovery of general laws in the climatic distribution of organic beings. From hence has a knowledge of the character of tropical countries, and of countries situated near the tropics, been brought by larger movements of masses of mankind, or by individual foreign settlers—a transplantation of scientific culture which has had a like beneficial influence on the intellectual existence and industrial prosperity both of colonies and parent states. And here we touch the point at which, in the commerce between mind and the world of sense, another form of enjoyment is associated to that which depends on excitement of the feelings—an enjoyment of nature which springs from ideas; the point at which, in the war of the conflicting elements, the orderly, the legitimate, is not merely surmised or suspected, but is positively known by force of reason: the point at which man, as the immortal poet has it—

Amidst fleeting phenomena, seeks the stable pole.(9)

To follow this variety of enjoyment, springing from ideas, to its source, we have only to cast our eye back upon the rise and progress of the history of the philosophy of nature; in other words, of the ancient doctrine of Cosmos.

An indefinite dread sense of the unity of the powers of nature, of the mysterious bond which connects the sensuous with the super-sensuous, is common even among savage communities; my own travels have satisfied me that this is so. The world which is revealed to man through the senses, blends, often without his consciousness, with the world which, in obedience to his internal promptings, he creates in the guise of a realm of wonders in his own interior. The latter, however, is nothing like a true reflection of the former; for however impotent the External be to dissever itself from the Internal, still creative fancy, and the disposition to represent in concrete shapes the significant in phenomena, proceed incessantly in their workings, even among the rudest nations. That which presents itself to single more gifted individuals as the rudiments of a natural philosophy, as an induction under the guidance of reason, acquires existence as the product of instinctive susceptibilities among whole tribes of men. In this way, out of the depth and activity of blind feeling, is also eliminated the first impulse to adoration, the sanctification of the preserving as of the destroying powers of nature; and, if man, in passing through the different phases of his progress, now feels himself less fettered to the earth, and rising by degrees to mental freedom, he can be satisfied no longer with a mere indefinite feeling, an obscure suspicion of the unity of the natural forces. The faculty of thought, with its attributes of analysis and arrangement, now asserts its rights, and growing in the same measure as the human kind improves, in presence of the plenitude of life that flows throughout creation, the eager desire to penetrate more deeply into the causal connection of phenomena is experienced.

It is extremely difficult to obtain speedy and, at the same time, certain satisfaction to such a desire. From imperfect observations, and still more imperfect inductions, erroneous views of the character of the natural forces arise; views which, embodied and fixed in significant words and phrases, distribute themselves, a common inheritance of fancy, through all classes of a nation. By the side of the scientific system of nature, another is then seen growing with an equal growth—a system of unproven, and, in part, entirely mistaken empirical knowledge. Embracing but few particulars, this kind of empiricism is the more presuming, because of its utter ignorance of the facts by which it is assailed. Shut up within itself, it is unchanging in its axioms, and arrogant, like every thing else that is restricted; whilst enlightened natural science, inquiring, and therefore doubting, goes on separating the firmly established from the merely probable, and perfects itself daily through the extension and correction of its views.

The crude heap of physical dogmas which one age transmits to and forces upon another, is not merely injurious because it cherishes individual errors, because it obstinately presents indifferently observed facts for acceptance; it does more than this, it opposes every thing like grand or comprehensive views of the fabric of the universe. Instead of investigating the medium point about which, despite the apparent unfettered aspect of nature, all phenomena oscillate within narrow limits, it takes cognizance of the exceptions only to the law; it seeks for other wonders in phenomena and forms than those of regulated and progressive development. It is ever disposed to presume the train of natural sequence interrupted, to overlook in the present all analogy with the past, and, trifling with the subject, to discover the cause of some fancied disturbance now in the depths of the vault of heaven, now in the interior of the globe we inhabit. It leads away from that comparative geognosy which Ritter's great and masterly work has shown can only acquire any thing like completeness when the whole mass of facts, which have been collected in all the climates of the earth, comprehended at a glance, stands marshalled at the disposal of the combining intellect.

It is one of the objects of these discourses upon nature, to correct a portion of the errors which have sprung from rude and imperfect empiricism, and continue to live on among the upper classes of society, associated frequently with distinguished literary tastes and acquirements, and thereby to increase the relish for nature by giving a clearer, a deeper, insight into her constitution. The want of such an ennobled relish for nature is generally felt; for a peculiar character of the age we live in, is proclaimed in the tendency among all the educated classes to enhance the pleasures of existence by adding to the store of ideas. The lively interest which is taken in these prelections bears witness to the prevalence of such a disposition.

I cannot, therefore, yield any place in my mind to the solicitude to which either a certain narrowness of understanding, or a kind of sentimental dulness, appears to lead—the solicitude, namely, that nature loses aught of her magic, of her charms in respect of mysteriousness and grandeur, by inquiries into the intimate constitution of her forces. The forces of nature, indeed, only operate magically, in the legitimate sense of the word, shrouded, as it were, in the gloom of some mysterious power, when their workings lie beyond the boundaries of generally ascertained natural conditions. The observer who determines the diameters of the planets with a heliometer, or a prism of double refracting spar(10), who measures the meridian altitudes of the same star for a series of years, who discovers telescopic comets amidst thickly aggregated nebulous spots, does not, probably, feel his fancy more excited than the descriptive botanist, whilst he is counting the divisions in the calyx and corolla of a flower, or is ascertaining, in the structure of a moss, the state of distinctness or coalescence of the teeth that surround the seed capsule; but measurements of angles, and the development of numerical relations, the careful observation of the Individual, prepares the mind for the loftier knowledge of nature as a whole, and leads to the discovery of the laws that rule the universe. To the natural philosopher, who, like Young, and Arago, and Fresnel, measures the undulations of unequal length, the interferences of which strengthen or weaken the ray of light; to the astronomer, who, by the space-piercing power of his telescope, studies the satellites of Uranus on the outermost verge of our system, or, like Herschel, South, and Struve, detects glimmering points of brighter light in the coloured double stars; to the initiated eye of the botanist, who perceives the circular movements of the sap-globules so conspicuous in the Charas, in almost all vegetable cells, and who finds unity of formation, in other words, enchainment of forms, in species and natural families—these cultivated intellects surely look into the depths of heaven, as they survey the flower-clad surface of the earth, with a grander eye, than the observer whose intellectual vision is not yet sharpened by any apprehension of the enchainment of phenomena. We cannot, therefore, assent to the proposition of the eloquent Burke, when he says, that "out of the uncertainty of the nature of things alone, do admiration and the feeling of sublimity arise."

Whilst vulgar sense conceives the stars inlaid in a crystalline vault, the astronomer actually extends the bounds of space; for if he circumscribes the cluster of stars, of which our sun is one, it is only that he may show others and others, a countless multitude of groups of suns, the infinite depths of space, till vision fails, still studded with astral systems like our own. The feeling of the sublime, in so far as it seems to spring from the simple contemplation of infinite space, is closely allied to that rapt mood of the mind which, in the realm of the spiritual, in abstract converse with our own consciousness, arises from the meditation of the endless and the free. Upon this affinity, this relationship of sensuous impressions, depends the magic, the feeling of infinitude, which we experience when we are gazing over the shoreless ocean, surrounding some isolated mountain peak, or are penetrating the depths of heavenly space with the telescope, and resolving nebulous specks into their constituent stars; nothing impresses the cultivated imagination more powerfully than spectacles like these.

One-sided treatment of the physical sciences, endless accumulation of the raw material, might indeed appear to countenance the now almost superannuated objection, that scientific knowledge must of necessity chill the feelings, quench the creative light of fancy, and so interfere with the enjoyment of nature. But he who countenances this idea, in the stirring times in which we live, very certainly misunderstands the joys of that higher intelligence which is the appanage of the general progress of human society—of that tendency of the mind which resolves multiplicity into unity, and loves especially to dwell with the General and the Exalted. To taste, to enjoy this Exalted, it is imperative that the individualities which have been the prize of the carefully cultivated field of special natural forms and natural phenomena be carefully kept in the background; he who has himself most clearly seen their importance, and whom they have most safely led to loftier views, must more especially hold them in reserve.

To the groundless fears for the loss of an unfettered enjoyment of nature, under the influence of reflective surveys, or scientific scrutinies of her domains, may be associated those which are derived from alarm lest a due measure of this knowledge, or an adequate conception of its bearings, prove unattainable to the mass of mankind. In the wonderful tissues of organized beings, in the eternal tendencies and workings of the living powers, each new and deeper inquiry seems but to lead to the entrance into a new labyrinth. But this very multiplicity of untrodden and intricate paths excites a kind of joyful amazement on each successive grade of science. Each natural law which reveals itself to the observer leads to the inference of one yet higher and unknown; for Nature, as Carus well says(11), and as the word itself was understood by the ancient Greeks and Romans, "is the Ever-becoming, the Ever-engaged in fashioning and evolving." The circle of organic types extends the wider the more the earth is searched over, in travels by land and voyages by sea; the more living organic forms are compared with the remains of those that are extinct, the more the microscope is improved, and adds to the empire of the eye. In the multiplicity and changes of organic forms, in consonance with climatic influences, the prime mystery of all formation is incessantly reproduced; it is the problem of metamorphosis, so happily developed by Goethe, upon the grandest scale, and proclaims the necessity for an ideal reference of organic forms at large to certain elementary types. With an extension of knowledge, the feeling of the immeasurableness of the life of nature is still increased, and we perceive that, neither in the solid crust of the globe, nor in the aerial covering that invests the solid, neither in the depths of the ocean, nor in the depths of heaven, will the bold scientific conqueror(12) lack scope for his inquiries for thousands of years to come.

General views of the Fashioned, be it matter aggregated into the farthest stars of heaven, be it the phenomena of earthly things at hand, are not merely more attractive and elevating than the special studies which embrace particular portions of natural science; they further recommend themselves peculiarly to those who have little leisure to bestow on occupation of the latter kind. The descriptive natural sciences are mostly adapted to particular circumstances: they are not equally attractive at every season of the year, in every country, or in every district we inhabit. The immediate inspection of natural objects, which they require, we must often forego, either for long years, or always in these northern latitudes; and if our attention be limited to a determinate class of objects, the most graphic accounts of the travelling naturalist afford us little pleasure if the particular matters, which have been the special subjects of our studies, chance to be passed over without notice.

As universal history, when it succeeds in exposing the true causal connection of events, solves many enigmas in the fate of nations, and explains the varying phases of their intellectual progress—why it was now impeded, now accelerated—so must a physical history of creation, happily conceived, and executed with a due knowledge of the state of discovery, remove a portion of the contradictions which the warring forces of nature present, at first sight, in their aggregate operations. General views raise our conceptions of the dignity and grandeur of nature; and have a peculiarly enlightening and composing influence on the spirit; for they strive simultaneously to adjust the contentions of the elements by the discovery of universal laws, laws that reign in the most delicate textures which meet us on earth, no less than in the Archipelagos of thickly clustered nebulæ which we see in heaven, and even in the awful depths of space—those wastes without a world. General views accustom us to regard each organic form as a portion of a whole; to see in the plant and in the animal less the individual or dissevered kind, than the natural form, inseparably linked with the aggregate of organic forms. General views give an irresistible charm to the assurance we have from the late voyages of discovery undertaken towards either pole, and sent from the stations now fixed under almost every parallel of latitude, of the almost simultaneous occurrence of magnetic disturbances or storms, and which furnish us with a ready means of divining the connection in which the results of later observation stand to phenomena recorded as having occurred in bygone times; general views enlarge our spiritual existence, and bring us, even if we live in solitude and seclusion, into communion with the whole circle of life and activity—with the earth, with the universe.

Who—to select a particular instance from the realms of space—who, that has paid any attention to scientific events in the course of the last few years, can perceive, without a general knowledge of the ordinary orbits of comets, how pregnant with results is Encke's discovery, that a comet, which, in its elliptical orbit, never leaves our planetary system, reveals the existence of a fluid controlling its centrifugal force? With the recent spread of a kind of half-education, which attracts scientific conclusions into the circle of social amusement and conversation, but so commonly distorts them, we have seen the old solicitude revived about a collision between the heavenly bodies, threatening danger or destruction to all, and cosmic influences, in an altered and therefore more deceitful guise, quoted to account for presumed deteriorations of climates, and the like. Clear conceptions of nature, though they may not be more than historical, preserve us from the presumptions of dogmatizing fancy. They assure us that Encke's comet, which completes its revolution in 1200 days, by reason of the form and position of its orbit, must ever be harmless to the inhabitants of the earth—as harmless as Halley's comet, the great comet of 1759 and 1835, with its period of 76 years; but that another comet, of shorter period, Biela's, to wit, with its course of six years, actually crosses the orbit of the earth, though it can only approach us nearly when its perihelion falls at the time of our winter solstice.

The quantity of caloric which one of the planets receives, and the distribution of which determines the grand meteorological processes of the atmosphere, is modified by the light-evolving power of the sun—the property of its surface, and the relative position of the sun and the planet; but the cyclic changes which the form of the earth's orbit, and the obliquity of the ecliptic, undergo, in conformity with the general laws of gravitation, are so slow, and confined within such narrow limits, that their influence will scarcely be perceptible to such instruments as we now possess for measuring temperature in the course of several thousand years. Cosmic causes of diminished temperature, of lessened fall of rain, and of epidemic diseases, which were much canvassed in the middle ages, and of which mention has again been lately made, are consequently seen to be entirely beyond the pale of actual experience.

If I would quote other instances from physical astronomy, which could excite no interest without a general knowledge of what has been already observed, I would refer to the numerous instances of differently coloured double stars which move in ellipses round one another, or rather around their common centre of gravity; to the periodical rarity of spots in the sun; to the regular appearance of innumerable falling stars, which have now been the subject of observation for so many years, and which are in all probability planetary in their nature, circulating 'round the sun, and crossing the earth's orbit, in their course on the 12th or 13th of November, and also, according to later observation, on the 10th or 11th of August.

In the same way, general views of Cosmos will alone enable us to perceive the connection betwixt the theory of the pendulum swinging in air, and the internal density—I might say, the degree of congelation or solidification—of our globe, a theory happily completed by the acuteness of Bessel; betwixt the production of crystalline rocks in stratified streams of lava upon the acclivities of still active volcanoes, and the endogenous granitic, porphyritic, and serpentine rocky masses, which, forced up from the interior of the earth, have burst through the floetz formations, and produced various effects upon them—hardening or silicifying them, converting them into dolomite, producing drusy cavities, filled with crystals, &c.; betwixt the elevation of islands and conical mountains, through elastic forces, and the uplifting of mountain chains and entire continents—a connection which has been acknowledged by the greatest geologist of our age, Leopold von Buch, and illustrated by a series of admirable observations. Such upheavings of granular mountain masses and floetz strata, as have even lately been witnessed over a vast extent of the coast of Chili, in connection with an earthquake, show us how possible it is that the marine shells, which Bonpland and I discovered on the slopes of the Andes, at an elevation of 14,000 feet above the level of the sea, were brought thither, raised from the bed of the ocean by volcanic forces, not by any general flood that overspread the surface as it now presents itself to us.

By Plutonism, or Vulcanism, taking either word in its most general sense, and using it not only with reference to the earth, but also to its satellite, the moon, I mean the reaction which the interior of a planet exerts upon its crust. He who is unacquainted with the observations that have been made on the gradual rise of temperature, as the crust of the earth is penetrated more deeply (observations which have led distinguished naturalists to conclude that at the depth of five geographical miles below the surface, a temperature adequate to keep granite in a state of fusion prevails[13]), is not prepared to appreciate many recent observations on the simultaneousness of the eruptions of volcanoes, separated by vast extents of country, on the limits of the circles within which earthquakes are likely to be felt, on the permanence of the temperature of hot mineral springs, and on the difference of temperature of the water in Artesian wells of different depths. And yet this knowledge of the internal temperature of the earth throws a feeble light upon the primary history of our planet. It proclaims the possibility at a former epoch of the general diffusion of a tropical climate over the surface of the globe, as a consequence of heat inherent, and of clefts pouring forth heat, in the lately concreted and oxidated crust of the earth. It reminds us of a state of things in which the temperature of the atmosphere may have been more intimately connected with the reaction of the interior upon the exterior, than with the position of the axis of revolution of our planet to the great central mass of our system, the sun.

Numerous productions of the tropics are now dug up by eager geologists from their tombs in the temperate and cooler regions of the earth: coniferous vegetables, trunks of palm trees, erect as when they grew, arborescent ferns, goniatites, and fishes with rhomboidal pearly scales, in the old coal formations([14]); skeletons of colossal crocodiles, long-necked plesiosaurians, the scales of planulites, and the stems of cicadeæ, in the Jura limestone; polythalamians and bryozoa in chalk, in several instances identical with species still existing in our seas; vast agglomerations of infusory animalcules, as brought to light by Ehrenberg's all-animating microscope, in beds of tripoli, semiopal and siliceous sinter (?) (*Kieselguhr*); bones of hyenas, lions, and elephantine pachydermatous animals, lying exposed in caverns, or covered merely with a layer of sand or mud. With a competent knowledge of other natural phenomena, these productions do not remain objects of mere idle curiosity and wonder; they become more worthily the occasion of much varied and interesting reflection.

In the multiplicity of objects which I have thus cursorily enumerated, the question presents itself: whether general views of nature can be brought to anything like precision without deep and earnest study of the several departments of natural science—natural history, natural philosophy, and physical astronomy? Here it is proper to distinguish carefully betwixt the teacher, who makes selections and delivers an account of results, and the pupil, who receives the account as something presented to him not investigated for himself. For the former, the most intimate knowledge of specialities is indispensably necessary; he must have long familiarised his mind with the several sciences, he must himself have taken the length and the breadth of things, observed and made experiments, before he can, with any confidence or propriety, venture on a picture of nature as a whole. The entire bearings of the problems whose investigation lends such attractions to the physical history of the world are perhaps scarcely to be comprehended in all their clearness where special preliminary knowledge is wanting; although, without it, the greater number of the propositions can still be satisfactorily discussed. If the great picture of nature be not presented with its outlines equally clear and sharp in every part, it will still be found sufficiently true and attractive to enrich the mind with ideas, and to arouse and fructify the imagination.

It has been made matter of reproach—and perhaps with some propriety—that the scientific works in our language do not sufficiently separate the General from the Particular—the review of actually established facts from the narrative of the means by which the results have been obtained. This imputation has led the greatest poet of our age([15]) humorously to say, that "the Germans possess the faculty of making the sciences inaccessible." But the scaffold left standing, we are hindered from obtaining a clear view of the building. And who will doubt, that the physical law in the distribution of the continental masses, which assume a pyramidal shape towards the south, whilst towards the north they spread out into vast bases—a law by which the division of climates, the prevalence of particular winds, the extension of tropical vegetable forms into the temperate northern zones, is explained in the most satisfactory manner—can be understood without reference to the trigonometrical surveys, and the astronomical determinations of precise geographical positions, by which the dimensions of the pyramids referred to have been ascertained? In the same way, we learn from physical geography, that the equatorial axis of our planet is greater than the polar axis by a certain number of miles, that the southern hemisphere is not flattened in a greater degree than the northern hemisphere, &c., without its being necessary to narrate at length how, by measurements of degrees of the meridian, and experiments with the pendulum, the figure of the earth has been finally determined to be that of an irregular spheroid of revolution in an ellipsis; and how this figure is reflected in the motions of our satellite, the moon.

Our neighbours on the other side of the Rhine possess an immortal work, Laplace's "Système du Monde," in which the results of the most profound mathematico-astronomical investigations of the phenomena of past centuries are luminously presented, freed from the individualities of the demonstration. The structure of the heavens there presents itself as the simple solution of a great problem in mechanics. Yet no one has ventured to charge the "Exposition du Système du Monde" with want of depth, because of its form. The separation of the Dissimilar in views, of the General from the Special, is not merely useful in facilitating the acquisition of knowledge; it farther gives an elevated and earnest character to the treatment of natural science. As from a higher station we overlook larger masses at once, so are we pleased mentally to grasp what threatens to escape the powers of our senses. If the successful cultivation of every branch of natural science in recent times, appear especially calculated to extend the study of particular departments—the chemical, the physical, the physiological, &c.—the progress made in each will nevertheless contribute in an eminent degree to abridge and render easy the way to the attainment of general principles.

The more deeply we penetrate into the essence of the natural forces, the more do we perceive the connection of phenomena, which, severally and superficially regarded, seemed long to resist every attempt at co-ordination and arrangement; the more do we see simplicity and brevity possible.

It is a certain indication of the extent and value of the discoveries which were to be looked for in any science, when the facts present themselves as still unconnected, almost, as it seems, without any thing like mutual reference, and when several of them, the fruit of the same degree of careful observation, even appear contradictory or subversive. We stand at this time in a state of lively expectation in regard to meteorology, to some of the departments of optics, and especially, since Melloni and Faraday came upon the stage, to the radiation of heat and electro-magnetism. The field of brilliant discovery here, has certainly not yet been exhausted, although a very remarkable connection of electrical, magnetical, and chemical phenomena has undoubtedly been developed in the voltaic pile. And who shall guarantee us that the entire number of the vital forces efficient in the universe has been fathomed?

In my mode of considering the scientific treatment of a general description of creation, I make no question of that unity which is arrived at by induction from a few fundamental principles supplied by reason. What I entitle a PHYSICAL HISTORY OF CREATION—in other words, a comparative natural history of the earth and heavens—consequently, makes no pretensions to the rank of a RATIONAL SCIENCE OF NATURE; it is a simple consideration of the phenomena that are known empirically, or by experience, as a natural whole. With the entirely objective constitution of my mind, it is under such restrictions alone that the history of creation falls within the scope of the inquiries which have exclusively occupied me in the long course of my scientific life. I do not venture upon a field that is strange to me, and that will probably be cultivated to better purpose by another. The unity attainable in such a history of creation as I propose to exhibit, is no more than that which historical representations in general can hope to achieve. Details, whether as to the form or arrangement of natural things, no more than in reference to the struggles of man with the elements, or the wars of one nation against another—all, in short, that falls within the sphere of mutability and true accident—cannot be derived or built up from *à priori* conceptions. The natural history of the earth, and universal history, consequently, stand on the same grade of the empirical ladder; but a luminous treatment of either, a rational arrangement of natural phenomena and of historical incidents, impresses us deeply with a belief in an old inherent necessity, which rules all the operations both of the spiritual and material forces within circles eternally reproduced and only periodically contracted or enlarged. This necessity, indeed, is the very essence of nature; it is nature herself, in the two spheres of her being—the material and the spiritual—and it leads to clearness and simplicity of view, to the discovery of laws which, in experimental science, present themselves as the ultimate term in human inquiries.

The study of every new science, especially of one which embraces the infinite field of creation, the universe at large, may be compared to a journey into a foreign country. Before undertaking such an expedition in company, we inquire as to its feasibility; we measure our own powers of endurance, and we look with a suspicious eye at the powers of our intended companions, with the perchance unjust anxiety lest they prove impediments in the way. But the times in which we live diminish the difficulties of the enterprise, and my confidence in ultimate success is based on the brilliant position now occupied by natural science itself, whose increasing stores may now be said to add less to the amount than to the enchainment of observation. The GENERAL RESULTS, which so powerfully interest every cultivated mind, have been wonderfully augmented since the end of the eighteenth century. Facts now stand less insulated; numerous gaps between different orders of beings and phenomena have been filled up; points which had remained inexplicable to the inquiring spirit at home, within the narrower circle of experience accessible to it, are frequently made clear by journeys undertaken into the remotest regions of the earth. Vegetable and animal forms that long appeared isolated, now appear connected by intermediate links or transition forms. A general concatenation, not in simple linear directions only, but in reticulate or more intricate modes, according to the higher development or the arrest of certain organs, according to relative preponderance in the several parts or systems, now presents itself to the mind of the enlightened naturalist. Appearances of stratification in trachytic syenite or porphyry, in green stone and serpentine, which are doubtful in Hungary, so rich in gold and silver, in the platina districts of the Ural chain, or deeper into Asia, in the south-western Altai, are unexpectedly cleared up by geological observations in the lofty plateaus of Mexico and Antioquia, and in the valleys of Choco. The materials which universal geography employs are not indiscriminately accumulated. In the present times, in virtue of the tendency which their individual character impresses upon them, it is admitted that new facts are only pregnant with future good, when the traveller is familiar with the actual state and requirements of the science whose boundary he pretends to widen; when ideas, in other words, insight into the spirit of nature, guide the taste for observation and collection.

Through this direction of the study of nature, through the happy, but, at the same time, often too readily satisfied taste for general results, can a very considerable portion of natural science be made the common property of cultivated humanity, and this with a full sense of the import and form, of the grandeur and worth of the subject, altogether different from that popular science which was held sufficient for the world at large up to the end of the last century. Let him, therefore, whom circumstances permit to escape from time to time from the narrow circle of his every-day occupations, lament that he has "remained so long a stranger to nature, unconscious of her charms," and learn, that in the contemplation of her grandeur and freedom, there dwells the purest delight which exalted intelligence can obtain for man. The study of general natural science, indeed, awakens organs in our interior that have long slumbered. We enter upon a new and more intimate intercourse with the external world, and are brought to feel a larger sympathy with that which proclaims at once the industrial progress, and the intellectual improvement of mankind.

The clearer the insight we obtain into the connection of phenomena, the more readily do we emancipate ourselves from the error of believing that every department of natural knowledge is not equally important in the culture and welfare of mankind, whether it be that department which measures and describes, or chemical inquiries, or the investigation of the generally diffused physical forces of matter. In the observation of a phenomenon which seems at first to stand isolated and alone, there frequently lies the germ of a great discovery. When Galvani stimulated the nerves of sensation by the contact of two dissimilar metals, his most intimate friends and contemporaries could never have expected that the voltaic pile, with its electricity of contaction, would one day show us a brilliant metal in the alkalis, silvery in its appearance, readily inflammable, and so light as to float upon the surface of water; that the same arrangement would by and by become the most powerful instrument in analytical chemistry, and prove at once a thermoscope and a magnet. When Huyghens began to investigate the optical properties of double refracting spar, no one imagined that the phenomena of coloured polarization would lead one of the singularly clear-sighted natural philosophers of our day(16) to discover in the fragment of a mineral a means of knowing whether the light of the sun proceeded from a solid mass or from a gaseous canopy; whether comets have the power of emitting light in themselves, or merely reflect the light they receive from other sources.

A like respect for every department of the study of nature is, however, especially necessary in the present times, when the material wealth and the increasing welfare of the nations is so closely connected with a more diligent use of natural productions and natural forces. The most superficial glance at the condition of Europe in these days, assures us that with the struggle against serious odds, any relaxation of effort would be followed, first by diminution, and then by annihilation of national prosperity; for in the destiny of nations it is as in nature, in which, as Goethe (17) says, finely, "there is neither rest nor pause, but ever movement and evolution, a curse still cleaving to standing still." Nothing but serious occupation with chemical, mathematical, and natural studies, will defend any state from evils assailing it on this side. Man can produce no effect upon nature, can appropriate none of her powers, if he be not conversant with her laws, with general relations according to measure and number. And here, too, lies the power of popular intelligence. It rises and falls with this. Science and information are the joy and the justification of mankind; they are portions of the wealth of nations, sometimes a substitute for material wealth, which nature has in many cases distributed with so partial a hand. Those nations which have remained behind in general manufacturing activity, in the practical application of the mechanical arts, and technical chemistry, in the transmission, growth, or manufacture of raw materials, nations among whom respect for such activity does not pervade all classes, must inevitably fall from any prosperity they may have attained; and this by so much the more certainly and speedily, as neighbouring states, instinct with powers of youthful renovation, in which science and the arts of industry co-operate or lend each other assistance mutually, are seen pressing forward in the race.

The taste for manufacturing industry, and for those portions of natural science which bear upon it more immediately—a characteristic of the present age—can in nowise be prejudicial either as regards philosophy, antiquities, or history, nor quench the all-animating flame of fancy, in the direction of the liberal arts. Where all the offshoots of civilization are permitted to expand in vigour, under the protection of wise laws and free institutions, no effort of mind in any one direction will be found to interfere with its aspirations in another quarter. Each presents its own peculiar fruit to the commonwealth: one, the means of maintenance and comfort to the citizen, another, the product of creative fancy, which, more durable than material wealth, transmits the name and fame of the community to the latest posterity. The Spartiates, despite the austerity of the Doric mind, prayed "the Gods to vouchsafe them the beautiful associated with the good."(18)

As in those higher circles of ideas and feelings—in the study of history, of philosophy, and of oratory—so in all the departments of natural science, the first and highest aim of intellectual activity is one that is INTERNAL; namely, the discovery of natural laws, the establishment of co-ordinate members in the images, the perception of necessary connection between all the changes that happen in the universe. So much of this science as flows over, and mingles with the industrial life of communities, elevating manufacturing industry, does so in virtue of the happy connection in human things, by which the true, the exalted, and the beautiful, mix unintentionally, as it seems, but certainly, with the useful, and co-operate with it in bringing about results. The improvement of agriculture by the hands of freemen, and on lands of moderate extent; the flourishing condition of manufactures, emancipated from oppressive restrictions; the extension of commercial relations, and the unimpeded progress of mankind in mental development as well as in their social institutions, are all inseparably connected, and severally and powerfully advance each other. The impressive picture of the late history of the world forces this faith upon the minds even of those that most eagerly oppose it.

Such an influence of natural science upon the welfare of the nations, and on the present condition of Europe, can receive nothing more than a passing allusion in this place. The course we have to complete is so vast in itself, that it would not become me to depart from the main object we have in view, namely, THE SURVEY OF NATURE AS A WHOLE, and intentionally to widen the field of our inquiries. Accustomed to wanderings in distant lands, I have, perhaps, without this, indicated the path to my fellow-travellers, as more distinctly traced and more attractive than they will find it in fact. This is ever the way with those who take pleasure in guiding others to the tops of mountains: they praise the view, though perchance large tracts of the country lie hidden in mist. They know that even in this concealment there dwells a certain mysterious charm; that the misty horizon calls up the image of the sensuous infinite in the mind, a picture which, as I have already observed, is reflected in grave and grand tints in the mind and affections. From the lofty stand, too, from which we propose to make our general survey of nature on the basis of science, all that is requisite cannot be commanded. In natural science, much yet lies but ill defined, and much—and shall I not gladly own to this in entering on a field so vast?—will appear indefinite and incomplete only because every thing like embarrassment becomes doubly detrimental to the speaker, who feels himself indifferently at ease in his subject, when separated from its individualities.

The purpose of this Introduction was not to present a picture of the importance of natural science, a thing universally admitted; it was rather to show how, without detriment to the deepest study of the several special departments of natural science, a higher position for physical scientific inquiry may be won, from which all the forms and powers of things shall be seen to reveal themselves, in the guise of a natural whole, actuated by intrinsic aptitudes. Nature is no dead aggregate; she is, "to the inspired inquirer," (as Schelling grandly expresses himself, in his admirable Discourse on the Fine Arts), "the holy, the eternally creative prime mover of the universe, engendering and evolving all things out of her pregnant self." The hitherto imperfectly seized idea of A PHYSICAL HISTORY OF THE EARTH expands, under more enlarged views and the comprehension of all created things in earth and heaven, into the idea of a PHYSICAL HISTORY OF THE

UNIVERSE. The latter of these titles is fashioned from the former. But it is the history of the universe, or the doctrine of COSMOS, as I conceive it; by no means an encyclopædic exposition of the most general and important results derived from particular natural historical, natural philosophical, and astronomical books. Such results will only be introduced incidentally into my description, and be used as materials only in so far as they illustrate the connection and co-operation of the forces of the universe, the production and limitation of natural phenomena. The study of the distribution of organic types according to soil and climate, the geography of plants and animals, is as dissimilar from descriptive botany and zoology, as geological knowledge of the crust of the Earth is different from oryctognosy. A physical history of the universe, consequently, must not be confounded with an encyclopædia of the natural sciences. In our survey of the Universe, the Individual will only be regarded in its relations to the General, and the higher the point of outlook now indicated is assumed, the more will this survey be made susceptible of especial treatment, and of interesting discussion.

THOUGHT and LANGUAGE, however, stand in most intimate and old relationship to one another. When speech adds grace and clearness to ideas, when its picturesqueness of derivation and organic structure favour our efforts sharply to define natural phenomena as a whole, it scarcely fails at the same time, and almost unconsciously to us, to infuse its animating power into the fulness of thought itself. The WORD is, therefore, more than the mere sign and form, and its mysterious influence still reveals itself most strikingly where it springs among free-minded communities, and attains its growth upon native soils. Proud of our fatherland, whose intellectual unity is the prop and stay of every manifestation of mental power, we turn our eyes with joy upon this privilege of our native country. Highly-favoured, indeed, may we call him who draws, in his accounts of the phenomena of creation, from the depths of a language, which, through the force aad unfettered application of intellect, in the regions of creative fancy, no less than in those of searching reason, has for centuries influenced so powerfully all that affects the destinies of man.

NOTES TO INTRODUCTION.

1 (page 4.)—This expression is borrowed from a fine description of a forest in Bernardin de St.-Pierre's Paul and Virginia.

2 (p. 5.) — These comparisons are only approximations. The more accurate elements (heights above the sea-level) are for the SCHNEE- or RIESEN-KOPPE, in Silesia, 824 toises, according to Hallaschka; for the RIGI, 923 t., assuming the surface of the Lake of Lucerne to be 223 t. (Eschmann's Results of Trigonometrical Measurements in Switzerland in 1840, p. 230); for MOUNT ATHOS, 1060 t. (Capt. Gaultier); for MOUNT PILATUS, 1180 t.; for ETNA, 1700·4 t., or 10,874 English feet, after Capt. Smyth. According to Sir John Herschel's barometric measurements, communicated by him to me in 1825, it is 10,876 Eng. ft. = 1700·7 t.; and, according to Cacciatore, from angular measurements, and, assuming the terrestrial refraction to be = 0·076, it is 10,898 Eng. ft., or 1,704 t. For the SCHRECKHORN, 2,093 t.; the JUNGFRAU, 2,145 t. (Tralles); for MONT BLANC, according to the results discussed by Roger, 2,467 t. (*Bibl. Univ.* May 1828, pp. 24—53); whilst Carlini determined it, from Mont Colombier, in 1821, at 2,460 t.; and Austrian engineers, operating from Trelod and the Glacier d'Ambin, fixed it at 2,463 t. The actual height of the Swiss snowy mountains varies, according to M. Eschmann, about 3½ t., owing to the variable thickness of the coating of snow. For CHIMBORAZO, my trigonometrical measurements give 3,350 t. (Humboldt, *Rec. d'Obs. astr.* vol. i. p. LIII.); for DHAWALAGIRI, 4,390 t. All these mountain-heights are given in *toises*, of six Paris feet each. As Blake and Webb's determinations differ by 70 t., I must here remark that the measurements of *Dhawalagiri* (or White Mountains, from the Sanscrit *dhwala*, white, and *giri*, mountain), cannot pretend to equal accuracy with those of JAWAHIR (4,027 t. = 24,160 Par. ft. = 25,749 Eng. ft. = 7,848 mètres), founded on a complete trigonometrical operation (*vide* Herbert and Hodgson, in Asiat. Res. vol. xiv. p. 189; and Supp. to Encycl. Brit. vol. iv. p. 643). I have shown in another place (*Ann. des Sciences nat.* Mars 1825), that the height of Dhawalagiri (4,391 t. = 26,345 Par ft. = 28,077 Eng. ft.) simultaneously depends on several imperfectly settled elements of astronomical positions and azimuths (Humboldt, *Asie cent.* vol. iii. p. 282). Still more unfounded is the surmise that some snowy peaks of the Tartarian chain, in the north of Tibet, near the Kuenlun chain, rise to the elevation of 30,000 Eng. ft. (4,691 t., nearly twice that of Mont Blanc), or at least to 29,000 Eng. ft. or 4,535 t. (*vide* Capt. Alexander Gerard and John Gerard's Journey to Boorendo Pass in 1840, vol. i. pp. 143 & 311). Chimborazo is styled "only one of the highest points of the Andes," since the learned and able traveller, Mr. Pentland, in 1827, during his memorable expedition to Upper Peru, or Bolivia, measured two mountains east of Lake Titicaca; namely, SORATA (3,948 t. = 23,688 Par. ft.) and ILLIMANI (3,753 t. = 22,518 Par. ft.), which far exceed Chimborazo (3,350 t. = 23,100 Par. ft.) in height, and nearly approximate to Jawahir (4,027 t.), the highest of the hitherto accurately measured Himalayan mountains. Mont Blanc (2,467 t. = 14,802 Par. ft.) is, therefore, 883 t. lower than Chimborazo, and Chimborazo 598 t. lower than Sorata, which is 79 t. lower than Jawahir, but probably 443 t. lower than Dhawalagiri. The measurements in this note may be taken as more accurate from being given in various scales, since false reductions of these scales have led to erroneous numerical statements in Modern maps and profiles. Pentland's more recent measurements of Illimani, in 1838, gives 7,275 mèt. = 3,732 t. for its height, differing only 21 t. from the measurements of 1827.

3 (p. 5.)—The absence of Palms and arborescent ferns in the temperate zones of the Himalaya is shown in Don's *Flora Nepaliensis* (1825), as also in the lithographed and remarkable catalogue of Willich's *Flora Indica*,—a catalogue which contains the enormous number of 7,683 almost entirely phanerogamous Himalayan species, although not yet sufficiently examined and classified. We as yet know of only one species of palm, Chamærops Martiana, *Wall.* (Plant. Asiat. vol. iii. p. 5, t. 211) in Nepaul (lat. 26½°—27¾°), 5,000 feet above the sea, in the shady valley of Bunipa. The splendid arborescent fern, Alsophila Brunoniana, *Wall.* of which the British Museum has had a stem, 45 feet long, since the year 1831, does not come from Nepaul, but from the mountains of Silhet, north-east of Calcutta, lat. 24° 50′. The Nepaul fern, Peranema cyathoïdes, *Don*, formerly Sphæropteris barbata, *Wall.* (op. cit. vol. i. p. 42, t. 48), is nearly related to the Cyathea, of which I saw a species, 30 feet high, in the South American Missions of Caripe; bu it was still no tree, properly so called.

4 (p. 5.) — Ribes nubicola, R. glaciale, R. grossularia. In spite of a declaration of the ancients on "Eastern Asia" (Strabo, lib. xi. p. 510, Cas.), the vegetation of the Himalayas is characterized by 8 species of Pinus, 25 oaks, 4 birches, 2 species of Aesculus (the 100 feet high wild chesnut-tree of Cashmir is inhabited up to 33° N. lat. by a great white ape with a black face—Charles von Hügel, Kashmir, 1840, part ii. p. 249), 7 maples, 12 willows, 14 roses, 3 strawberry species, 7 Alpine roses (Rhododendra), one of which is 20 feet high, and many other Northern forms. Amongst the Coniferæ we find the Pinus Deodwara, or Deodara (properly *dêwadâru*, god-timber,) nearly related to Pinus Cedrus. Near the eternal snows the Gentiana venusta, G. Moorcroftiana, Swertia purpurescens, S. speciosa, Parnassia armata, P. nubicola, Pæonia Emodi, Tulipa stellata, display their large blossoms. Even next to the peculiar Hindoo mountainous species of European *orders*, we find eight genuine European species, as Leontodon taraxacum, Prunella vulgaris, Galium Aparine, Thlaspi arvense. The heath, already mentioned by Saunders, in Turner's Journey, and which has even been confounded with Calluna vulgaris, is an Andromeda—a fact of great importance for the geography of Asiatic plants. If, in this note, I make use of the unphilosophical expression, "European forms, or European species, growing wild in Asia," it is a consequence of the ancient botanical language, which very arbitrarily subjects the idea of the distribution, or rather of the coexistence of organic forms, to the historical hypothesis of an immigration, even premising a movement from west to east, out of prejudice to European cultivation.

5 (p. 5.)—The snow-line of the southern declivity of the Himalayan chain is 2,030 t. = 12,180 ft. above the sea-level, whilst on the northern side, or rather on the peaks which rise, in 30½° to 32° lat., above the Tartaro-Tibetan table-land, it is 2,600 t. = 15,600 ft., the snow-line being at the height of only 2,470 t. = 14,820 ft. under the equator in the Quito Andes. I have deduced this result from comparing together several observations of Webb, Gerard, Herbert, and Moorcroft. *Vide* my two *Mémoires sur les montagnes de l'Inde* of 1816 and 1820 in the *Ann. de Chimie et de Physique*, tom. iii. p. 303; tom. xiv. pp. 6, 22, 50. The eternal snow-line on the Tibetan declivity is a consequence of the radiation of heat by the near table-land, of the serenity of the sky, and of the scanty formation of snow in very dry cold air (Humboldt, *Asie cent.* tom. iii. pp. 281—326). The conclusion, in regard to this line on both sides of the Himalayas, which I proposed as the more probable one, had the sanction of Colebrooke's great authority. "I find," as he wrote to me in June 1824, "that the height of the eternal snows, according to the materials which I possess, is 13,000 English feet (= 2,033 t.). On the southern declivity, under the parallel of 31°, Webb's measures would give me 13,500 Eng. ft. (= 2,111 t.), or 500 feet more than Captain Hodgson's observations. Gerard's measurements perfectly confirm your announcement, that the snow-line is higher on the northern than on the southern side." Only in this year (1840) have we at length received, through Mr. Lloyd, a copy of the entire journal of both the brothers, Gerard (Narrative of a Journey from Caunpoor to the Borrendo Pass, in the Himalaya, by Captain Alexander Gerard and John Gerard, edited by George Lloyd, vol. i. pp. 291, 311, 320, 327, and 341). A great deal on single localities is comprised in the "Visit to the Ghatool, for the purpose of determining the line of perpetual snow on the southern face of the Himalaya, in August 1822;" but, unfortunately, the travellers always confound the height where accidental snow falls, with the maximum height at which the snow-line rises over the Tibetan plateau. Captain Gerard distinguishes the peaks in the middle of the plateau, the eternal snow-line of which he fixes at from 18,000 to 19,000 Eng. ft. (= 2,815 to 2,971 t.), and the northern declivities of the Himalayan chain, which limit the passage of the Sutlej, and where the plateau is deeply furrowed, with, of course, little radiation. The village Tangno is placed only at 9,300 Eng. ft. (= 1,454 t.), whilst the plateau about the sacred lake, Manasa, is said to be 17,000 Eng. ft. (= 2,658 t.) high. Capt. Gerard finds at the break in the chain, that the snow is 500 Eng. ft. (= 78 t.) *lower* on the northern declivity than on the southern towards India; on which latter face the snow-line is estimated by him at 15,000 Eng. ft. (= 2,346 t.).

The botanical relations offer the most striking differences between the Tibetan table-land and the southern aspect of the Himalayan chain. In the latter, the harvest (and the corn is often cut green) extends to 1,560 t. only; the upper woody limit, with tall oaks and Dewadaru firs, to 1,870 t., low dwarf birches to 2,030 t. On the plateau, Capt. Gerard saw pastures up to 2,660 t.; cereals prosper up to 2,200 t., and even to 9,900 t.; tall birches to 2,200 t.; underwood, for fuel, to 2660 t., that is, 200 t. higher than the eternal snow-line under the equator at Quito. It is most desirable that travellers, accustomed to general views, should re-determine the *mean* altitude of the Tibetan table-land, which I assume to be 1,800 t., between the Himalaya and Kuen-lün, as also the relative glacial heights on the northern and southern declivities. Hitherto estimates have been often confounded with actual measurements, and the heights of some prominent peaks, with that of the table-land wherefrom they rise (compare Carl Zimmermann's acute hypsometric remarks in his "Geographical Analysis of the Map of the Interior of Asia," 1841, p. 98). Mr. Lord directs our attention to a contrast between the heights of eternal snow on both declivities of the Himalaya and the Apine chain, Hindoo Koosh. "In the latter," he says, "we find the table-land in the south, and the altitude of the snow-line is consequently greater on the southern declivity: the reverse of the Himalaya, which is bounded by warm plains on the north, as the Hindoo Koosh is on the south." However considerable the critical corrections that may be required for these several details, it is still an indisputable fact, that the wonderful configuration of a portion of the earth's surface in the interior of Asia allows to the human race the possibility of propagation, food, fuel, and colonization, at a height above the sea-level, which, in almost every other district of both continents (excepting the parched, snow-free Bolivia, where Pentland found the snow-line under 16°—17¾° S. lat. at the mean height of 2,450 t. in 1838,) is eternally covered with ice. The probable differences of the north and south declivities of the Himalaya range, in regard to the eternal snow-line, have been amply confirmed by the barometric measurements of Victor Jacquemont, who so early became the victim to his noble and untiring zeal (*vide* his "Correspondance pendant son Voyage dans l'Inde, 1833, tom. i. p. 299; and "Voyage dans l'Inde pendant less années 1828 à 1832, livr. 23, pp. 290, 296, 299). "The eternal snows," says Jacquemont, "descend lower on the southern than on the northern declivity of the Himalaya, and their limit constantly rises as we advance to the north of the border-chain of India. On the Kioubrong peak, 5581 mètres high (2863 t.), according to Captain Gerard, I was still considerably beneath the limit of the eternal snows, which in this part of the Himalaya I believed (certainly too great—*Humboldt*) to be at 6,000 mètres = 3078 t." The same traveller observes, that, to whatever height we rise on the southern declivity, the climate retains the same character, the same division of seasons, as in the plains of India. "The summer solstice brings the same showers of rain, which uninterruptedly last until the autumnal equinox. Only at Kashmir, which I have found to be 5,350 Eng. ft. high," (=837 t., therefore nearly that of the cities Merida and Popayan,) "begins a new and distinct climate." —*Jacquem. Corresp.* tom. ii. pp. 58 and 74. Leopold von Buch accurately remarks that the monsoons do not impel the moist and warm sea-air of the Indian lowlands across the Himalayan barrier to the tramontane Tibetan district of Ladak and Lhassa. Carl von Hügel estimates the height of the valley of Kashmir above the sea-level, from observations of the boiling point of water (Part ii. p. 155, ant.; Journal of the Geog. Soc. vol. vi. p. 215) at 5,818 Eng. ft. (= 910 t.). In this perfectly calm and almost tempest-free valley, under 34° 7′ lat., the snow lies many feet deep from December to March.

6 (p 5.)—See generally my "Essai sur la Géographie des Plantes et Tableau Physique des Régions equinoxiales," 1807, pp. 80—88; on the diurnal and nocturnal oscillations of temperature in the ninth plate of my "Atlas géog. et phys. du nouveau Continent," and the tables to my work, "De distributione geographica plantarum secundum cœli temperiem et altitudinem montium," 1817, pp. 90—116; the meteorological portion of my "Asie centrale," tom. iii. pp. 212—214; lastly, the more recent and accurate account of the height-decreasing temperature among the Andes in Boussingault's "Mémoire sur la profondeur à laquelle on trouve la couche de température invariable sous les tropiques," *Ann. de Chimie et de Phys.* 1833, tom. liii. pp. 225—247). The essay last quoted contains the determination of the height and mean temperature of 128 points, from the sea-level to the declivity of Antisana, at 2,800 t. height, between the aerial temperatures of 27°5 and 1°7 Cent. (= 81°5 and 35° Fahr.).

7 (p. 6.)—"On the Kawi Language in the island of Java, with an introduction on diversities in the structure of language, and their influence on the mental development of the human race, by William v. Humboldt," 1836, vol. i. pp. 5—310.

8 (p. 6.)—Respecting the proper Madhjadêça, *vide* Lassen's excellent *Indische Alterthumskunde*, vol. i. p. 92. The Chinese term South Bahar *Mo-kie-thi*, meaning the part lying south of the Ganges.—*Vide* Chy-Fa-Hian's *Foe-koue-ki*, 1836, p. 256. *Djambu-dwipa* is entire India, sometimes comprehending one of the four Buddhist continents.

9 (p. 6.)—Schiller's Elegy, Der Spaziergang, or the Walk, which first appeared in 1795, in the *Horen*:—

> Within his silent chamber, casting circles
> Pregnant with meaning, sits the thoughtful sage—
> Creative mind compelling new results:—
> Testing the forces that inhere in matter,
> Proving the magnet's wondrous hate and love,
> Pursuing sound through the air, the ray of light
> Through ether, still intent on finding laws
> Amidst the incongruous in what seems chance,
> Intent on making out the stable pole
> Amidst the flight of mere phenomena.

10 (p. 7.)—Arago's ocular micrometer, a happy improvement upon Rochon's prismatic or double-refraction micrometer, *vide* M. Mathieu's note in Delambre, "Hist. de l'Astr au 18me siècle," 1827, p. 651.

11 (p. 8.)—Carus on the Elementary Parts of the Osseous and Crustaceous Frame-work of Animals, 1821, p. 6.

12 (p. 8.)—Plut in vit. Alex. Magn. cap. vii.

13 (p. 8.)—The melting-points of difficultly fusible substances usually assumed are too high. Mitscherlich's always accurate researches limit the melting-point of granite to 1,300° C. = 2,372° F.

14 (p. 9.)—Louis Agassiz's classical work on fossil fishes, "Rech. sur les Poissons fossiles," 1834, vol. i. p. 38; vol. ii. pp. 3, 28, 34, Addit. p. 6. The entire species Amblypterus, *Agass.* nearly related to Palæoniscus (Palæothrissum), is buried beneath the Jura, in the old coal formation. Scales, which, in single layers, are formed like teeth, and are covered with enamel, from the Lepidoid family (Order *Ganoïdes*), belong, after Placoïdes, to the oldest forms of fossil fishes, whose now living representatives are found in two species, Bichir (Nile and Senegal) and Lepidosteus (Ohio).

15 (p. 10.)—Goethe's "Aphorismen über Naturwissenschaft" (Works, small edit. 1833, vol. L. p. 155.)

16 (p. 11.)—Arago's discovery in 1811 (Delambre, op. cit. p. 652).

17 (p. 12.)—Goethe's "Aphoristisches über die Natur" (op. cit. vol. L. p. 4).

18 (p. 12.)—Pseudo-Plato, Alcib. ii. p. 148, ed. Steph.: Plut. Instituta laconica, p. 253, ed. Hutten.

LIMITATION AND SCIENTIFIC TREATMENT OF A PHYSICAL HISTORY OF CREATION.

In the general views with which I have opened my prolegomena to a survey of universal nature, I have sought to explain, and, by examples, to illustrate, how the enjoyment of nature, diverse in its intimate sources, may be enhanced through clear ideas of the connection of her phenomena, and of the harmony that reigns among her actuating forces. It will now be my endeavour to enunciate more particularly the spirit and leading idea of the following scientific inquiry; carefully to separate from it all that is foreign; and with comprehensive brevity to convey the scope and contents of the doctrine of the Cosmos as I have apprehended and worked it out, after long years of study in various climates of the globe. Let me flatter myself with the hope that such an exposition will bear me out in the bold title I have given my work, and free me from the charge of presumption. My prolegomena comprise, under four divisions, and in consonance with my introductory remarks on the foundation of the laws of the universe, 1st. The conception and limitation of physical cosmography, as a separate and distinct science.

2d. The objective contents, the comprehensive empirical survey, of nature at large, in the scientific form of a general picture.

3d. The reflex action of nature upon the imagination and feelings, as stimulating to its study, through animated descriptions of remote countries, landscape poetry (a branch of modern literature), beautiful landscape painting, the cultivation and contrasted grouping of exotic plants, &c.

4th. The history of creation—in other words, an account of the gradual development and extension of the idea of the Cosmos as a natural whole.

The higher the point of view from which the phenomena of nature are contemplated, the more distinctly must the science, the foundations of which are now to be laid, be bounded, and marked off from all allied departments of natural knowledge. Physical Cosmography embraces the description of all that is created, of all that exists in space, both natural things and natural forces, as a simultaneously existing coordinate whole. It divides itself for man, the inhabitant of the earth, into two principal divisions; one telluric, another sidereal or uranological. To confirm the scientific independence of physical cosmography, and show its relations to other departments—to physics or natural philosophy, to natural history or the special description of natural objects, to geognosy and comparative geography, or the description of the earth—we shall first pause over the telluric portion of our subject. Even as little as the history of philosophy consists in a crude arrangement side by side, or in sequence, of the various philosophical opinions that have been entertained, so little is the telluric portion of cosmography any encyclopædic aggregate of the natural sciences enumerated above. The lines of demarcation between branches so intimately allied as these, are the more confused in consequence of the custom which has prevailed for centuries, of designating by specific titles certain groups of experimental knowledge, which are now too narrow, now too comprehensive for the matters comprised, and which, in times of classical antiquity, and in the languages from which they were borrowed, had a totally different signification from that now attached to them. The titles of particular natural sciences, such as anthropology, physiology, natural philosophy, natural history, geognosy, and geography, arose and became universally current before mankind had attained to any clear conception of the diversity of objects embraced by these several sciences, and the precise line of demarcation between each—that is to say, of the grounds of separation themselves. In the language of one of the most polished nations of Europe, natural philosophy (*physics*) is scarcely distinguished from medicine (*physic*); whilst technical chemistry, geology, and astronomy, treated in an entirely empirical manner, are jumbled together, and papers on all are published under the joint title of Philosophical Transactions, by a Society whose fame is justly as wide as the world.

Alterations of old, often ill chosen, but generally well understood names, for newer titles, have been repeatedly attempted, but always, as yet, with indifferent success, by those who have turned their attention to the classification of the several departments of human knowledge, from the Margarita Philosophica (a great Encyclopædia) of the Carthusian monk Gregory Reisch(1), to Bacon; from Bacon to d'Alembert, and, not to forget the very latest times, to the acute geometrician and natural philosopher, Ampère(2). The unfelicitous choice of a fantastical nomenclature has perhaps been more prejudicial to every attempt of the kind, than the excessive number of divisions and subdivisions that have been introduced.

Physical cosmography, whilst it embraces the world "as an object of the external senses," requires, it is true, the association of general physics and natural history as auxiliary sciences; but the consideration of corporeal things, under the guise of a natural whole, moved and actuated by inherent forces, has an entirely special character as a distinct science. Physics occupies itself with the general properties of matter: it is an abstraction from the manifestations of force by matter; and in the very place where its first foundations, as a science, are laid, viz., in the eight books of the Physics of Aristotle(3), all the phenomena of nature are represented as vital manifestations of a general cosmic force. The telluric portion of physical cosmography, to which I willingly concede the old title, Physical History of the Globe, treats, among other matters, of the distribution of magnetism over our planet, with reference to intensity and direction; not of the laws of magnetical attraction and repulsion, nor of the means of exciting electro-magnetical effects, now of a more passing, now of a more permanent character. Physical cosmography displays, in bold outlines, the partitionings of continents and the distribution of their masses in either hemisphere—points that influence climate and the more important meteorological processes in the most remarkable manner; it goes farther—it indicates the prevailing characters of the several great mountain ranges, their extension in more continuous and even chains, or their connections in the manner of a grating, and their association with the several epochs and systems of formation; it determines the mean height of continents above the present level of the sea; the points of the centres of gravity of their volumes; the relations of the higher peaks of extensive chains to their acclivities, to neighbouring seas, and to the mineral nature of their constituent rocks; it informs us how these mountain masses, now active and moving, breaking through a superimposed crust, now passive and moved, present their strata under every variety of inclination—level, sloping, perpendicular; it considers the succession or isolation of volcanoes; the indications of their manifestations of activity, the extent of the circles they severally shake, and which in the course of centuries enlarge or contract. It farther informs us, to select a few examples from the conflict of the fluid with the solid, of the points of resemblance between all mighty streams in one part or another of their course: how they are liable to bifurcate, either in their superior or inferior channels; how at one time they cut across colossal mountain chains at right angles, at another, run in lines parallel to them, whether this be near the declension of the chain, or at some considerable distance from it, as a consequence of the influence which an elevated mountain system has exerted upon the surface of entire districts of country, and on the saline bottoms of neighbouring plains. Only the chief results of comparative orography and hydrography belong to the science which I here circumscribe, not minute descriptions of mountain masses; of volcanoes that are now active; of the volume of waters of particular rivers, &c.: all this, according to my views, belongs to special or descriptive geography, and will be comprised in the notes which illustrate my work. The enumeration of similar, or closely-allied, natural relations, the general survey of terrestrial phenomena with reference to their distribution in space, or their relations to particular zones of the Earth, is not to be confounded with the consideration of the individual things of Nature, to wit, terrestrial substances, animated organisms, physical phenomena; a consideration which would only lead to a systematic arrangement of objects, according to their intimate analogies.

Special geographical descriptions are, it is true, the most available material for a general physical geography; but the most painstaking accumulation of such descriptions would as little convey to the mind the characteristic idea of terrestrial nature at large, as the mere coordination of all the individual floras of the earth would give a notion of the geography of plants. It is the business of the combining intellect, out of the individualities of organic forms (morphology, the doctrine of the external forms of plants and animals), to extract the common in climatic distribution; to fix the numerical laws—the proportions in the number of certain forms of natural families, to the entire number of plants or animals of the more perfect types; to determine in what zone each of the principal forms attains its maximum in point of numbers of kinds and organic development, and even to show how the impression made upon the mind by a landscape at different distances from the equator, in so far as this is connected with the vegetable growths that cover the surface of our planet, is mainly dependent on the laws of vegetable geography.

Those systematically arranged catalogues of organic forms, which in former times were designated by the somewhat ostentatious titles of Systems of Nature, present a wonderful enchainment in reference to similarity of form (structure), to the conception of a gradual unfolding or evolution of leaf and calyx into coloured blossom and fruit, but not any concatenation with reference to distribution in space, that is to say, to climate, elevation above the level of the sea, and to temperature, to which the whole surface of the globe is exposed. The highest aim of physical geography, however, as already observed, is the recognition of unity in multiplicity, the investigation of the Common and Intimately-connected in all terrestrial phenomena. Where individualities are indicated, no more is done than may help to bring the laws of organic arrangement into unison with those of geographical distribution. The mass of living forms, in this point of view, appears to be arranged rather according to the zones of the earth, or to the course of isothermal lines, than in conformity with internal relationship, or the principle of gradation and individualizing development of organs inherent in the whole of nature. The natural sequence of vegetable and animal forms will therefore be here assumed from our ordinary descriptive botany and zoology. It is the province of physical geography to investigate the mysterious generical relations in which, with an apparent dispersion of families and species over the surface of the earth, the most dissimilar forms still stand to one another; to show how the various organisms constitute a natural whole; how they modify the atmosphere by the slow processes of combustion and assimilation that go on in their interior; and how, influenced by promethean light in their evolution, in their very being, despite their inconsiderable mass, they act upon the whole life of the globe.

The mode of presenting the subject which I here propose as alone appropriate to physical cosmography, gains in simplicity when we apply it to the uranological portion of the Cosmos, to the physical history of heavenly space, and of the heavenly bodies. If we distinguish physics, or natural philosophy, as used formerly to be done, but as deeper and clearer views of nature allow us no longer to do—physics, or the general consideration of matter, of force, and of motion, from chemistry, or the consideration of the different natures of matter, its combinations and changes through admixture, not through affinities in virtue of the simple relations of mass, we then perceive, in the telluric region, physical and chemical processes existing together. Besides that fundamental property of all matter, attraction at a distance (gravitation), other forces affect us here upon earth, which come into operation at infinitely small distances, or upon immediate contact between material particles(4), forces which are designated chemical affinities, and which, called into action variously by electricity, caloric, and even simple contact, are incessantly efficient in inorganic nature, as well as in living organisms. In the celestial spaces we have as yet no apprehension of any other than physical processes, affections of matter which depend on mass alone, and which are subjected to the dynamic laws of a pure doctrine of motion. Such affections are regarded as independent of all qualitative differences—of heterogeneousness or specific difference of matter.

The inhabitants of the earth are brought into relation with the matter dispersed over space, only by the phenomena of light and the influence of general gravitation (attraction according to mass). The influences of the sun and moon upon the periodical variations of terrestrial magnetism, are still buried in obscurity. We have no immediate knowledge or experience of the qualitative nature of the matter which circulates in, or perhaps fills, the universe, unless, perchance, it be through the fall of aerolites, if these heated masses, involved in vapour, be assumed as constituting small planetary bodies which have come within the sphere of the earth's attraction in their course through space; an assumption which the direction and extraordinary centrifugal force of the bodies in question appears to render probable. The familiar aspect of their constituent elements, and the identity in nature of these with such as we have in abundance among the mineral masses of the earth, are very striking. They may serve, on analogical grounds, to lead us to conclusions in regard to the nature of such planets as belong to the same group, and have been formed, under the dominion of one central body, by precipitation from revolving rings of vaporous matter. Bessel's pendulum-experiments, which bear the impress of such accuracy as has never yet been attained, have given a renewed faith in the truth of the Newtonian axiom, that bodies of the most dissimilar constitution—water, gold, quartz, granular limestone, aerolites—experience a perfectly similar acceleration of motion through the attraction of the earth. Many purely astronomical results, indeed, for example the almost equal mass of Jupiter, in consequence of the influence of the planet on his satellites, on Encke's comet, on the small planets Vesta, Juno, Céres, and Pallas, assure us that every where it is the quantity of matter alone which influences its power of attraction(5).

This exclusion of every appreciable circumstance referrible to diversity of material, simplifies the mechanism of the heavens in a remarkable manner; it brings the infinite realms of space under the sole dominion of the laws of motion; and the astrognostic portion of physical cosmography draws from established theoretical astronomy, in the same way as the terrestrial portion draws from physics, chemistry, and organic morphology. The departments of science just mentioned, indeed, embrace phenomena so intricate, and at times so opposite to mathematical views, that the terrestrial portion of the doctrine of the Cosmos cannot boast of the same certainty and simplicity of treatment as the astronomical portion. In the distinction now indicated lies undoubtedly the reason wherefore, in the earlier periods of the Greek civilization, the Pythagorean philosophy of nature was rather directed to the heavens than to the earth; wherefore it became fruitful, with reference to our solar system, in a much higher degree, through Philolaus, and, in later times, through Aristarchus of Samos, and Seleucus the Erythrean, than the Ionic natural philosophy could prove in regard to the physics of our globe. More indifferent as to the specific nature of that which filled space, as to qualitative differences of matter, the forces of the Italic school were directed with Doric earnest upon regulated formations, on shape, on form and measure alone(6); whilst the Ionic physiologists occupied themselves especially with the consideration of species of matter, with their supposed transmutations and generic relations. It was reserved for the powerful, truly philosophic, and, at the same time, thoroughly practical mind of Aristotle, to plunge with equal delight into the world of abstraction, and into the measureless abundance of material diversity in organic forms.

Several, and these very excellent works upon physical geography, comprise an astronomical section in their introduction, in which the earth is first considered in its planetary dependence, or in its relations to the rest of the solar system. This plan is the very opposite of that which I have chalked out for myself. In a system of cosmography, the astronomical portion, which Kant entitled the Natural history of the heavens, must not appear as subordinate to the telluric portion. In the Cosmos, as the old Copernican philosopher, Aristarchus of Samos, said, the sun with his attendants is a star amongst innumerable stars. A general survey of creation must consequently begin with the heavenly bodies that occupy space, with a graphic delineation, a kind of map of the celestial universe, such as the bold hand of the elder Herschel first ventured to design. If we see that, despite the relative insignificance of our planet, the terrestrial portion still occupies the largest space in the history of the universe, and is most fully handled, this only happens in respect of the unequal mass of that which is Known to the inequality of that which is Empirically accessible. This subordination of the uranological portion we already find in the great geographer, Bernhard Varenius, in the middle of the 17th century(7). He distinguishes with much acumen between the General and Special description of the earth, and subdivides the former, into the absolutely terrestrial and the planetary, according as the relations of the surface of the earth in different zones, or the sol-lunar life of the earth—the relations of our planet to the sun and moon—are considered. It is a great and enduring honour to Varenius, that the realization of this plan of a General and of a Comparative Geography attracted Newton's attention in a very decided manner; but owing to the imperfect state of the accessory sciences from which Varenius drew, the way in which the idea could be carried out was not in accordance with the grandeur of the conception. It was reserved for our own times to see comparative geography, in the widest sense of the expression, even in its reflex on the history of mankind—the influence which the figure of continents has had on the course of the great migrations of the human family, and the progress of civilization, worked out in the most masterly manner(8).

The enumeration of the various rays which unite as in a focus in the natural sciences considered as a whole, may serve as an apology for the title of the work which I venture to produce in the late evening of my life. This title is perhaps even bolder than the undertaking itself, considering the limits which I have prescribed myself. In the special departments, I had hitherto avoided as much as possible the use of new names for the indication of new conceptions. Where I attempted any extension of our nomenclature, it was always confined to individual objects in zoology and botany. The term, Physical Cosmography, which I here employ, is imitated from the phrase, Physical Geography, which has long been familiar to all. The great extent of the subject embraced, the purpose of surveying nature at large, from the remote nebulous specks in the heavens, to the climatic distribution of the organic tissues that colour the face of our rocks, make the introduction of a new term necessary. And however completely our old and usual terms EARTH, and WORLD, blend together, as we see them in the familiar phrases of, a voyage round the world, a map of the world, the new world, &c., this is a mere consequence of the former more limited knowledge of mankind; the scientific distinction between the world, or universe at large, and the earth we inhabit, is now felt to be a matter of common necessity. The grander and more correct expressions, UNIVERSE, FABRIC OF THE UNIVERSE, CREATION, and NATURE,* employed to designate the conception and origin of all matter, terrestrial as well as that of the farthest stars, seem to approve the propriety of this distinction. To make this more definite, I might say more solemn and impressive, and also to recur to the antique name, I have placed the word COSMOS (ΚΟΣΜΟΣ) at the head of my work; this term, in the Homeric times, having been used to indicate beauty and order, but by and by employed as a philosophical expression to indicate the harmony or arrangement of the world, even of the entire mass of matter filling space, of the universe at large.

The difficulty of distinguishing the normal—the regular and legitimate—amidst the ceaseless changes of earthly phenomena, appears at an early period to have directed the mind of man, in an especial manner, to the uniform and harmonious movements of the heavenly bodies. According to Philolaus, and the concurring testimony of the whole of antiquity(9), Pythagoras was the first who employed the word Cosmos as synonymous with creation, with the order and arrangement of the earth and heavenly bodies. From the Italic philosophical school, the word passed into the language of the poets of nature, Parmenides and Empedocles; and by and by it was adopted by the prose writers. It is beyond my purpose to expatiate in this place on the various particular applications of the term, according to Pythagorean views—now to the planets that revolve around the focus of the world, now to groups of stars in the firmament; or to explain that Philolaus, on one occasion, distinguishes between Olympus, Kosmos, and Uranus. In my plan of a cosmography, as this was understood in times posterior to Pythagoras, and as the term is used by the unknown author of the book, De Mundo, which was so long ascribed to Aristotle, Cosmos is used to designate the conception of the heavens and earth—of the whole of the material universe. The Romans, in the spirit of imitation, and when they came to pay a tardy attention to philosophy, adopted the word Mundus, which originally signified ornament, never order, for the designation of the universe. The introduction of the technical term into the Latin tongue, the literal translation of the Greek Kosmos, used in a double sense, is probably to be ascribed to Ennius(10), a disciple of the Italic school, and the translator of the Pythagorean philosophical speculations of Epicharmus, or of one of his imitators.

As a physical history of the world, in the widest sense of the word, were the materials accessible for such an undertaking, would pass in review the changes which the Cosmos undergoes in the lapse of time, from the new stars which suddenly make their appearance in the heavens, and the nebulæ which either dissolve and disappear, or become condensed in their centres, to the most insignificant vegetable tissue that first covered the cold crust of the earth, or that gradually and progressively overspreads the coral reef which rises from the bosom of the ocean, so would a physical description of the world, on the other hand, portray the coexistent in space, the simultaneous agency of the natural forces, and of the concrete forms that are the product of these forces. The Existing, however, in our conception of nature, is not to be absolutely distinguished or separated from the Coming into Existence; for it is not the organic alone that is to be conceived as ceaselessly involved in coming into being and ceasing to be; the whole life of the globe, in each stage of its existence, refers us to earlier conditions that have been successively passed through. The various superimposed strata, of which the outer crust of our earth consists in principal part, inclose the remains of a creation that has almost entirely disappeared; they give us to wit of a series of formations, which, in groups, have successively supplanted one another; they disclose to the eye of the observer the aggregate faunas and floras of bygone millenniums. In this sense, the Description of Nature, and the History of Nature, are not entirely to be dissevered. The geologist cannot

* Weltgebäude, Weltkörper, Weltschöpfung, Weltraum, German; literally Worldfabric, Worldbodies, Worldcreation, Worldspace.

apprehend the present without understanding the past. Each penetrates the other, and blends in a natural picture of the globe; just as in the vast domain of language, the etymologist finds reflected in various states of grammatical forms, in their rise and progressive development, the whole of the present in the past. But this reflection of what has been, is by so much the clearer in the material world, as we now see several products forming themselves under our eyes. Among mountain masses, to choose an example from geology, trachytic cones, basalt, layers of pumice and amygdaloidal scoriæ, enliven the landscape in a remarkable manner. They work upon our imagination like tales from antiquity; their form is their history.

Existence in its whole extent and intimacy is first completely known as a something that has become. To this original blending of conceptions, classic antiquity bears witness in the use of the word *History*, both by Greece and Rome. If not included in the definition which Verrius Flaccus(11) gives of the term, History is used in the zoological writings of Aristotle to signify a narrative of things investigated, of matters recognized by the senses. The description of the World of the elder Pliny bears the title Historia Naturalis; in the letters of the nephew, it is more worthily designated "a History of Nature." In the times of classical antiquity, the early historian makes little distinction between descriptions of countries and the narrative of events of which these countries were the theatre. Physical geography and history continued long to present themselves pleasantly mingled together, until increasing political interests, and deeper movements in civic existence, pushed aside the former element, which then took its place as a separate department of human science.

To embrace the multiplicity of the phenomena of the Cosmos in unity of thought, in the form of a purely rational series, is not, as I conceive, possible in the present state of our empirical knowledge. The sciences of experiment are never complete; the realm of the impressions of sense is not to be exhausted; no generation of men will ever have it in their power to boast, that they have surveyed the whole of the world of phenomena. It is only where phenomena can be grouped, and separated from one another, that we recognize in the individual groups the empire and agency of grand and simple natural laws. The more the physical sciences improve, the wider also does the boundary of this empire extend. Brilliant instances of the truth of this have been afforded by recent views of the processes going on in the solid crust of the globe, as well as in the atmosphere, which depend on electro-magnetic forces, on radiant heat, and the propagation of pulses of light; brilliant examples, too, are supplied by the late insight gained into the laws of organic evolution, where all that is to be, is indicated beforehand, where the continuous growth and progressive development of cells give rise to all the varied tissues of plants and animals. In this generalization of laws; which at first seemed only to comprise much narrower circles, mere isolated groups of phenomena, there are numerous grades. The empire of recognized laws gains in extent, that of ideal connection in clearness, so long as inquiries are pursued in what may be called analogous and allied masses. But where our dynamic views, which are based on figurative atomic premises, no longer suffice us, because the specific nature of matter, and its heterogeneousness come into play, we find ourselves striking suddenly upon reefs that rise from fathomless depths, when we strive after unity of comprehension. Here the operation of a new kind of force is unfolded. The law of definite proportions, or numerical relations, which the genius of modern chemistry has recognized, and has applied so happily, so brilliantly, but still under an antique vesture, in the symbols of atomic representative expressions, has yet remained isolated, has not been brought under the dominion of the laws of pure dynamics.

The individualities to which all the immediate perceptions of the mind are limited, can be logically arranged into classes and families. Such arrangements lead, as I have already had occasion to remark, in so far as Nature is concerned, to the high-sounding titles of Systems of Nature. They facilitate the study, it is true, of organic forms and their linear enchainment with one another; but as catalogues, they present a mere formal enumeration; they introduce more of unity into the exposition than into the knowledge itself. As there are degrees in the generalization of natural laws, according as they comprise larger or smaller groups of phenomena, wider or narrower circles of organic forms and members, so are there also grades in empirical inquiry. It begins with isolated views, which are separated and ordered according to their kinds. From observation it goes on to experiment, to evocation of phenomena under determinate conditions, according to guiding hypotheses; in other words, according to the presentiment of the intimate connection of natural things and natural forces. What is attained through observation and experiment, leads, on grounds of analogy and induction, to the knowledge of empirical laws. These are the phases through which observing intellect must pass, and which indicate, at the same time, particular epochs in the history of natural science among men.

Two forms of abstraction dominate the entire mass of our knowledge: one, QUANTITATIVE, indicative of relationship according to number and volume; the other, QUALITATIVE, relationship in reference to material constitution. The former, and more accessible form, belongs to the mathematical, the second to the chemical sciences. In order to subject phenomena to calculation, matter is assumed as composed of molecules, or atoms; the number, form, position, and polarity of which give occasion to phenomena. All myths about imponderable matters and special vital forces inherent in organized beings, only render views of nature perplexed and indistinct. Under great variety of conditions and forms of apprehension, the heavy burthen of our accumulated, and still accumulating knowledge, is moved lazily and reluctantly. Reason, boldly and with increasing success, now seeks to break down the ancient forms, by means of which, as with mechanical contrivances and symbols, man has still been wont to strive to obtain mastery over rebellious matter.

We are still far from the time when it will be possible to concentrate all perceptions of sense, into unity of conception of Nature. It may even be said to be problematical whether this time will ever come. The complicated character of the problem, and the infinity of the universe, seem almost to render vain the hope that it ever will. But though the complete solution of the problem may remain unattainable, its partial solution may still be anticipated; the effort, indeed, to understand the phenomena of the universe is still the highest, as it is the eternal goal of all natural investigation. Faithful to the character of my early writings, as to the nature of my occupations, which have still been devoted to experiments, to measurements, to the minute examination of facts, I limit myself in my present undertaking to the empirical, or experimental method. It supplies the only ground upon which I feel that I can move with less of uncertainty. But this treatment of an empirical science, or rather of an aggregate of empirical knowledge, does not preclude arrangement of the conclusions come to, in harmony with leading ideas, the generalization of the special, the ceaseless search after empirical natural laws.

Knowledge acquired under the guidance of thought, the attainment of a rational comprehension of the universe, holds out yet a higher object. I am far from blaming efforts in which I have myself made no trial of my strength, because their fruits still remain subject of doubt. Greatly misunderstood, and much against the views and the counsel of the powerful thinkers whom these, the special matters that engaged antiquity, have again attracted, systems of what was called the Philosophy of Nature, threatened, for a time, to lead men away from the study of the mathematical and physical sciences, so important in themselves, so intimately connected with the material welfare of mankind. The intoxicating delirium of possession obtained by toil, a peculiarly adventurously symbolical language, a schematic discipline, narrower than ever the middle age of humanity forced itself into, have, in the youthful misapplication of noble powers, been the features that distinguished the brilliant, but short-lived Saturnalia of this purely ideal natural science—I repeat the expression, misapplication of powers; for the sober spirits dedicated at once to philosophy and to observation, continued strangers to these excesses. The conception of Experimental Science in general, and of a Philosophy of Nature complete in all its parts, if such perfection can ever be obtained, cannot stand in contradiction or opposition to one another, if the Philosophy of Nature, true to its promise, be the rational comprehension of the phenomena of the universe. Where contradiction shows itself, the blame lies in the hollowness of the speculation, or in the arrogance of empiricism, which thinks it gains more from experience than experience warrants

And here the realm of the Spiritual might be opposed to the Natural; as if the spiritual, too, were not contained within the concept of nature as a whole! Or Art may be opposed to Nature, by Art being implied something more than the idea of the spiritual faculty of producing which is inherent in man. Yet these opposites must not lead to such a separation of the physical from the intellectual as would make the physics of the universe sink down into a mere heap of empirically collected individualities. Science begins at the point where mind dominates matter, where the attempt is made to subject the mass of experience to the scrutiny of reason; science is mind brought into connection with nature. The external world exists to us only when we receive it into our interior, when it has fashioned itself within us into a natural perception. Mysteriously indivisible, as are mind and language, as are thought and the fructifying word, even so and to us all consciously, does the external world blend with the interior in man, with thought and with emotion. "External phenomena," says Hegel, in his Philosophy of History, "are thus translated into internal conceptions." The external or objective world, conceived by us, reflected in us, is then subjected to the eternal, necessary, and all-influencing forms of our spiritual existence. Our intellectual activity then exercises itself upon the material that has been taken in through perceptions of sense. There is, therefore, a tendency to philosophical ideas even in the infancy of human society, in the simplest views that can ever be taken of nature. This impulse is various, more or less lively, according to the temper of the mind, to national peculiarity, and to the state of intellectual culture among communities. The work of the mind begins so soon as thought, impelled by internal necessity, takes up the material of sensible impressions.

History has preserved us records of the oft and variously repeated attempt to comprehend the world of physical phenomena in its multiplicity, to get at the knowledge of a peculiar penetrating, moving, compounding, and decompounding power pervading the universe. These attempts, in classical times, constituted the physiologies and doctrines of the primeval matter of the Ionic school, in which, by the side of a poorly arranged empiricism, a scanty display of facts, ideal efforts, or efforts to explain nature upon grounds of pure reason, prevailed. But the more the material of certain empirical knowledge accumulated, under the influence of a brilliant extension of all the natural sciences, the more did the impulse cool which led men to seek to comprehend the essence of phenomena, and to discover their unity as a natural whole, by the construction of systems prompted by pure reason. In times that have but recently gone by, the mathematical portion of natural philosophy has had to rejoice in a great and noble development. The methods and the instrument (Analysis), have advanced towards perfection simultaneously. And what was elicited in such a variety of ways—by a judicious application of atomical premises, by a more general and more immediate contact with nature, by the invention and improvement of new instruments, is now, as of old, the common inheritance of mankind, and ought not to be lost to the freest operations of philosophy, however changing in her forms. Hitherto, indeed, the inviolability of the material has run certain risks in the process of reconstruction; and in the ceaseless change of idealistic views, it is little to be wondered at, if, as finely observed by Bruno([12]), "Many regard philosophy as susceptible of no more than a sort of meteoric existence, so that even the larger and more remarkable forms in which she has revealed herself to mankind share the fate of comets, which are not regarded as belonging to the imperishable and eternal works of nature, but are merely reckoned among the number of fiery vapours."

Misuse or misdirection of the mental energies, however, must not lead to any conclusions tending to degrade intellect; as if the world of thought were, from its very nature, the realm of phantasms and deceptions; as if the precious stores of empirical knowledge, which have been accumulating for centuries, were threatened by philosophy as by some hostile power! It becomes not the spirit of these times to reject, as groundless hypothesis, every generalization of ideas, every attempt, based upon analogy and induction, to investigate the concatenation of the phenomena of nature; and, among the noble faculties with which nature has so wonderfully furnished man, to condemn at one time reason, inquiring, searching every where for causal connections; at another imagination, the active, the exciting, the indispensable to all invention, to all discovery.

NOTES TO PRECEDING SECTION.

1 (p. 17.)—The "Margarita philosophica" of the Carthusian prior of Freiburg, Gregorius Reisch, first appeared under the title of "Aepitome omnis Philosophiæ, alias Margarita philosophica tractans de omni genere scibili," *vide* the Heidelberg edition of 1486, and that of Strasburg of 1504. In the Freiburg edition of that year, and in the twelve following editions, which appeared in the short interval till 1535, the first part of the title was omitted. This work exercised a great influence on the diffusion of mathematical and physical knowledge at the beginning of the 16th century; and Chasles, the learned author of the "Aperçu historique des méthodes en géométrie" (1837), has shown now important is Reisch's Encyclopedia for the mathematical history of the middle ages. I have endeavoured, by means of a passage of the "Margarita philosophica," and which only occurs in the edition of 1513, to unravel the important relations of Hylacomilus (Martin Waldseemüller) the geographer of St. Dié, who first (1507) named the New Continent *America*, with Amerigo Vespucci, with René King of Jerusalem and Duke of Lorraine, and with the celebrated editions of Ptolemy of 1513 and 1522. *Vide* my "Examen critique de la géographie du nouveau Continent, et des progrès de l'astronomie nautique aux 15e et 16e siècles," tom. iv. pp. 99—125.

2 (p. 17.)—Ampère, "Essai sur la Phil. des Sciences," 1834, p. 25. Whewell's Inductive Philos., vol. ii. p. 277; Park's Pantology, p. 87.

3 (p. 17.)—All changes of state in the material world are reduced to *motions*. Aristot. Phys. ausc. iii. 1 and 4, pp. 200—201; Bekker, viii. 1, 8, and 9, pp. 250, 262, 265; De gener. et corr. ii. 10, p. 336; Pseudo-Aristot. de mundo, cap. vi. p. 398.

4 (p. 19.)—Respecting the question raised by Newton of the difference between mass-attraction and that of molecules, *vide* Laplace's "Exposit. du syst. du monde," p. 384, and in the "Supplément au livre x. de la mécanique cél." pp. 3, 4.—(Kant's Metaphysical Elements of Natural Philosophy, in collective Works, 1839, vol. v. p. 309; Peclet's Physique, 1838, tom. i. pp. 59—63.)

5 (19.)—Poisson, in Conn. des tems pour l'année 1836, pp. 64—66; Bessel, in Poggendorff's Annalen der Physik, vol. xxv. p. 417; Encke, in Berlin Academy's Transactions, 1826, p. 257: Mitscherlich's Man. of Chemistry, 1837, vol. i. p. 352.

6 (p. 19.)—Compare Otfried Müller's Dorians, vol. i. p. 365.

7 (p. 19.)—"Geographia generalis in qua affectiones generales telluris explicantur." The oldest Amsterdam edition (Elzevir) is of 1650; the second of 1672, and the third of 1681, were edited by Newton. This all-important work of Varenius is a Physical Geography in its proper sense. Since the excellent description of the New Continent by the Jesuit, Joseph da Acosta (Historia natural de las Indias, 1590), never had the telluric phenomena been so generally contemplated. Acosta is richer in individual observations; Varenius embraces a greater circle of ideas—his residence in Holland, then the centre of the commerce of the world, having connected him with many intelligent travellers. "Generalis sive universalis Geographia dicitur, quæ tellurem in genere considerat atque affectiones explicat, non habita particularium regionum ratione." Varenius's Universal Geography (*Pars absoluta*, cap. i.—xxii.) is altogether a *comparative* one, although the author uses the term *Geographia comparativa* (cap. xxxiii.—xl.) in a much more restricted meaning. The remarkable parts are the enumeration of mountain-systems and reflections, or the relations of their directions with the whole continents (pp. 66—76, ed. Cantab. 1681); the list of the active and extinct volcanoes; the conjunction of results on the division of islands and island groups (p. 220); on the depth of the ocean compared with the height of the coast (p. 103); on the equal levels of the surface of all open seas (p. 97); on the currents as dependent on the prevailing winds, the unequal saltness of the sea, and the configuration of the coasts (p. 139); the directions of the wind as resulting from differences of temperature, &c. Excellent likewise are the considerations on the general equinoctial current, from east to west, as the cause of the gulf-stream which begins at Cape St. Augustine and breaks forth between Cuba and Florida (p. 140). The directions of the current along the Western African coast, between Cape Verd and the island of Fernando Po in the gulf of Guinea, are most accurately described. Varenius considers sporadic islets to be "the elevated ocean-bed;"—"magna spiritum inclusorum vi, sicut aliquando montes e terra protrusos esse quidam scribunt" (p. 215). The edition of 1681, by Newton (*auctior et emendatior*), unfortunately has no additions by this great man. There is no mention of the spheroidal flattened figure of the earth, although Richer's pendulum experiments were published nine years before the Cambridge edition, but Newton's "Principia mathematica philosophiæ naturalis" was only communicated in manuscript to the Royal Society in 1686. There is much uncertainty about the native country of Varenius. According to Jöcher, he was born in England; according to the "Biographie Universelle" (tom. xlvii. p. 495), in Amsterdam: but the dedication of the Universal Geography to the burgomasters of this city shows that both assertions are equally false. Varenius expressly says that he had fled to Amsterdam, "his native town having been burnt to ashes and completely destroyed in the long war." These words appear to refer to Northern Germany, and the ravages of the 30 Years' War. Varenius likewise remarks, in the dedication of his "Descriptio Regni Japonicæ" (Amst. 1649) to the Hamburg Senate, that he had made his first studies at the Hamburg Gymnasium. It is probably incontrovertible that this acute geographer was a German, and, moreover, of Lüneburg. (Witten's Mém. Théol. 1685, p. 2142; Zedler's Universal Lexicon, 1745, part xlvi. p. 187.)

8 (20.)—Charles Ritter's Geography in relation to Nature and the History of Man, or general comparative geography.

9 (20.)—Κόσμος, in its original and proper meaning, signified *ornament* (for men, women, and horses); figuratively, *order*, εὐταξία, and *ornament of speech*. The ancients unanimously assure us that Pythagoras was the first to employ this word in the sense of *order of the world*, or *world* itself. Not having written himself, the earliest proofs are in the fragments of Philolaus (Stob. Eclog. pp. 360, 460; Heeren's Philolaos, by Boeckh, pp. 62, 90). We do not cite Timæus of Locrus, his authenticity being doubtful. Plutarch (de plac. phil. ii. 1) decidedly says that Pythagoras was the first to call the whole universe *Cosmos*, by reason of the order observed therein: (likewise Galen, hist. phil. p. 429). In its new meaning, the word passed from the philosophical school to the poets of nature and the prosaists. Plato continues to call the celestial bodies Uranos; but he still styles the order of the world Cosmos: and, in the Timæus (p. 30, B.), the universe is called a soul-endowed animal (κόσμος ζῶον ἔμψυχον). Compare, on the immaterial world-arranging spirit, Anaxagoras Claz. (ed. Shaubach, p. 111) and Plutarch (op. cit. ii. 3). With Aristotle (de Cælo, i. 91), Cosmos is, "World and its Arrangement;" is is also considered as specially divisible into the sublunary world, and the higher above the moon (Meteor. I. ii. 1, and I. iii. 13, pp. 339, *a*, and 340, *b*, Bekk.). The definition of Cosmos, cited by me in the text, is from the "Pseudo-Aristoteles de Mundo," (cap. ii. p. 391), namely: κοσμος ἐστι σύστημα ἐξ οὐρανοῦ καὶ γῆς καὶ τῶν ἐν τούτοις περιεχομένων φύσεων. Λέγεται δὲ καὶ ἑτέρως κόσμος ἡ τῶν ὅλων τάξις τε καὶ διακόσμησις, ὑπὸ θεων τε καὶ δία θεων φυζαττομένη. Most passages of the Greek writers on Cosmos are collected—1. In Richard Bentley's polemical pamphlet against Charles Boyle (Opuscula philologica, 1781, pp. 347, 445; Dissertation upon the Epistles of Phalaris, 1817, p. 254) on the historical existence of Zaleucus, the Locrian legislator: 2. In Nœke's excellent Sched. crit. 1812, pp. 9—15: and 3. In Theop. Schmidt ad Cleom. cycl. theor. met. I. i. pp. ix. 1, 99. The closer meaning of Cosmos was likewise used in the plural (Plut. i. 5), as, either every star (celestial body) was so called (Stob. i. p. 514; Plut. ii. 31), or many singular systems (world-islands) were assumed in infinite space, each having a sun and moon (Anaxag. Claz. fragm. pp. 89, 93, 120; Brandis's History of Græco-Roman Philosophy, vol. i. p. 252). As each group became a Cosmos, the universe τὸ πᾶν receives a higher signification distinct from Cosmos (Plut. ii. 1). The last word is used for the Earth only a long time after the Ptolomaic age. Böckh has communicated inscriptions in praise of Trajan and Hadrian (Corp. Insc. Græc. tom. i. Nos. 334, 1306), wherein κόσμος is used for οἰκουμένη, just as we often understand by *world* only the *earth*. The above-mentioned strange threefold division of space into Olympus, Cosmos, and Uranos, (Stob. i. p. 488; Philolaos, pp. 94—102) refers to the different regions which surround the *hearth of the universe*, the Pythagorean

'Εστία τοῦ παντός. The inner region, between the earth and moon, the realm of the *variable*, is termed Uranos in the Fragment. The middle portion, that of the unchangeable orderly circulating planets, is exclusively termed Cosmos, after a very partial view. The exterior region, a fiery one, is the Olympus. "If," says that profound diver into the affinities of language, Bopp—"if we derive *κόσμος* from the Sanscrit root *s'ud'*, *purificari*, as Pott has done (Etymol. Researches, part i. pp. 39, 252), we must regard, in respect to the sounds—1. that the Greek *κ* (in *κόσμος*) has proceeded from the palatal *s*, (expressed by Bopp with an *s'* and Pott with a *ç*,) like *δεκα*, *decem*, Gothic *taihun*, from the Indian *das'an*; 2. that the Indian *d'* regularly corresponds (Compar. Gramm. § 99) to the Greek *θ*, whence we clearly ascertain the relation of *κόσμος* (for *κοθμος*) to the Sanscrit root *s'ud'*, whence *καθαρος*. Another Indian word for *World* is *g'agat* (pronounce *dschagat*), properly meaning the *going*, as a participle from *g'a-gâmi*, I go (from the root *gâ*)." In the inner circle of Hellenic etymology, *κόσμος* is (according to Etym. M. p. 532, 12) nearest connected with *κάζω*, or rather *κάινυμαι*, whence *κεκασμέγος* or *κεκαδμεγος*. Herewith Welcker (eine Cretische Col. in Theben, p. 23) connects the name *Κάδμος*, as in Hesychius *κὰδμος* denotes a Cretan suit of armour. When the Romans introduced the philosophical technical language of Greece, they similarly employed the word *mundus*, originally used like *κόσμος* for female ornament, to express the world or universe. Ennius appears to have been the first to venture on this innovation; he says, in a fragment preserved to us by Macrobius (Sat. vi. 2), in his strife with Virgil, "mundus cæli vastus constitit silentio;" like Cicero, "quem nos lucentem mundum vocamus" (Timæus s. de univ. cap. 10). The Sanscrit root *mond*, whence Pott (Etym. Res. part i. p. 240) deduces the word *mundus*, unites both meanings of *shining* and *adorning*. *Lôka* signifies *world* and *men* in Sanscrit, like the French *monde*, and, according to Bopp, is derived from *lôk*, to *see* and *illuminate*: similarly the Slavonian *swjet* (Grimm's German Gramm. vol. iii. p. 394) is *light* and *world*. This word *Welt*, which the Germans now use, old High German *wëralt*, old Saxon *worold*, Anglo-Saxon *vëruld*, originally denotes, according to Jacob Grimm, only "the idea of time, *sæculum* (age of man), not the spacial *mundus*." Amongst the Tuscans, the open *mundus* meant an inverted dome, which turned its cupola towards the world below, and imitated the heavenly vault.—(Otf. Müller's Etruscans, part ii. pp. 96, 98, 143.) In its narrower telluric signification, the world appears in the Gothic language as the sea-(*marei*, *meri*) surrounded horizon, as *merigard*, a sea-garden.

10 (20.)—*Vide* about Ennius, Leopold Krahner's acute researches in his "Grundlinien zur Geschichte des Verfalls der Römischen Staats-Religion," 1837, pp. 41—45. Probably Ennius did not draw from the Epicharmic pieces, but from poems which went by the name of Epicharmus, and were written according to his system.

11 (p. 21.)—Gellius, Noct. Att. v. 18.

12 (p. 23.)—Schelling's Bruno on the Divine and Natural Principle of Things, p. 181.

PICTURE OF NATURE.

GENERAL SURVEY OF NATURAL PHENOMENA.

When the human mind essays to dominate matter—in other words, to comprehend the world of physical phenomena—when we strive, in thoughtful contemplation of existing things, to penetrate the life of Nature in its ample fulness, and to unveil the empire of her various forces, we feel ourselves raised to an eminence, whence, in the wide-spread horizon around, individualities present themselves gathered into groups, and surrounded with a kind of vaporous haze. This figurative language is used to give some idea of the point of view from which we shall here attempt to survey the universe, and to present it for contemplation in both of its spheres—the celestial and the terrestrial. The boldness of such an attempt I do not conceal from myself. Of all the kinds of representation to which these pages are dedicated, that of the General Picture of Nature is by far the most difficult. Here we do not condescend upon the minutiæ of individual forms; we only pause upon the grander masses, whether in the world of fact or of idea. By separation and subdivision of phenomena, by a kind of foreboding penetration of the play of obscure forces, by liveliness of representation, in which the impression made on the senses is reflected true to nature, may we hope to grasp and to describe the Infinite All (τὸ πᾶν), in a way that shall become the grand word Cosmos, in its sense of Universe, Order of Creation, Beauty of Arrangement. May the infinite diversity of elements that crowd into a picture of Nature so vast, not disturb the harmonious impression of repose and unity, which it is the last purpose of every literary and artistical composition to convey!

We begin with the depths of space, and the region of the farthest nebulæ; we descend, step by step, through the stratum of stars to which our solar system belongs, and at length set foot on the air- and sea-surrounded spheroid we inhabit, discussing its form, its temperature, and its magnetical tension, till we reach the LIFE, that, under the stimulus of light, is evolved upon its surface. A picture of the universe, therefore, worked with a few grand touches, comprehends the immeasurable depths of space, as well as the microscopic organisms of the vegetable and animal kingdom that live in our stagnant waters, and cling to the weatherworn faces of our rocks. All that the most careful study of nature, in its present direction, up to the passing hour, has discovered, constitutes the material in harmony with which the canvass is to be filled; it includes within itself the evidence of its truth and endurance. A descriptive natural picture, however, such as we would indicate it in these prolegomena, must not present all the individual, all the single; it needs not, to be complete, an enumeration of all the forms of life, of every natural thing and natural process. Striving against the tendency to endless subdivision of the Known and the Collected, the thinker who orders and arranges must rather seek to escape the danger of empirical overabundance. A considerable mass of the qualitative forces of matter, or, to speak in the language of the philosophers of nature, of its qualitative manifestations of force, is certainly still unknown. The discovery of unity in totality must, therefore, and on this account, remain imperfect. Beside the joy, mixed as it were with wo, which we feel in knowledge possessed, there dwells in the eager spirit, unsatisfied with the present, the longing after yet untrodden, yet unimagined, regions of knowledge. But such a longing only knits more firmly the bond which, in virtue of ancient laws, controlling the very core of the world of thought, binds the Sentient with the Supersentient; it vivifies the commerce between that "which the mind receives from the world without, and that which, from its own depths, it gives back."

If nature, therefore, or the conception formed of natural things and natural phenomena, considered in its boundary and contents, be infinite, so is she also, with reference to the intellectual powers of man, an incomprehensible, and, in the general causal co-operation of her forces, an unresolvable problem. Such an avowal is proper where existence and evolution (Being and to Be) are only subjected to immediate scrutiny, in circumstances where the empirical path, and the strictly inductive method, cannot be quitted for a moment. But if the ceaseless longing to comprehend nature in its totality remain unsatisfied, the history of human progress in contemplating nature, which is reserved for another section of the prolegomena, teaches us, on the other hand, how, in the course of centuries, mankind have gradually attained to a partial insight into the relative dependence of phenomena. It is my duty to pass in review the contemporaneously known, according to the measure and the boundaries of the present. In all that is mobile, changeable in space, mean numerical values are the ultimate object—they are the expression, indeed, of physical laws; they show us the stable in the change and in the flight of phenomena. The progress of our modern measuring and weighing physics is particularly distinguished by the attainment and correction of the mean values of certain quantities or masses; and here, as dwelt on by the old Italic school, but in a wider sense, we find those wide-spread, hieroglyphic signs, numbers, coming into play as powers in Cosmos.

The serious inquirer rejoices in the simplicity of numerical relations, by which are indicated the dimensions of the celestial spaces, the magnitude of the bodies they enclose, and the periodic perturbations which these suffer; the threefold elements of terrestrial magnetism; the mean pressure of the atmosphere, and the quantity of heat which the sun dispenses in the course of every year, and in each division of the year, over the several points of the solid or liquid surface of our planet. The poet of nature is less satisfied with such results; the appetite for the marvellous, inherent in the many, is less appealed to by them. The poet complains that science has made a desert of nature; the vulgar find many questions returned to them with doubtful solutions, or declared unanswerable, which formerly were met without misgivings. In her graver form, in her less ample garments, she is robbed of that seducing grace by which the dogmatic and symbolic physics of former times sought to deceive the reason, to occupy the imagination. Long before the discovery of the New World, it was thought that land could be seen in the West from the Canaries and the Azores. These were phantasms, not produced by any extraordinary refraction of rays of light, but merely by a longing for the distant, for that which lies beyond the present. The natural philosophy of the Greeks, and the physics of the middle ages, and even of much later centuries, presented swarms of such fantastic forms to the imagination. The mental eye still essays to pass the horizon of limited knowledge, even as the material eye endeavours to pierce the natural horizon from an island height or shore. Faith in the unusual and wonderful gives definite outlines to every product of imagination, and the realm of fancy, a strange land of cosmological, geognostical, and magnetic dreams, is incessantly blended with the world of reality.

Nature, in the manifold significance of the term, now as implying entireness of that which is, and is becoming; now as an inherent actuating force; and again, as the mysterious prototype of all phenomena, reveals itself to the simple sense and feeling of mankind as something more especially terrestrial, as something that is near akin to them. We seem at first to recognize our proper home in the living circle of organic formation. Where the bosom of the earth is adorned with fruits and flowers, where it supports and nourishes innumerable kinds of animals, there does the image of nature come up in living tints before the soul. We are more immediately connected with the earth, with the terrestrial; the canopy of heaven, inlaid with shining stars, the boundless realms of space, belong to a picture, the magnitude of whose elements—hosts of suns, glimmering nebulous specks, infinity of space—arouse our wonder and amazement, indeed, but still remain foreign to our mind and feelings, through a sense of desolation, and a total want of immediate impression through the presence of organic life. To mankind at large, therefore, the heaven and the earth have still remained distinct, as the above and the below in space, in consonance with the earliest notions entertained on the subject. Were a picture of nature at large, then, solely intended to meet the requirements of sense, it would have to be begun with a representation of our proper home for a foundation. It would first portray the earth in its dimensions and configuration, in its increasing density and temperature as its centre was approached, in its solid and fluid superposed strata; it would exhibit the severance of sea and land; the life which in both is evolved as cellular tissue in plants and animals; and the atmospheric ocean, with its waves and currents, from the bottom of which wood-crested mountain-chains emerge like reefs and shoals. After this exhibition of purely terrestrial relations, the eye would rise to the celestial spaces; the earth, the well-known seat of organic formative processes, would now be contemplated as a planet. It would fall into the series of bodies which circulate around one of the innumerable host of self-effulgent stars. This sequence of ideas indicates the path pursued in the first contemplation of nature by the senses; it still reminds us of the "sea-surrounded disc of earth," which supported heaven: it sets out from the station of simple perception, from the known and the near, to the unknown and the far removed. It corresponds with the method observed in our elementary astronomical works, which pass from the apparent to the true motions of the heavenly bodies.

In a work, however, which undertakes to speak of the actually known, of that which, in the present state of science, is held for certain, or which, in various degrees, is looked upon as probable, but which does not propose to give the details upon which results are founded, another course of procedure appears advisable. Here we do not set out from the subjective point of view, from that which regards human interests. The terrestrial can only appear as a part in the whole, and as subordinate to this. The view taken of nature must be general, it must be grand and free, not contracted by notions of vicinity, of affection, of relative usefulness. A physical cosmography, or true picture of the universe, cannot, therefore, commence with the terrestrial; it must needs begin with the contents of heavenly space. But as the spheres of contemplation, in reference to space, contract, the amount of individual details, the variety of physical phenomena, knowledge of the qualitative heterogeneousness of matter, augment. From regions in which we can only distinguish the empire of the universal laws of gravitation, we descend to our planet, to the intricate play of forces that constitute the life of the globe. The natural descriptive method now sketched out is opposed to that which establishes conclusions. The one enumerates what the other demonstrates.

Man assumes the external world into his interior by means of certain organs. The phenomena of light make us aware of the existence of matter in the farthest depths of heaven. The eye is the organ by which the universe is perceived, and the discovery of telescopic vision some century and a half ago has conferred a power upon later generations whose limits have not yet been reached. The first and most general consideration in Cosmos is that of the contents of space, the contemplation of division in matter, of Creation, as we are accustomed to designate all that is or is about to be. We perceive matter here aggregated into revolving and circulating masses of most dissimilar density and magnitude; there diffused in the shape of self-luminous clouds or vapours. If we first turn our attention to these NEBULÆ (WORLD-MISTS, separating into determinate forms), we discover that they are in course of suffering change in their state of aggregation. They present themselves to our eyes apparently of small dimensions, as rounded or elliptical discs, single or in pairs, occasionally connected by a luminous streak; of larger size they are variously shaped—elongated or shooting out into several branches; or they look fan-shaped; or they form sharply defined rings with dark included centres. These nebulæ are believed to be in process of various and progressive changes, according as the star-dust or vapour composing them is becoming condensed, in harmony with the laws of attraction, around one or several nuclei. The number of these unresolvable nebulæ—nebulæ in which the most powerful telescope does not enable us to distinguish a single star—that have been reckoned, and their position in space determined, now amounts to about one thousand five hundred.

The genetic evolution, the ceaseless, progressive formation that appears to be going on in these portions of infinite space, has led reflective minds to the analogy of organic phenomena. As in our woods we observe the same kind of tree in every stage of growth at the same time, and from this view, this co-existence, derive the impression of progressive vital development; so do we, in the mighty garden of the universe, recognise different stages in the progressive formation of stars. The process of condensation, indeed, which Anaximenes and the Ionic school once taught, seems here to proceed, as it were, under our eyes. This object of inquiry and conjecture is peculiarly attractive to the imagination. That which, in the circles of life, and in all the internal impulsive forces of the universe, fetters us so unspeakably, is less the recognition of Being, than of what is About to be; even though the latter be nothing more than a new condition of matter already extant; for of proper creation as an efficient act, of a protogenesis of matter, of entity succeeding nonentity, we have neither conception nor experience.

It is not merely by a comparison of the various moments of development which are exhibited by nebulæ, in greater or less degrees of condensation of their interiors, that astronomers have inferred changes in their structure. We have now a series of observations made immediately upon particular nebulæ, on the one in Andromeda, on that which occurs in the ship Argo, and also in the flocky portion of that which presents itself in Orion, which led to the belief that actual changes in their form have been observed. Inequality of power of light in the instruments employed, however, different states of our atmosphere, and other optical conditions, it must be admitted, render a portion of these results questionable as true historical data.

The peculiar multiform nebulæ, the several parts of which have different degrees of brightness, and which, with a diminution of their areas, will perhaps become concentrated into stars, and those nebulæ that have been entitled planetary, the round or somewhat oviform discs of which shine in every part with a mild and equable light, are not to be confounded with nebulous stars. Here there is no appearance of a star projected accidentally, as it might seem, upon a remote nebulous ground; no, the vaporiform matter, the light-cloud, forms a single mass with the star which it surrounds. From the frequently very considerable magnitude of their apparent diameters, and the distances whence they glimmer, both planetary nebulæ and nebulous stars must possess enormous dimensions. New and acute considerations([1]) on the very different influence of distance upon the intensity of the light of a disc of measurable diameter, or of a single self-luminous point, make it not improbable that planetary nebulæ are extremely remote nebulous stars, in which the distinction between the central star and its hazy envelope has disappeared even to our telescopic vision.

The brilliant zones of the southern celestial hemisphere, between the parallels of 50° and 80°, are particularly rich in nebulous stars, and concentrated but unresolvable nebulæ. The Magellanic clouds which circulate round the starless, desolate south pole (especially the larger of the two), appear, according to the latest observations([2]), "as a wonderful mixture of groups of stars, of globular clusters of nebulous stars of different magnitudes, and of unresolvable nebulæ, which, producing a general brightness of the field of vision, form a kind of back-ground to the picture." The aspect of these clouds, of the light-streaming ship Argo, of the milky way between the Scorpion, the Centaur, and the Cross—the whole of the charming landscape presented by the southern heavens, has left an indelible impression upon my mind. The zodiacal light, which rises like a pyramid from the sun, and in its gentle radiance proves another of the eternal ornaments of the tropical night, is either an immense nebulous ring rotating betwixt the earth and Mars, or (but this is less probable) it is the outermost stratum of the sun's atmosphere itself. Besides these luminous clouds and nebulæ of determinate form, accurate and still coinciding observations seem to proclaim the existence and general diffusion of an infinitely rare, and apparently not self-luminous matter, which, OFFERING RESISTANCE, reveals itself by lessening the eccentricity, and shortening the period of revolution of Encke's, and perhaps also of Biela's comet. This impeding æthereal and cosmic matter may be conceived as in motion, despite its original tenuity as gravitating, as condensed in the vicinity of the great body of the sun, and even as increased in the course of myriads of years, by vapours thrown off from the tails of comets.

If we now pass on from the nebulous matter of the infinities of heavenly space (οὐρανοῦ χόρτος[3]), here scattered without form or boundary, a cosmic world-ether, there condensed into nebulous specks, to the conglobated solid portions of the universe, we approach a class of phenomena which are exclusively designated by the title of stars, or fixed stars. And here, too, the degree of solidity or density of the conglobated matter is different. Our own solar system presents us with every grade of mean density; in other terms, of difference betwixt the relations of volume and mass. When we compare the planets from Mercury to Mars with the Sun and with Jupiter, and Mars and Jupiter, again, with Saturn, we proceed in a descending scale of density; selecting familiar objects as standards of comparison, from matter of the density of antimony, to matter of the density of honey, of water, and of pine timber. In Comets, which, numerically speaking, constitute the largest portion of the individualized physical forms of our solar system, the most concentrated part, which we call nucleus or head, still allows the light of the stars to pass through it unrefracted. The mass of comets, perhaps, never exceeds the five-thousandth part of the mass of the earth: so variously do the formative processes meet us in original and perhaps progressive conglobations of matter. Setting out from what is most general, it was especially necessary to indicate this diversity, not as a thing possible, but as a reality—as a datum in universal space.

What Wright, Kant, and Lambert have deduced from the conclusions of pure reason, in regard to the construction of the universe, to the distribution of matter in space, has been established by Sir William Herschel upon the securer basis of observation and measurement. This great, inspired, and yet cautious observer, first cast the plumb-line into the depths of heaven, to determine the boundaries and the form of the separate cluster of stars which we inhabit; and he was the first who ventured to offer an explanation of the relations in point of position and distance, of remote nebulous specks to our own astral system. William Herschel, as the elegant inscription on his monument, at Upton, says so happily, "broke through the barriers of the heavens (*cælorum perrupit claustra*)." Like Columbus, he forced his way into an unknown ocean, and caught a glimpse of coasts and groups of islands whose true position it is reserved for future centuries to determine.

Considerations on the varying intensity of the light of the stars, and on their relative numbers—in other words, their numerical abundance or rarity in equal fields of the telescope—have led to inferences concerning the unequal distances and distribution in space of the strata which they compose. Such inferences, considered as leading to circumscription of the several portions of the universe, do not, however, admit of the same degree of mathematical certainty as is attained in all that concerns our own solar system; the revolutions of double stars, with unequal velocities, around a common centre of gravity, and the apparent or actual motions of the stars in general. We are almost disposed to compare the chapter in our physical cosmography which discusses the nebulous specks of heaven, with the mythological portion of general history. They both begin alike—the one in the twilight of remote antiquity, the other in the depths of illimitable space; and where reality threatens to disappear, fancy is doubly excited to draw from her own abundance, and to give form and endurance to the Indefinite and the Changeable.

If we compare the universe with one of the isle-studded oceans of our planet, we think that we can perceive matter distributed groupwise: now, collected into unresolvable nebulous specks of various age; now condensed around one, or several, nuclei, and again rounded into clusters of stars, or isolated sporades. The cluster of stars, the islet in the infinity of space, to which we belong, forms a lenticular, compressed, and everywhere distinct or separate layer, the longer axis of which has been estimated at from seven to eight hundred, and the shorter axis at some one hundred and fifty, distances of Sirius. Presuming that the parallax of Sirius is not greater than that of the bright star in the Centaur, which has been accurately ascertained (viz. 0″·9128), light would pass through one distance of Sirius from the Earth in three years, whilst, from Bessel's admirable earlier paper(*) on the parallax (0″·3483) of the remarkable star in Cygnus (the 61st), the very distinct proper motion of which must admit of a very close approximation, it follows, that the light of this star only reaches us after travelling through space for some nine years and a quarter. Our stratum of stars, a disc of relatively moderate thickness, is divided, through one-third of its extent, into two arms; and it is thought that we are placed somewhat near to this division—nearer to Sirius than to the constellation of the Eagle, almost in the middle of the material extension of the layer, in the line of its thickness, or lesser axis.

This position of our solar system, and the formation of the whole lens, are deduced by means of a process of what has been aptly designated gauging the heavens; *i. e.* reckoning the number of stars included in the same field of the telescope turned on every side around. The increasing, or decreasing, number of stars measure the depth or thickness of the layer in different directions. Precisely as the point at which the plummet strikes the bottom determines the length of the line that it is cast from the hand, do these soundings of the heavens give the lengths of the visual ray, when the bottom of the starry depths, or rather, and more correctly, as there is neither above nor below here, when the limits of starry space are attained. In the direction of the longer axis, and where the greatest numbers of stars lie one behind another, the eye perceives the farthest off thickly crowded together, connected, as it seems, by a milky glimmer (light-mist), and projected, in perspective, upon the visible vault of heaven in the form of a belt or girdle. This narrow belt of beautiful, but unequal radiance, for its continuity is broken by less luminous spaces, divides into two branches, and, save where it is interrupted for a few degrees, forms a great circle upon the hollow sphere of the heavens. This is in consequence of the position of our system, near the middle of the great astral group to which it belongs, and almost in the plane of the milky way itself. Were our planetary system placed far without the cluster, the milky way would present itself to the assisted eye as a complete ring, and at a still greater distance, as a resolvable disc-shaped nebula.

Amongst the many self-luminous bodies, erroneously designated fixed stars, for they are all in motion, which constitute our island in the universe, our sun is the only one which we know, through actual observation, as a central body in reference to the conglobated masses of matter, in the shape of planets, comets, and aërolitic asteroids, which revolve around, and immediately depend upon him. Among the multiple or double stars or suns, in so far as their nature has yet been studied, there does not appear to reign the same planetary dependence, in respect of relative motion and illumination, which characterizes our solar system. Two or more self-luminous stars, whose planets and moons—if any such exist—escape our present telescopic powers of vision, revolve unquestionably around a common centre of gravity; but this centre falls in a space that perchance is filled with unaggregated matter (world-mist), whilst with our sun it is always situated in the inner confines of a visible central body. When we consider our sun and earth, or our earth and moon, as double stars, and our whole planetary system as a multiple group of stars, the analogy with the proper multiple or double fixed stars, which such a designation presents to the mind, extends no farther than to motions connected with systems of attraction of different orders, quite independently of light evolving processes, and kinds of illumination.

In this generalization of cosmic views, which befits the sketch of a Picture of Nature or the Universe, the solar system to which our earth belongs may be considered in a two-fold relationship: immediately, to the several classes of individualized conglomerate matter—to the magnitude, the fashion, the density, and the distance of the bodies of the system; and, next, in its relations to other parts of our astral system, to the sun's change of place within the same.

The solar system, in other words, the very variously fashioned matter which circulates about the sun, consists, according to our present knowledge, of eleven principal planets, of eighteen moons or satellites, and of myriads of comets, three of which, called planetary comets, never quit the limited spheres of the proper planets. We may further, with no slight show of propriety, reckon as falling within the empire of our sun, as included within the sphere of his central force—1st, a ring of vaporous matter, revolving, in all probability, betwixt the orbits of Venus and Mars, certainly extending beyond the orbit of the earth(5), which is visible to us in a pyramidal form, and is known under the name of the zodiacal light; 2d, a host of very small asteroids, whose orbits either intersect the orbit of the earth, or approach it very nearly, and give occasion to the phenomena of aërolites and falling stars. When we direct our attention to the complexity of formations which circulate about the sun in orbits more or less excentric, unless, with the immortal author of the "Méchanique Céleste," we regard the greater number of comets as nebulous stars which sweep from one central system to another(6), we must confess, that the planetary system, strictly so called—the group of bodies which revolve, with their attendant satellites, in but slightly excentric orbits round the sun—constitutes but a small portion of the entire system, when the number, not the mass, of the individuals is made the basis of consideration.

The telescopic planets, Vesta, Juno, Ceres, and Pallas, with their mutually intersecting, much inclined, and more excentric orbits, have been viewed as constituting a kind of zone of separation between two divisions of our planetary system, and as forming in themselves a middle group. According to this view, the inner planetary group, comprising Mercury, Venus, the Earth, and Mars, presents several remarkable points of contrast with the outer group, consisting of Jupiter, Saturn, and Uranus(7). The inner planets, nearer to the sun, are of moderate dimensions, of greater density, turn more slowly upon their axes, and very nearly in the same period of time (twenty-four hours), are flattened towards their poles in a less degree, and, with one exception, are unaccompanied by moons. The outer, and, from the sun, more distant planets, are vastly larger, of but one-fifth of the density, more than twice as rapid in their periods of rotation about their axes, flattened towards their poles in a much greater degree, and attended by a far larger number of moons; in the ratio of 17 to 1, if Uranus have actually so many as six satellites.

These general observations on certain characteristic peculiarities of the two great groups, are not, however, precisely or in all respects applicable to the particular planets of each group; for example, to the ratios of their absolute magnitudes, to their distances from the central body, to their densities, to the times of their rotations on their axes, to their excentricities, and to the inclinations of their orbits and of their axes. We know as yet of no intimate necessity, of no mechanical natural law, like the beautiful law which connects the squares of the times of revolution with the cubes of the greater axes, which makes the six elements of the planets just indicated, and the form of their orbits, dependent on one another, or on their mean distances. Mars, more remote from the sun, is smaller than the Earth or Venus; he approaches Mercury—the nearest of all the known planets to the sun—most closely in his diameter; Saturn, again, is smaller than Jupiter, yet much larger than Uranus. The zone of the telescopic planets, so insignificant in point of volume, lies, in a series of distances setting out from the sun, immediately before Jupiter, the most considerable of all the planetary bodies; and yet these asteroids, several of whose discs can scarcely be measured, are barely one half more in their superficies than France, or Madagascar, or Borneo. Again, however remarkable the very small density of all the colossal planets that lie farthest from the sun, there is still nothing like a regular sequence among them(8). Uranus appears to be more dense than Saturn, even when Lamont's smaller mass, $\frac{1}{24605}$, is adopted; and although the differences in point of density of the inner group of planets (9), are insignificant, we still find Venus and Mars, on either side of the Earth, of less density than itself. The time of rotation decreases, it is true, with the distance from the sun; but for Mars it is relatively greater than for the Earth, and for Saturn it is greater than for Jupiter. The greatest excentricities in the elliptical orbits of any of the planets, occur in those of Juno, Pallas, and Mercury; the least in those of Venus and the

Earth, the two planets which follow each other immediately. Mercury and Venus present the same contrast in the excentricity of their orbits which we observe in the four so closely allied asteroids. The excentricities of Juno and Pallas, which are very nearly alike, are three times greater than those of Ceres and Vesta. It is the same with reference to the inclination of the planetary orbits to the plane of projection of the ecliptic, and to the position of the axes of rotation on their orbits, this position influencing climate, season, and length of day, still more than excentricity. The planets which have the most elongated elliptical orbits, Juno, Pallas, and Mercury, are also inclined in the greatest degree, although not in equal measure, to the ecliptic. The orbit of Pallas is almost comet-like, and its inclination is nearly twenty-six times greater than that of Jupiter; while the orbit of the little Vesta, which is so near to Pallas, scarcely exceeds the angle of inclination of the orbit of Jupiter six times. The positions of the axes of the four or five planets, whose axes of rotation are known with any degree of certainty, also offer nothing like regularity of series. Judging from the position of Uranus's satellites, two of which (the 2d and 4th) have recently been certainly seen again, we should say, that the axis of Uranus, the outermost of all the planets, was scarcely inclined 11° to the plane of his orbit; but Saturn, whose axis of rotation almost coincides with the plane of his orbit, revolves between Jupiter, whose axis is nearly perpendicular, and Uranus, where, as we have seen, it is but little inclined.

The world of planetary formations, in this brief enumeration of the relations of these bodies in space, is assumed as a fact, as a thing that exists in nature, not as an object of intellectual intuition, of internal causally-founded concatenation. The planetary system, in its relations of absolute magnitude and relative position of axis, of density, time of rotation, and different degree of excentricity of orbit, does not strike us as naturally more necessary, than is the measure of separation between the land and the sea on the surface of our planet, than are the outlines of its continents, or the heights of its mountain-chains. In this respect there is no general law discoverable either in celestial space, or in the inequalities of our earth's surface. The things that we meet with are facts in nature, which have proceeded from the conflict of multifarious forces in operation under former and unknown conditions. But in formation of the planets, man sees as accidental what he is incapable of explaining genetically. If the planets have been formed out of separate rings of vaporous matter circulating round the sun, differences in the density, the temperature, and the electro-magnetic tension of these rings, may have given rise to the most diverse fashions of the conglobated matter; in the same way as the amount of the velocity of projection, and trifling aberrations in the direction of the projection, may have given rise to manifold forms and inclinations of the elliptical orbits. The attraction of masses, and the laws of gravitation, have undoubtedly been at work here, as in the geognostic relations of continental upheavings; but we are not to draw conclusions from the present state of things, as to the entire series of conditions which have been passed through from their commencement. Even the law, as it has been styled, of the distances of the planets from the sun, the progression from the failing member in which Kepler was led to suspect the existence of a planet betwixt Mars and Jupiter, has been found incorrect numerically for the distances between Mercury, Venus, and the Earth, and because of a supposed first member, inapplicable to the idea of a regular series.

The eleven principal planets which have been discovered circulating round the sun, are accompanied by at least fourteen, and very probably by eighteen, secondary planets (satellites or moons). The primary planets are therefore, in their turn, central bodies with reference to subordinate systems. And here, in the structure of the universe, we recognize the same formative process which the evolution of organic life so often exhibits to us in the extremely complex groups of animals and plants, in the typical repetition of forms of subordinate spheres. The secondary planets, or moons, occur in larger numbers in the outer region of the planetary system, in connection with the three great planets that lie without the zone formed by the four telescopic planets. With the single exception of the earth, all the planets within this zone are moonless, and the satellite of the earth is relatively of very large dimensions, inasmuch as its diameter amounts to one-fourth of that of the earth; whilst the largest of all the secondaries known, the sixth of Saturn, is not more perhaps than the $\frac{1}{17}$th, and the largest of Jupiter's moons, the third, is not above $\frac{1}{28}$th the diameter of its primary. The planets which have the greatest number of moons are the most remote, and they are, at the same time, the largest, the least dense, and the most flattened at the poles. The late measurements of Mädler seem to indicate Uranus as the planet which is flattened towards the poles in the greatest degree, $\frac{1}{9.22}$. In the earth and her moon, whose mean distance from one another amounts to 237,000 English miles, the differences in the masses and the diameters of the two bodies are much smaller than we are accustomed to meet with them in the primary and secondary planets, and bodies of a different order in the solar system[10]. Whilst the density of the earth's satellite is $\frac{5}{9}$ths less than that of the earth itself, it would appear, supposing we can depend on the determinations that have been come to on the magnitudes and the masses of the satellites, that of the moons which attend upon Jupiter, the second is denser than the primary planet.

Of the fourteen satellites the relations of which have been determined with something like accuracy, the system of Saturn presents instances of the most remarkable contrast in the absolute magnitudes and distances from the primary. The sixth satellite of Saturn is probably not much smaller than Mars, whilst the earth's moon is only one-half the diameter of this planet. Next in order, in point of volume, to the two outermost satellites of Saturn (the sixth and the seventh), comes the third and brightest of the moons of Jupiter. On the other hand, the two innermost satellites of Saturn, which were discovered by Sir William Herschel, in 1789, with his great 40 foot telescope, and which have been again seen by Sir John Herschel at the Cape, by Vico at Rome, and by Lamont at Munich, belong, in common with the satellites of Uranus, to the smallest of the visible bodies that enter into the constitution of our solar system. These satellites, indeed, are only to be seen under peculiarly favourable circumstances, and with the most powerful telescopes. All determinations of the true diameters of satellites, deductions of these from measurements of the apparent magnitudes of small discs, are exposed to many optical difficulties; and physical astronomy, which calculates before-hand, and with such admirable precision, the motions of the heavenly bodies, as they are exhibited from our place of observation, the earth, is more concerned about motion and mass, than volume.

The absolute distance of a satellite from its primary, is greatest in the case of the outermost or seventh satellite of Saturn, which is half a million of geographical miles* remote, or ten times as far as the distance of our moon from the earth. In reference to Jupiter, the outermost or fourth satellite is no more than 260,000 geographical miles* from the planet; the fifth satellite of Uranus, however, if it actually exist, must be at the distance of 340,000 miles.

On comparing, in each of these subordinate systems, the volume of the primary planet, with the distance of the farthest orbit in which a satellite has been formed, we discover totally dissimilar numerical relations. Expressed in semidiameters of the principal planets, the distances of the farthest satellites of Uranus, Saturn, and Jupiter, are as 91, 64, and 27. Saturn's outermost satellite, therefore, is but a very little ($\frac{1}{15}$th) more remote from the centre of the primary than our moon is from the earth. The satellite that approaches its primary most closely, is undoubtedly the first or innermost of Saturn, which, in addition, presents the only instance of a revolution in less than 24 hours. The distance of this satellite from Saturn's centre, according to Mädler and Beer, expressed in semidiameters of the primary, is only 2·47, or 20,022 geographical miles.* This satellite cannot, therefore, be distant from the surface of its primary more than 11,870 g. miles; and from the outer edge of the ring, only 1,229 g. miles. One who has been a traveller readily forms an idea of so short a distance, the more so when he thinks of that bold seaman, Captain Beechey, having sailed over 18,200 geographical miles in the course of three years. Recurring to semidiameters of the primary as measures of distance, we find that the first or innermost satellite of Jupiter is no more than six semidiameters of the planet from his centre; our moon, on the contrary, is 60¼ semidiameters of the earth from its centre. The first satellite of Jupiter is, nevertheless, 6,500 miles farther from his centre, than our moon from the centre of the earth.

In the subordinate systems of the satellites, in other respects, all the laws of gravitation are reflected that have been established in connection with the sun and the primaries which revolve around him. The twelve satellites of Saturn, Jupiter, and the Earth, all revolve, like the primary planets, from west to east, and in elliptical orbits, which differ but little from circles. It is only the moon, and probably the first, or innermost satellite of Saturn (0·068), which have orbits, whose eccentricity surpasses that of Jupiter. Bessel's very accurate observations on the 6th satellite of Saturn show that the excentricity here (0·029), exceeds that of the Earth.

It is only in connection with the satellites of Uranus, on the extreme limit of the planetary system, at nineteen times the distance of the earth from the sun, and where his central force must be notably diminished, that we find any thing like contrasts to admitted laws. Instead of moving, like all the other satellites, in orbits but little inclined to the ecliptic, and from west to east, (the ring of Saturn, a kind of fused or undivided satellite, not excepted), the moons of Uranus revolve in planes nearly perpendicular to the ecliptic, and, as Sir John Herschel has found, after many years of observation, in retrograde courses from east to west. If the primary and secondary planets of our system have actually been formed out of rotating rings of vapour, by condensations of former solar and planetary atmospheres, there must have been strange, and to us altogether inconceivable conditions of retardation or counteraction among the vaporous rings that revolved around Uranus, to have brought about such a singular opposition to the motions of the central body as we observe in his 2d and 3d satellites.

It is highly probable, that the period of rotation of all the satellites is the same as their period of revolution, so that they still keep the same side turned towards their primaries. Inequalities, as a consequence of slight variations in the revolution, nevertheless, occasion oscillations of from 6 to 8 degrees—an apparent libration—both in longitude and latitude. We therefore actually see, in succession, more than one half of the surface of the moon; at one time more of her eastern and northern, at another more of her western and southern limb. By the libration(11) the annular mountain Malapert, which the south pole of the moon covers at times, is made more visible to us, and then we obtain a better view of the arctic landscape around the mountain-crater, Gioja, as also of the extensive grey level near Endymion, which surpasses the Mare vaporum in superficial extent. In spite of all this, however, three-sevenths of the moon's surface remain, and, unless some new and unexpected cause of perturbation interferes, will ever remain withdrawn from our eyes. These cosmic relations remind us, involuntarily, of a nearly similar position of things in the intellectual world, in the products of thought, where, in the deep investigation of the dark elaboratory of nature and the prime creative power, there are also regions turned from our ken, and that seem unattainable, though, in the course of thousands of years, mankind have, from time to time, caught a glimpse of some narrow stripe or margin, now in a true and steady, now in a more false and flickering light.

We have hitherto regarded the principal plan-

* The miles are always German geographical miles, 15 to a degree of the Equator.—TRANSLATOR.

ets, their satellites, and the concentric ring that belongs to at least one of the outermost of them, as products of a projectile force, and as connected with one another by intimate bonds of mutual attractions.

We have still to speak of COMETS, an innumerable host, which revolve around the sun in definite orbits, and from him derive their light. When we estimate the relative lengths of the orbits of these bodies, the boundaries of their perihelia, and the great likelihood of their remaining invisible to the inhabitants of the earth, by the rule of probabilities, we find that they must amount to such myriads as makes the imagination pause amazed. Kepler, with the liveliness of expression that distinguished him, says, that there are more comets in the depths of space, than there are fishes in the bosom of the ocean; and yet we have scarcely the accurately-computed orbits of some 150 of the six or seven hundred of these bodies, upon whose appearance and course through known constellations we have indications more or less rude. Whilst the classic nations of the west, the Ancient Greeks and Romans, occasionally give the place in the heavens where a comet was first seen, but never say a word of its apparent course, the ample literature of the Chinese, those accurate observers of nature and of individual things, contains circumstantial notices of the constellations through which each comet passed. These notices extend to more than five hundred years before the commencement of the Christian era, and many of them are used by astronomers at the present day(12).

Of all planetary bodies, comets are those which, with the smallest masses, occupy the largest fields of space. The particular observations that have hitherto been made upon them, indicate masses much under the $\frac{1}{5000}$th of that of the earth; yet have these bodies tails, which often extend over many millions of miles, both in length and breadth. The light-reflecting tail, or cone of vaporous matter which comets emit, has occasionally been observed to be as long as is the distance of the earth from the sun, a line which intersects the orbits of two of the planets, those of Mercury and Venus. This was the case with the remarkable comets of 1680 and 1811; and it is even probable that our atmosphere was mingled with the vapour of the comets' tails of the years 1819 and 1823.

Comets exhibit such variety of forms or appearances, often appertaining to the individual rather than to the kind, that a description of one of these travelling LIGHT-CLOUDS—for so they were called by Xenophanes and Theon of Alexandria, the contemporaries of Pappus—can only be applied with certain precautions to another. The feeblest telescopic comets are generally without any visible tail, and resemble the nebulous stars of Herschel. They appear as rounded, palely-glimmering nebulæ, with the light stronger or more concentrated towards the middle. This is the simplest type; but it is even as little a rudimentary or nascent type on this account, as it is a type of a planetary body grown old, and become exhausted by exhalation. In larger comets we distinguish a head, or nucleus, as it is commonly called, and a simple or compound tail, which the Chinese astronomers entitle, very characteristically, the brush (sui). In general the nucleus has no definite outline, although, in some cases, it has the splendour of a star of the first or second magnitude; and in the great comets of 1402, 1532, 1577, 1744, and 1843, it had such brilliancy, that it could be seen in bright sunshine(13). This last circumstance seems to testify to the existence, in some members of the family at least, of greater density and a highly reflective faculty in the mass. But no more than two comets have yet been seen, which, in Herschel's great telescope, presented well-defined discs(14); these two are the one of 1807, discovered in Sicily, and the magnificent one of 1811. The disc of the former appeared under an angle of 1″, that of the latter under an angle of 0″·77, from which an actual diameter of 134 and 107 miles respectively is obtained. The less precisely defined nuclei of the comets of 1798 and 1805, indicated diameters of no more than 6 or 7 miles. In several comets that have been accurately observed, particularly in the one of 1811, mentioned above, and that was seen so long, the nucleus, and the misty envelope which surrounded it, were wholly separated from the tail by a darker space. The intensity of the light of the nucleus does not go on increasing continuously towards the centre; bright zones are repeatedly separated by concentric misty envelopes. The tail, as stated, has appeared now single, now double; but rarely, although this was the case in the comets of 1809 and 1843, of very different lengths in the two branches; one comet, that of 1744, has appeared, which had six tails. The tail, again, is either straight or curved, now to both sides, now outwardly (1811), or convex to the side towards which the comet is tending (1618); occasionally the tail has been waving or flame-shaped. The tails of comets are always turned from the sun in such wise that their axes produced would pass through the centre of that luminary; a fact which Biot assures us was notified by the Chinese astronomers so long ago as the year 837, but which was first distinctly mentioned in Europe by Fracastorius and Petrus Apianus in the 16th century. These effusions may be regarded as conoidal envelopes, having thicker or thinner walls—a view upon which several very remarkable optical appearances may readily be explained.

The several comets, however, are not so characteristically distinguished by their mere forms or appearance—they are not in one case tailless, in another provided with a tail of 104 degrees in length, as was the third of the year 1618; we further observe them passing through a rapid succession of varying formative processes. This change of form was most accurately and ably observed by Heinsius, of St. Petersburgh, in the comet of 1744, and in Halley's comet, on its last appearance in 1835, by Bessel, of Königsberg, by whom it has been very carefully described. On the part of the nucleus which was turned towards the sun there was a kind of tufted emanation apparent. The rays of this that bent backwards went to form part of the tail. "The nucleus of Halley's comet, with its emanations, presented the appearance of a burning rocket, the train of which was deflected sideways by a current of air." The rays proceeding from the head were seen by Arago and myself from the Parisian observatory on successive nights with very different appearances([15]). The great Königsberg astronomer, from numerous measurements and theoretical considerations, concluded "that the outstreaming cone of light departed distinctly, both to the right and left, from the line of direction towards the sun; but always returned to this line again, to pass over to the opposite side; that the outstreaming cone of light, therefore, as well as the body of the comet itself, which engenders and throws it out, has a rotatory, or rather a vibratory motion in the plane of the orbit." He found, further, "that the ordinary attractive force of the sun which is exerted upon heavy bodies, is not adequate to account for these vibrations;" and is of opinion "that they proclaim a power of polarity in the comet, which keeps one semidiameter of the body turned towards, the other semidiameter turned from, the sun; that the magnetic property possessed by the earth may present something of an analogous nature; and should the opposites of the telluric polarity inhere in the sun, the influence of this might show itself in the precession of the equinoxes." This is not the place for a more particular development of the grounds upon which explanations that accord with the phenomena have been built; but observations so remarkable([16]), views of such magnitude in reference to the most wonderful class of bodies that belong to our solar system, could not be passed by unnoticed in this sketch of a general picture of nature.

Notwithstanding the rule according to which the tails of comets increase in size and brightness as the perihelion is approached, and are turned from the central body of our system, the comet of 1823 presented the remarkable example of two tails, which formed an angle of 160° with each other, and of which one was turned from the sun, as usual, whilst the other was turned towards him. Peculiar modifications of the polarity, and unequal distribution and conduction of this, may, in the rare instance just quoted, have occasioned a twofold and uninterrupted effusion of nebulous matter([17]).

In the natural philosophy of Aristotle, the phenomena of comets and the existence of the milky way may be brought into a most strange juxtaposition or connection. The countless multitude of stars which compose the milky way give off a self-igniting or luminous mass; and the nebulous streak that divides the vault of the heavens is therefore regarded by the Stagirite as a mighty comet, which ceaselessly reproduces itself([18]).

Occultations of the fixed stars by the head or nucleus of a comet, or its immediate vaporous envelope, might throw some light upon the physical constitution of these wonderful heavenly bodies; but we have no observations which give us unquestionable assurance that any occultation has been observed which was completely central([19]); for, as we have above observed, there are alternate concentric scales of dense and very rare vapour in the parts lying near the nucleus. On the other hand, there is no question of the fact, that on the 29th of September, 1835, the light of a star of the 10th magnitude passed through an extremely dense vapour, at the distance of 7·″78 from the central point in the head of Halley's comet, according to Bessel's very accurate measurements; and that the light of this star suffered not the slightest deflection from its rectilinear course at any moment of the passage through this vapour([20]). Such an absence of refractive power, if it actually extends to the centre of the nucleus, renders it difficult to imagine that the matter of comets is at all of the nature of a gasiform fluid. Or, is the absence of refringent power the mere result of an almost infinite rarity of a fluid of this description? or does a comet consist of segregated particles, forming a cosmic cloud, which affects the ray of light passing through it in no greater degree than the clouds of our atmosphere, which have no influence in altering the zenith distance of the fixed stars or the edges of the sun? A greater or less diminution of the light of a fixed star has indeed been remarked during the passage of a comet over it, but this has been ascribed, with great propriety, to the lighter ground from which the star appears to stand out during the occultation.

The most important and decisive observations which have yet been made upon the nature of the light of comets, are those of Arago on its polarization. The polariscope of this distinguished philosopher gives us information of the physical constitution of the sun as well as of that of comets; the instrument, in a word, informs us whether a ray of light that reaches us after travelling many millions of miles, is direct or reflected light, and whether, in the former case, the source of the ray is a solid, a liquid, or a gaseous body. The light of Capella, and that of the great comet of 1819, were examined by the same apparatus: the comet showed polarized and therefore reflected light; the brilliant star, as was to have been anticipated, proclaimed itself a self-luminous sun([21]). The existence of polarized light in connection with the comet, however, was not merely made known by the inequality of the images; on the reappearance of Halley's comet in the year 1835, it was still more distinctly indicated by the striking contrast of complementary colours, in accordance with the laws of chromatic polarization discovered by Arago, in 1811. But it still remains undetermined, even by the beautiful experiments just referred to, whether, besides the reflected sun-light, comets have not also a light proper to themselves. In some, at least, of the true planets, Venus for example, it appears to be extremely probable that there is an inherent independent capacity to evolve light.

The variable brightness of comets is not always to be explained from their position in their orbit, and their distance from the sun. It certainly points, in particular individuals, to internal processes of condensation, and of augmented or diminished power of reflecting borrowed light. In the case of the comet of 1618, as also of the one with a period of three years, Hevelius observed the nucleus to be lessened as the sun was approached, increased as he was quitted; and this remarkable phenomenon, so long neglected, has lately been again referred to and confirmed by Balz, the able astronomer of Nismes. The regularity in the alteration of volume according to the distance from the sun is particularly striking. The physical explanation of the phenomenon cannot well be sought for in any increased density of the layers of the world-ether at distances progressively nearer the sun; for it is difficult to conceive the vaporous envelope of the comet's nucleus as vesicular, and impenetrable to the ether that fills the universe(22).

The very dissimilar excentricities in the elliptical orbits of comets has led in recent times (1819) to brilliant additions to our knowledge of the solar system. Encke made the discovery of a comet of so short a period that it always remains within the limits of our planetary orbits; he found that the place of its aphelion or greatest distance from the sun lay between the orbits of the telescopic planets and that of Jupiter. The excentricity of this comet's orbit is 0·845, that of Juno (the greatest excentricity among all the planetary orbits) being 0·255. Encke's comet has repeatedly been seen with the naked eye, although it is not easily discovered; it was seen, however, in Europe in 1819, and, according to Rümker, in New Holland in 1822. The period of this comet is nearly 3⅓ years; but from careful comparisons of the times of its return to the perihelion, the remarkable fact has been discovered that its periods from 1786 to 1838 have been going on regularly contracting from revolution to revolution, viz., in the course of 52 years, by one and $\frac{3}{10}$ths of a day. So remarkable a circumstance has led to the admission of the very probable existence of a vaporiform matter diffused in planetary space, and capable of opposing a certain resistance to bodies in motion through it. Something of the kind, indeed, seems necessary in order to bring the most careful consideration of every source of planetary perturbation into harmony with the results of observation and calculation. The tangential force is diminished, and with it the greater axis of the cometary orbit. The value of the constant of resistance appears, moreover, to be somewhat different before and after the passage of the perihelion, which is perhaps to be ascribed to the altered form of the small nucleus, and to the effect of inequality in density of the layers of ether in the sun's vicinity(23). These facts, and their explanation, must be reckoned among the number of the most interesting results of modern astronomy. And then, if Encke's comet led us at an earlier period to subject the mass of Jupiter—always so important in every reckoning of perturbation—to a closer scrutiny, its course has subsequently obtained for us the first, although merely approximative, determination of an inferior mass for Mercury.

The first comet of short period, namely, Encke's, of 3⅓ years, was followed, in 1826, by another planetary one, the aphelion of which lies beyond the orbit of Jupiter, but much within that of Saturn. This, or Biela's comet, completes its revolution in 6¾ years. Its light is still more feeble than that of Encke's comet. The motion of both these comets is direct, whilst that of Halley's is retrograde—contrary to the motion of the planets properly so called.

Biela's comet presents the first certain instance of the orbit of a comet intersecting that of the Earth; its path is, therefore, one of possible danger—if we can regard as dangerous a phenomenon which has not been observed within the historical period, and of which the consequences are doubtful. Small masses, possessed of enormous velocities, may certainly exercise a notable force; but, though Laplace demonstrated the mass of the comet of 1770 to be less than the 1-5000th of that of the Earth, he supposes, with a certain degree of probability, that the average masses of the comets are much below the one hundred thousandth part of the Earth's (about the 1-1200th of the moon's) mass(24). We must not confound the passage of Biela's comet through our earth's orbit, with its proximity to, or absolute encounter with the Earth itself. When this passage took place on October 29th, 1832, the Earth was still a full month off from the point of intersection of the two orbits.

The orbits of these two comets of short period mutually intersect each other; and it has been correctly observed(25), that owing to the numerous perturbations which such small celestial bodies suffer from the planets, it is not impossible for them to encounter, and that, should this occur about the middle of the month of October, the inhabitants of the Earth might behold the extraordinary spectacle of a cosmical combat; in other words, of the mutual penetration of two comets, of their agglutination, or of their destruction, in consequence of exhaustive emanations. The immense ethereal expanse may have witnessed during millions of years several events of this kind, consequences of deviations produced by perturbing masses, or of originally intersecting orbits; still they are insulated phenomena, having as little general influence in modifying the form or state of the universe, as the appearance or extinction of a volcano in the limited sphere of the Earth.

A third planetary comet of short period was discovered by Faye on November 22d, 1843, at the Paris Observatory Its elliptical orbit approximates more nearly to a circle than that of any other known comet, and is included between the paths of Mars and Saturn. This comet (which Goldschmidt says stretches beyond the orbit of Jupiter), is therefore one of the few known which has its perihelion beyond the orbit of Mars. It accomplishes its revolution in 7·29 years, and probably owes the present form of its orbit to its great proximity to Jupiter at the close of 1839.

When we consider comets in their closed elliptical orbits as members of our solar system, with reference to their major axes, their excentricities, and their periods of revolution, it seems probable that in the last particular the three planetary comets (Encke's, Biela's, and Faye's) are immediately succeeded by Messier's of 1766 (supposed by Clausen to be identical with the third comet of 1819), and by the fourth of 1819, discovered by Blanpain, which Clausen considers identical with that of 1743, but which, as well as Lexell's, has suffered great orbital changes from the proximity and attraction of Jupiter. These two comets appear to have a period of from five to six years, and their aphelia fall in the neighbourhood of the orbit of Jupiter. From 70 to 76 years are occupied in their revolutions, by Halley's comet (so important in a theoretical point of view, of which the last appearance, in 1835, was less brilliant than its former ones had led astronomers to expect it would prove), by Olbers's comet of March 6, 1815, and by Pons's comet of 1812, the elements of which were calculated by Encke. Both of the latter were invisible to the naked eye. The great comet of Halley has already greeted us for the ninth known time; Laugier's computations([26]) having recently demonstrated that it is identical with the comet of 1378, recorded in Ed. Biot's Chinese Catalogue of Comets. From 1378 to 1835 its period has varied between 74·91 and 77·58 years, the mean having been seventy-six years.

Contrasted with the celestial bodies above mentioned, we behold another series of bodies requiring millenniums for their barely determinable periods. Thus, Argelander says that the splendid comet of 1811 requires 3065 years for its revolution, whilst Encke fixes 8800 years for the awfully grand one of 1680. These bodies, therefore, recede respectively 21 and 44 times farther from the Sun than Uranus; that is, 8400 and 17,600 millions of miles. The Sun's attractive force extends therefore even to this enormous distance; but then, whilst the comet of 1680, at its perihelion, travels at the rate of 53 miles (above 1,300,000 English feet) per second, or 13 times faster than the Earth, its velocity hardly attains 10·8 E. feet per second at its aphelion. The last-mentioned rate is only thrice greater than the velocity of water in our most sluggish European rivers, and but half the velocity which I observed in the Cassiquiare, a branch of the Orinoko. Amongst the immense number of uncomputed or undiscovered comets, there are most probably many which have a major orbital axis far exceeding that of the comet of 1680. In order to give some idea, if not of the extent of the sphere of attraction, at least of the spacial distance of a fixed star, or other sun, from the aphelion of the comet of 1680 (the most distant traveller of all the celestial bodies of our system, according to our present knowledge), I need only remind the reader that the most recent estimates of parallax still make the nearest fixed star 250 times farther from the sun than the aphelion of this comet, which is only 44 times as remote as Uranus, whilst the star *α* Centauri is 11,000, and the star 61 Cygni (after Bessel's very accurate observations) is 31,000 times more distant than the planet.

After this consideration of the greatest elongations of comets from the central body of the solar system, let us glance at those which have approached it most nearly. The instance of the greatest known proximity of a comet to the earth occurred with that of Lexell and Burkhardt, celebrated for the perturbations it suffered from Jupiter; this comet was only six times the distance of the moon from us on June 28th, 1770. In 1767 and 1779, the same comet twice traversed the system of Jupiter's satellites, without causing the slightest perceptible derangement in their orbits; orbits which have been so thoroughly investigated by physical astronomers. But the great comet of 1680, when at its perihelion, was from eight to nine times nearer to the surface of the sun than Lexell's was to the earth. On December 17th, the sun and the comet of 1680 were only one-sixth of the diameter of the former body apart; in other words, seven-tenths of the moon's distance from us. Owing to the feebleness of the light of distant comets, perihelia beyond the orbit of Mars are rarely observable by man; the comet of 1729 is, in fact, the only one of those hitherto computed which has its perihelion between the orbits of Pallas and Jupiter, and which has been observed beyond the path of the latter planet.

Since scientific acquirements, some solid, by the side of much superficial learning, have penetrated in wider circles into social life, the fears of the possible evils wherewith comets threaten us have increased in weight, and their direction has become more definite. The certainty of there being several periodical comets within the known planetary orbits, visiting us at short intervals; the considerable perturbations which Jupiter and Saturn cause in their paths, whereby apparently harmless wanderers of the sky may be converted into peril-fraught bodies; the orbit of Biela's comet passing through that of the earth; the existence of a cosmical ether, that resisting and retarding fluid which tends to contract the orbits of all the planetary bodies; the individual differences in the bodies of comets which permit us to suspect considerable gradations in the quantity of the mass of the nucleus; all these circumstances amply replace, in multiplicity of grounds, the dread which, in former centuries, was entertained of flaming swords, and an universal conflagration to be lighted up by fiery stars.

As the grounds for confidence derivable from the doctrine of probabilities only operate on the understanding, are only of avail among the reflecting, and produce no effect on gloomy apprehension and imagination, modern science has been charged, not altogether without reason, with seeking to allay the fears which it has itself created. It is a principle laid deeply in the desponding nature of man, in his inherent disposition to view things on the dark rather than on the bright side, that the unexpected, the extraordinary, excites fear, not hope or joy([27]). The strange aspect of a mighty comet, its pale nebulous gleam, its sudden appearance in the heavens, have in all countries, and almost at all times, been held as portentous indications of change or dissolution of the old-established order of things. And then, as the apparition is never more than short lived, arises the belief that its significance must be reflected in contemporaneous or immediately succeeding events. And such is the enchainment of events, that some particular incident scarcely fails to turn up which can be fixed upon as the calamity prognosticated. It is only in these times that a spirit of greater hopefulness, in connection with the appearance of comets, has shewn itself among the people. In the beautiful valleys of the Rhine and the Moselle, ever since the appearance of the brilliant comet of 1811, comets have been regarded as exerting a favourable influence on the ripening of the grape; nor have various years of indifferent vintage, along with the appearance of other comets, instances of which have not been wanting, been able to shake the faith of the wine-growers of the north of Germany in their beneficial influences.

I now pass from comets to another and yet more enigmatical class of agglomerated matter, to the smallest of all asteroids, which, in their fragmentary condition, and when they have arrived in our atmosphere, we designate by the name of Aërolites, or Meteoric Stones. If I dilate at greater length on these bodies than I have done on comets, and accumulate those individual features which should otherwise be excluded from a general survey of nature, it is done with a purpose. The very remarkable characteristic diversities of comets have been long known. From the little that has yet been learned of their physical condition, it is difficult, in an exposition such as is here required, to seize the Common, and to separate the Necessary from the Accidental, in phenomena observed with very different degrees of accuracy. The measuring and calculating astronomy of comets has alone made marvellous progress. In this state of our knowledge, a scientific consideration must be limited to physiognomical differences in the fashion of the nucleus and tail; to examples of close approximations to other planetary bodies; to extremes in orbits with reference to space, and in periods of revolution to time. Natural truth in these, as in the phenomena that are immediately to be spoken of, is only to be attained by a delineation of the Individual, and by the animated and contemplative expression of reality.

Shooting Stars, Fire-balls, and Meteoric Stones, are, with great appearance of probability, regarded as small masses moving with planetary velocity in conic sections round the sun, in harmony with the laws of universal gravitation. When these masses encounter the Earth in their course, and, attracted by it, become luminous on the verge of our atmosphere, they frequently let fall stony fragments, heated in greater or less degree, and covered on their surface with a black and shining crust. By careful analysis of all that has been observed at different epochs when great numbers of shooting stars have fallen, as at Cumana in 1799, in North America in 1833 and 1834, &c., it seems no longer proper to separate fire-balls from shooting stars. Both phenomena are not only frequently contemporaneous and intermingled, but they also pass into one another, and this whether we pay particular attention to the dimensions of the discs, to the sparks or trains of fire which they emit, or to the velocities of their respective motions. Whilst there are fire-balls that have the apparent diameter of the moon, that explode and emit smoke, and possess such brilliancy that they can be seen at noon-day([28]), there are, on the other hand, shooting stars in countless multitudes, of such small dimensions that they only present themselves to the eye in the form of moving points or of phosphorescent lines([29]). But whether or not among the many luminous bodies that shoot through the sky in the form of falling stars and meteors, there are not several of different natures, remains to be shown. Occupied, shortly after my return home, with the impression which the phenomena of shooting stars had left upon my mind, and remembering that I had observed them in greater numbers, of brighter colours, and more commonly accompanied by long and brilliant trains, both on intertropical plains just raised above the level of the sea, and on mountains at the height of twelve and even fifteen thousand feet above its surface, than in the temperate and frigid zones, I soon perceived that the ground of the more vivid impression lay in the glorious transparency of the tropical atmosphere itself([30]). There one sees deeper into space. Sir Alexander Burnes, too, speaks of the magnificent and constantly recurring spectacle of coloured shooting stars, which he enjoyed in Bokhara, and which he attributes to the purity of the atmosphere.

The connection of meteoric stones with the grander and more brilliant phenomena of fire-balls—that stones actually fall from these fire-balls, and penetrate ten or fifteen feet into the ground, has been shown, among many other instances of the kind, by the well-known fall of aërolites at Barbotan, in the department Des Landes, on the 24th July, 1790, at Lima on the 16th of June, 1794, at Weston, in Connecticut, on the 14th of December, 1807, and at Juvenas, in the department of Ardèche, on the 15th of June, 1821. Other phenomena connected with the fall of aërolites are those where the masses have descended, shaken, as it were, from the bosom of a small dark cloud, which had formed suddenly in the midst of a clear sky, accompanied with a noise that has been compared to the report of a single piece of artillery. Whole districts of country have occasionally been covered with thousands of fragments of stones, of very dissimilar magnitudes, but like constitution, which had been rained down from a progressive cloud of the kind described. In rarer instances, as in that which occurred at Klein-wenden, not far from Mühlhausen, on the 16th of September, 1843, large aërolites have fallen amidst a noise like thunder, when the sky was clear and without the formation of any cloud. The close affinity between fire-balls and shooting stars is also shown by the fact of instances having occurred, of the former throwing down stones, though they had scarcely the diameter of the balls that are projected from our fireworks called Roman candles. This happened notably at Angers on the 9th of June, 1822.

With regard to the form-producing forces, the physical and chemical processes in these phenomena, we are still completely in the dark. We know not whether the particles which form the compact mass of the aërolite lay originally apart from one another, in the shape of vapour, as in comets, and first contracted and ran together when they began to lighten within the gleaming ball; we know nothing of what takes place in the black cloud, where it sometimes continues to thunder for minutes before the stones descend; neither are we aware whether from the smaller shooting stars there be any precipitation of solid matter, or only an attenuated dry haze, or a ferruginous and nickeliferous meteoric dust([31]). We, however, know the immense, the wonderful and perfectly planetary rapidity of shooting stars, fire-balls, and meteoric stones; we recognise the General in reference to them, and in this Generality perceive uniformity of phenomena only, nothing of genetical cosmic process, the consequence of change. If meteoric stones revolve already consolidated into dense masses[32] (less dense, however, than the mean density of the Earth), then must they form very insignificant nuclei to the fire-balls, surrounded by inflammable vapours or gases, from the interior of which they shoot, and which, judging from their height and apparent diameters, must have actual diameters of from 500 to 2600 feet. The largest meteoric masses of which we have information, those to wit of Bahia and Otumpa in Chaco, which Rubi de Celis has described, are from 7 to 7½ feet in length. The meteoric stone of Aegos Potamos, so celebrated through the whole of antiquity, and which is even mentioned in the Marble Chronicle of Paris, is described as having been of the magnitude of two millstones, and of the weight of a wagon load. Despite the vain attempts of the African traveller, Browne, I have not yet abandoned the hope that this great Thracian meteoric stone, which must be so difficult of destruction, though it fell more than 2300 years ago, will again be discovered by one or other of the numerous Europeans who now perambulate the East in safety. The enormous aërolite which fell in the beginning of the 10th century in the river at Narni, projected a whole ell above the surface of the water, as we are assured by a document lately discovered by Pertz. It is to be observed, however, that none of these aërolites, whether of ancient or modern times, can be regarded as more than principal fragments of the mass which was scattered by the explosion of the fire-ball or murky cloud whence they descended.

When we duly consider the mathematically determined enormous velocities with which meteoric stones fall from the outer confines of our atmosphere to the earth, or with which, as fire-balls, they speed for long distances through even the denser fields of air, it seems to me more than improbable that the metalliferous mass, with its internally disseminated and very perfect crystals of olivine, labrador, and pyroxene, could have run together in so short an interval into a solid nucleus from any state of gas or vapour. The mass that falls, besides, even in cases where the chemical constitution varies, has always the particular characters of a fragment; it is commonly of a prismatoidal or irregular pyramidal form, with somewhat arched surfaces and round edges. But whence this figure, first observed by Schreibers, of a mass detached from a rotating planetary body? Here, too, as in the circle of organic life, all that has reference to the history of evolution is hidden in obscurity. Meteors begin to lighten and to burn at elevations which we must look upon as almost perfect vacuums, or that cannot contain 1-100,000th of oxygen. Biot's new researches on the interesting crepuscular phenomenon([33]), reduce the line very notably which, somewhat hardily perhaps, is frequently spoken of as the *limits* of our atmosphere; but luminous phenomena take place independently of the presence of oxygen, and Poisson has admitted the combustion of aërolites, or meteors, as occurring far beyond the confines of our atmosphere. It is only in so far as calculation and geometrical admeasurement can be applied to meteoric stones, as to the greater bodies of the solar system, that we feel ourselves proceeding on surer grounds. Although Halley had already pronounced the great fire-ball of 1686, the motion of which was in opposition to that of the earth, a cosmic phenomenon([34]), Chladni was the first (1794) who, in the most general terms, and most clearly recognized the connection betwixt fire-balls and the stones that fall from the atmosphere, as well as the correspondence between the motions of these bodies and those of the planetary masses at large([35]). A brilliant confirmation of this view of the cosmic origin of such phenomena has been supplied by Denison Olmsted, of New-Haven, Connecticut, in his observations on the showers of shooting stars and fire-balls which made their appearance in the night from the 12th to the 13th of November, 1833. On this occasion, all these bodies proceeded from the same quarter of the heavens—from a point, namely, near the star γ Leonis, from which they did not deviate, although the star, in the course of the lengthened observation, changed both its apparent elevation and its azimuth. Such an independence of the rotation of the earth proclaimed that the luminous bodies came *from without*—from outer space into our atmosphere. According to Encke's calculations of the entire series of observations that were made in the United States of North America, between the paralells of 35° and 42°, the whole of the shooting stars came from the point in space towards which the earth was moving at the same epoch([36]). In the subsequent American observations on the shooting stars of November 1834 and 1837, and the Bremen ones of 1838, the general parallelism of their courses, and the direction of the meteors from the constellation Leo, were perceived. As in the November periodical recurrence of shooting stars, a more decided parallel and particular direction has been noted than in the case of those that appear sporadically at other seasons, so in the August phenomenon it has also been believed that the bodies came for the major part from a point between Perseus and Taurus, the point towards which the earth is tending about the middle of the month of August. This was particularly remarked in the summer of 1839. This peculiarity in the phenomenon of falling stars, the direction of retrograde orbits in the months of November and August, is especially worthy of being either better confirmed or refuted by the most careful observations upon future occasions.

The altitudes at which shooting stars make their appearance, by which must be understood the periods between their becoming visible and their ceasing to be so, are extremely various; in a general way, they may be stated as varying between four and thirty-five geographical miles. This important result, as well as the extraordinary velocity of the problematical asteroids, was first arrived at by Benzenberg and Brandes, by means of a series of contemporaneous observations and determinations of parallax, at either extremity of a base line 46,000 feet in length([37]). The relative velocity of the motion was from four and a quarter to nine miles per second; it was therefore equal to that of the planets([38]). Such a velocity of movement, as well as the frequently observed course of shooting stars and fire-balls in a direction the opposite of that of the earth, has been used as a principal element in combating that view of the origin of aërolites, in which they were presumed to be projected from still active volcanoes in the moon. The supposition of any volcanic power, of greater or less energy, inherent in a small planetary body surrounded by no atmosphere, is, indeed, in the nature of things, and numerically considered, extremely arbitrary. It is not difficult, indeed, to conceive the reaction of the interior of a planet against its crust, as ten or even a hundred times greater than that which we now observe in connection with the volcanoes of the earth. The direction of the masses, too, which could be projected from a satellite moving from west to east, might appear retrograde, in consequence of the earth, in its orbit, arriving later at the point of its path where the masses fall. But, then, if the entire circle of relations, which I felt myself compelled to specify, even in this general picture of nature, to escape the suspicion of making unfounded assertions, be surveyed, it will be found that the hypothesis of a lunar origin of meteoric stones([39]) is dependent on a majority of conditions, the accidental association of which could alone give to the barely possible, the form and substance of reality. The admission of the original existence of small planetary masses circulating in space, is simpler, and seems more in harmony with what we know or infer with reference to the formation of the solar system.

It is highly probable that a great proportion of these cosmic bodies pass undestroyed in the vicinity of our atmosphere, and only suffer a certain deflection in the excentricity of their orbits by the attraction of the earth. We may conceive that the same bodies only become visible to us again after the lapse of several years, and when they have made many revolutions round their orbit. The ascent of some fire-balls and shooting stars (which Chladni endeavoured to explain, not very happily, by a reflection produced by a body of greatly condensed air) appears, at first sight, to be a consequence of a mysterious projectile force throwing off the meteors from the earth; but Bessel has shown on theoretical grounds, and indeed proved, by means of Feldt's very accurate calculations, that in the absence of perfect agreement in point of time, of the disappearances recorded, there is not one amongst the whole of the observations published which impresses the assumption of an ascent, with a character of probability, none which does not allow us to regard it as an effect of observation([40]). Whether the explosion of shooting stars, and of the smoking and flaming fire-balls which do not always move in straight lines, may force the meteors upwards in the manner of rockets, or otherwise influence the direction of their path, in certain cases, as Olbers supposes, must remain matter for further observation.

Shooting stars fall either singly and rarely, and at all seasons indifferently, or in crowds of many thousands (Arabian writers compare them to swarms of locusts), in which case they are periodical, and move in streams generally parallel in direction. Amongst the periodic showers, the most remarkable are those that occur from the 12th to the 14th of November, and on the 10th of August; the "fiery tears" which then descend, are noticed in an ancient English church-calendar, and are traditionally indicated as a recurring meteorological incident([41]). Independently of this, however, precisely in the night from the 12th to the 13th of November, 1823, according to Klöden, there was seen at Potsdam, and in 1832, over the whole of Europe from Portsmouth to Orenburg on the river Ural, and even in the southern hemisphere, in the Isle of France, a great mixture of shooting stars and fire-balls of the most different magnitudes; but it appears to have been more especially the enormous fall of shooting stars, which Olmsted and Palmer observed in North America between the 12th and 13th of November, 1833, when they appeared in one place as thick as flakes of snow, and 240,000 at least were calculated to have fallen in the course of nine hours, that led to the idea of the periodic nature of the phenomenon, of great flights of shooting stars being connected with particular days. Palmer of New Haven recollected the fall of meteors in 1799, which Ellicot and I first described([42]), and from which, by the juxtaposition of observations which I had given, it was discovered that the phenomenon had occurred simultaneously over the New Continent from the equator to New-Hernhut in Greenland (N. Lat. 64° 14′), betwixt 46° and 82° of Longitude. The identity in point of time was perceived with amazement. The stream, which was seen over the whole vault of heaven between the 12th and 13th of November, 1833, from Jamaica to Boston (N. L. 40° 21′), recurred in 1834, in the night between the 13th and 14th of November, in the United States of North America, but with something less of intensity. In Europe, its periodicity since this epoch has been confirmed with great regularity.

A second, even as regularly recurring shower of shooting stars as the November phenomenon, is the one of the month of August—the feast of St. Lawrence phenomenon—between the 9th and the 14th of the month. Muschenbroeck([43]) had already called attention in the middle of the preceding century to the frequency of meteors in the month of August; but their periodic and certain return about the time of the feast of St. Lawrence was first pointed out by Quetelet, Olbers, and Benzenberg. In the course of time other periodically recurring showers of shooting stars([44]) will very certainly be discovered—perhaps from the 22d to the 25th of April; from the 6th to the 12th of December, and, in consequence of the actual fall of aërolites described by Capocci, from the 27th to the 29th of November, or about the 17th of July.

However independent all the phenomena of falling stars yet witnessed may have been of polar elevation, temperature of the air, and other climatic relations, there is still one, although perhaps only accidental, accompanying phenomenon which must not be passed by unnoticed. The Northern Lights showed themselves of great intensity during the most brilliant of all these natural incidents, that, namely, which Olmsted has described (Nov. 12–13, 1833). The same thing was also observed in Bremen in 1838, where, however, the periodic fall of meteors was less remarkable than at Richmond, in the neighbourhood of London. I have also referred, in another work([45]), to the remarkable observation of Admiral Wrangel, which he has confirmed to me verbally oftener than once, that during the appearance of the Northern Lights, on the Siberian shores of the Icy Sea, certain regions of the heavens which were not illuminated, became inflamed and continued to glow whilst a shooting star passed through them.

The different meteor-streams, each of them made up of myriads of little planets, probably intersect the orbit of our earth in the same way as Biela's comet does. Upon this view we may imagine these shoot-star asteroids as forming a closed ring, and pursuing their course in the same particular orbit. The smaller telescopic planets between Mars and Jupiter, with the exception of Pallas, present us, in their closely connected orbits, with a similar relationship. It is impossible as yet to decide whether alterations in the epochs at which the stream becomes visible to us, whether retardations of the phenomenon, to which I long ago directed attention, indicate a regular recession or change of the nodes (the points of intersection of the earth's orbit and the ring), or whether from unequal clustering or very dissimilar distances of the little bodies from each other, the zone is of such considerable breadth, that the earth only passes through it in the course of several days. The lunar system of Saturn likewise shows us a group of most intimately associated planetary bodies of amazing breadth. In this group, the orbit of the 7th or outermost satellite, is of so considerable a diameter, that the earth, in her orbit round the sun, would take three days to pass over a space of like extent. Now, if we suppose that the asteroids are unequally distributed in the course of one of the closed rings which we picture to ourselves as forming the orbits of the periodic currents, that there are but a few thickly congregated groups such as would give the idea of continuous streams, we can understand wherefore such brilliant phenomena as those of November 1799 and 1833 are extremely rare. The acute Olbers was inclined to announce the return of the grand spectacle, in which shooting stars mixed with fire-balls should fall like a shower of snow, for the 12th–14th of November, 1867.

Hitherto the current of the November asteroids has only been visible over limited portions of the earth's surface. It appeared, for example, with great splendour in England in the year 1837, as a meteoric shower; whilst an experienced and very attentive observer at Braunsberg, in Prussia, saw nothing more than a few scattered shooting stars in the course of the same night, from seven o'clock in the evening till sun-rise, the sky having continued uninterruptedly clear the whole of the time. Bessel concluded from this, "that a group of the great ring which is occupied by these bodies, of but limited extent, had approached the earth over England, whilst districts to the east passed through a relatively empty portion of the ring"([46]). Should the idea of a regular precession or variation of the nodal lines, occasioned by perturbations, acquire greater likelihood, the discovery of older observations of the phenomenon would become a matter of particular interest. The Chinese annals, in which, beside the appearance of comets, there are also notices of great showers of shooting stars, go back beyond the time of Tyrtæus, or the second Messenic war. They describe two streams occurring in the month of March, one of which is 687 years older than the commencement of the Christian era. Edward Biot has already remarked, that among the fifty-two appearances which he finds recorded in the Chinese annals, the most frequently recurring were those that fell near the date from the 20th to the 22d of July (old style), which may very possibly be the now advanced stream occurring about the time of the feast of St. Lawrence([47]). If the great fall of shooting stars which Bogulawski, jun., finds recorded in Benessius de Horowic's "Chronicon Ecclesiæ Pragensis," as having been seen in full day light on the 21st of October, 1366 (old style), corresponds with our present November fall, the precession in the course of 447 years informs us that this shoot-star system (that is to say, its common point of gravity), describes a retrograde course about the sun. It also follows, from the views now developed, that when seasons pass by in which neither of the streams as yet observed—that, namely, of November and that of August—is seen in any part of the earth, the reason of this lies either in the interruption of the ring—in other words, in the occurrence of gaps or vacancies between the clusters of asteroids that follow each other — or, as Poisson will have it, in the influence which the larger planets exercise upon the form and position of the ring([48]).

The solid, heated, although not red-hot, masses which are seen to fall to the earth from fire-balls by night, from small dark clouds by day, accompanied with loud noises, the sky being generally clear at the time, show, on the whole, a very obvious similarity, in point of external form, in the character of their crust and the chemical composition of their principal ingredients. This they have maintained through centuries, and in every region of the earth in which they have been collected. But so remarkable and early asserted a physiognomical equality in these dense meteoric masses is subject to many individual exceptions How different are the readily forged masses of iron of Hradschina, in the district of Agram, or that of the banks of the Sisim, in the government of Jenesiesk, which have become celebrated through Pallas, or those which I brought with me from Mexico([49]), all of which contain 96 per cent. of iron, from the aërolites of Siena, which scarcely contain 2 per cent. of this metal, from the earthy meteoric stone of Alais (Dép. du Gard), which crumbles when put into water, and from those of Jonzac and Juvenas, which, without metallic iron, contain a mixture of oryctognostically distinguishable, crystalline and distinct constituents! These diversities have led to the division of the cosmical masses into two classes—nickeliferous meteoric iron, and fine or coarse-grained meteoric stones. Highly characteristic is the crust, though it be but a few tenths of a line in thickness, often shining like pitch, and occasionally veined([50]). So far as I know, it has only been found wanting in the meteoric stone of Chantonnay, in La Vendée, which, on the other hand—and this is equally rare—exhibits pores and vesicular cavities like the meteoric stone of Juvenas. In every instance the black crust is as sharply separated from the clear gray mass, as is the dark-coloured crust or varnish of the white granite blocks which I brought from the cataracts of the Orinoko(51), and which are also met with by the side of other cataracts in different quarters of the globe—those of the Nile, the Congo, &c. It is impossible to produce anything in the strongest heat of the porcelain furnace which shall be so distinct from the unaltered matter beneath, as is the crust of aërolites from their general mass. Some, indeed, will have it that here and there indications of penetration of fragments, as if by kneading, appear; but in general the condition of the mass, the absence of flattening from the fall, and the not very remarkable heat of the meteoric stone, when touched immediately after its fall, indicate nothing like a state of fusion of the interior during the rapid passage from the limits of the atmosphere to the earth.

The chemical elements of which meteoric masses consist, upon which Berzelius has thrown so much light, are the same as those which we encounter scattered through the crust of the earth. They consist of eight metals (iron, nickel, cobalt, manganese, chrome, copper, arsenic, and tin); five earths; potash and soda; sulphur, phosporus, and carbon; in all, one-third of the entire number of simple substances at present known. Despite this similarity to the ultimate elements into which inorganic bodies are chemically decomposable, the appearance of meteoric masses has still something that is generally strange to us; the kind of combination of the elements is unlike all that our terrestrial mountain and rocky masses exhibit. The native iron, which is met with in almost the whole of them, gives them a peculiar, but not therefore a lunar character; for, in other regions of space, in other planetary bodies besides the moon, water may be entirely wanting, and processes of oxidation may be rare.

The cosmic gelatinous vesicles, the nostoc-like organic masses, which have been attributed to shooting stars ever since the middle ages, and the pyrites of Sterlitamak (westward from the Ural Mountains), which have been said to be composed of hail-stones in the interior, belong to the fables of meteorology(52). It is only the finely granular texture, only the mixture of olivine, augite, and labrador spar(53), of some aërolites, of the doloritic-looking mass of Juvenas in Ardèche, for example, that gives them somewhat more of an indigenous character, as G. Rose has shown. These aërolites, indeed, contain crystalline substances exactly similar to those of the crust of our Earth; and in Pallas's Siberian mass of meteoric iron, the olivine is only distinguished by the absence of nickel, which is there replaced by oxide of tin(54). As meteoric olivine, like that of our basalt, contains from 47 to 49 per cent. of magnesia, and this earth, according to Berzelius, generally constitutes one-half of the earthy ingredients of aërolites, we must not be astonished at the large quantity of silicate of magnesia which we find in these cosmic masses. If the aërolite of Juvenas contains separable crystals of augite and labrador, it is at least probable, from the numerical relations of the ingredients, that the meteoric mass of Chateau-Renard is a diorite composed of hornblende and albite, and those of Blansko and Chantonnay of hornblende and labrador. The indications of a telluric or atmospheric origin of aërolites, which have been derived from the oryctognostic resemblances just mentioned, do not appear to me of any great weight. Wherefore should not—and here I might refer to a remarkable conversation between Newton and Conduit at Kensington(55)—wherefore should not the matter belonging to a particular cluster of celestial bodies, to the same planetary system, be for the major part the same? Why should it not be so, when we feel at liberty to surmise that these planets, like all larger and smaller conglobated masses which revolve about the sun, have separated from particular and formerly much more widely-expanded sun-atmospheres, as from vaporous rings, and which originally held their courses round the central body? We are not, I believe, more authorized to regard nickel and iron, olivine and pyroxene (augite), which we find in meteoric stones, as exclusively terrestrial, than I should have been had I indicated the German plants which I found beyond the Obi, as European species of the flora of northern Asia. If the elementary matters in a group of planetary bodies of various magnitudes be identical, why should they not also, in harmony with their several affinities, run into determinate combinations—in the polar circle of Mars, into white and brilliant snow and ice; in other smaller cosmic masses into mineral species that contain crystalline, augite, olivine, and labrador? Even in the region of the merely Conjectural, the unbridled caprice that despises all induction must not be suffered to control opinion.

The extraordinary obscurations of the sun which have occasionally taken place, during which the stars became visible at mid-day (as in the three days' darkness of the year 1547, about the time of the fateful battle near Mühlberg), and which are not explicable on the supposition of a cloud of volcanic ashes, or of a dense dry-fog, were ascribed by Kepler, at one time, to a materia cometica, at another to a black cloud, the product of sooty exhalations from the sun's body. The observations of shorter periods of darkness—of three and six hours, in the years 1090 and 1203—Chladni and Schnurrer have explained by the passage of meteoric masses. And since the stream of shooting stars from the direction of its orbit has been regarded as forming a closed ring, the epochs of these mysterious celestial phenomena have been brought into a remarkable connection with the regularly recurring showers of shooting stars. Adolph Erman has, with great acuteness, and after a careful analysis of all the data collected up to the present time, directed the attention of philosophers to the coincidence of the conjunction with the sun, as well of the August asteroids (7th of February) as of the November asteroids (12th of May), at the epoch which coincides with the popular belief in the celebrated *cold days* of Mamertius, Pancratius, and Servatius([56]).

The Greek natural philosophers, little disposed in general to observation, but incessantly, inexhaustibly addicted to speculation on the manifold import of half-seen truths, have left views behind them on shooting stars and meteoric stones, several of which chime in most remarkably with those at present so commonly entertained of the cosmic nature of the phenomenon. "Shooting stars," says Plutarch([57]), in the Life of Lysander, "according to the opinion of some naturalists, are not excretions and emanations of the ethereal fire, quenched in the air immediately after their ignition; neither are they any kindling and combustion of the air, produced by those which have become dissolved in quantities in the upper regions; they are rather a fall of celestial bodies, occasioned by a certain abatement of the centrifugal force, and the impulse of an irregular motion, and are cast down, not only upon the inhabited earth, but also beyond it into the ocean, on which account they are not then found." Diogenes of Apollonia([58]) speaks still more clearly on the subject. According to his view, "along with the visible stars, others move that are invisible, and therefore are unnamed. These last frequently fall to the earth and are extinguished, as was the case with the stony star which descended in fire at Aegos Potamos." The Apollonian, who also regards all the other stars (the luminous ones) as pumice-like bodies, probably founded his opinions of the nature of shooting stars and meteoric masses upon the doctrines of Anaxagoras, of Clazomenæ, who maintained that all the heavenly bodies were "mineral masses, which the fiery ether, in the power of its revolution, had torn from the earth, had ignited and converted into stars." In the Ionic school, according to the statement of Diogenes of Apollonia, and as it has come down to us, aërolites and the heavenly bodies were placed in one and the same class; both are alike terrestrial in their original production; but only in the sense that the earth, as the central body, had formerly([59]) fashioned all around her; in the same way as our present ideas lead us to conceive that the planets of a system arise from the extended atmosphere of another central body—namely, the sun. These views, consequently, are not to be confounded with that which speaks familiarly of meteoric stones, as of telluric or atmospheric origin, nor yet with the extraordinary conjecture of Aristotle, to the effect that the enormous mass of Aegos Potamos had been raised by a tempestuous wind.

The presumptuous skepticism which rejects facts without caring to examine them, is, in many respects, even more destructive than uncritical credulity. Both interfere with rigour of investigation. Although, for fifteen hundred years, the annals of various nations have told of the fall of stones from the sky—although several instances of the circumstance are placed beyond all question by the unimpeachable testimony of eye-witnesses—although the Bætylia formed an important part of the meteor-worship of the ancients, and the companions of Cortes saw the aërolites in Cholula, which had fallen upon the neighbouring pyramid—although Caliphs and Mongolian princes have had sword blades forged from meteoric masses that had but lately fallen, and men have even been killed by stones from heaven (a certain monk at Crema, on the 4th September, 1511; another monk in Milan, 1650; and two Sweedish sailors on ship-board, 1674), so remarkable a cosmical phenomenon remained almost unnoticed, and, in its intimate relationship with the rest of the planetary system, unappreciated, until Chladni, who had already gained immortal honour in physics by his discovery of phonic figures, directed attention to the subject. But he who is penetrated with the belief of this connection, if he be susceptible of emotions of awe through natural impressions, will be filled with solemn thoughts in presence, not of the brilliant spectacles of the November and August phenomena only, but even on the appearance of a solitary shooting star. Here is a sudden exhibition of movement in the midst of the realm of nocturnal peace. Life and motion occur at intervals in the quiet lustre of the firmament. The track of the falling star, gleaming with a palely lustre, gives us a sensible representation of a path long miles in length across the vault of heaven; the burning asteroid reminds us of the existence of universal space every where filled with matter. When we compare the volume of the innermost satellite of Saturn, or that of Ceres, with the enormous volume of the Sun, all relation of great and small vanishes from the imagination. The extinction of the stars that have suddenly blazed up in several parts of the heavens, in Cassiopea, in Cygnus, and in Ophiucus, leads us to admit the existence of dark or non-luminous celestial bodies. Conglobed into minor masses, the shooting-star asteroids circulate about the sun, intersect the paths of the great luminous planets, after the manner of comets, and become ignited when they approach or actually enter the outermost strata of our atmosphere.

With all other planetary bodies, with the whole of nature beyond the limits of our atmosphere, we are only brought into relationship by means of light, of radiant heat, which is scarcely to be separated from light([60]), and the mysterious force of attraction which distant masses exert upon our earth, our ocean, and our atmosphere, according to the quantity of their material parts. We recognize a totally different kind of cosmic, and most peculiarly material relationship, in the fall of shooting-stars and meteoric stones, when we regard them as planetary asteroids. These are no longer bodies, which, through the mere excitement of pulses, influence us from a distance by their light or their heat, or which move and are moved by attraction; they are material bodies, which have come from the realms of space into our atmosphere, and remain with our earth. Through the fall of a meteoric stone, we experience the only possible contact of aught that does not belong to our planet. Accustomed to know all that is non-telluric solely through measurement, through calculation, through intellectual induction, we are amazed when we touch, weigh, and subject to analysis a mass that has belonged to the world beyond us. Thus does the reflecting, spiritualized excitement of the feelings work upon imagination, in circumstances where vulgar sense sees nothing but dying sparks in the clear vault of heaven, and in the black stone that falls from the crackling cloud the crude product of some wild force of nature.

If the crowd of shooting asteroids, upon which we have paused so long with pleasure, be assimilated in some respects, in their small masses and in the variety of their orbits, with comets, they are still essentially distinguished from these bodies in this—that we first become aware of their existence almost in the moment of their destruction, when fettered by the earth they become luminous, and ignite. But to embrace everything that belongs to our solar system, which has now become so complex, so rich in variety of forms, by the discovery of the telescopic planets, of the inner comets of short period, and the meteoric asteroids, we have still to speak particularly of the ring of Zodiacal Light, to which we have already alluded incidentally oftener than once. He who has lived for years in the zone of the palms, retains a delightful recollection of the mild radiance with which the zodiacal light, rising like a pyramid from the horizon, illumines a portion of the unvarying length of the tropical night. I have seen it occasionally more intensely luminous than the milky way in Saggitarius ; and that not only in the thin and dry atmosphere of the summits of the Andes, at the height of twelve or fourteen thousand feet above the level of the sea, but also in the boundless grassy plains (Llanos) of Venezuela, as well as on the coasts of the ocean under the ever-serene sky of Cumana. Of most peculiar beauty was the phenomenon, when small fleecy clouds appeared projected upon the light, and stood out picturesquely from the luminous back-ground. A leaf of my journal, during the sea voyage from Lima to the western coast of Mexico, preserves the memorial of this air-picture : "For the last three or four nights (between 10° and 14° N. lat.) I see the zodiacal light with a splendour such as I have never observed before. In this part of the Pacific, judging from the brilliancy of the stars, and the distinctness of the nebulæ, the transparency of the air is wonderfully great. From the 14th to the 19th of March, very regularly for three-quarters of an hour after the disc of the sun has dipped into the sea, there is no trace of the zodiacal light, although it is by this time completely dark ; but, an hour after sun-set, it suddenly becomes visible, of great brilliancy, between Aldebaran and the Pleiades ; and on the 18th of March having an altitude of 39° 5′. Long narrow stripes of cloud show themselves, scattered over the beautiful blue, and deep on the horizon in front of a kind of yellow screen. The higher clouds are playing from time to time with variegated tints. It seems as if the sun were setting for the second time. On this side of the vault of heaven, the brilliancy of the night appears to be increased, almost as it is in the first quarter of the moon. Towards ten o'clock, the zodiacal light, in this part of the Pacific, was usually extremely faint ; about midnight I could merely perceive a trace of it. On the 16th of March, when the phenomenon presented itself in its greatest splendour, there was a counter-blush of mild light apparent in the east." In our misty northern temperate zone, as it is called, the zodiacal light is only to be distinctly seen in the early spring, after the evening twilight, in the western, and towards the end of autumn before the morning twilight, in the eastern horizon.

It is difficult to comprehend how a natural phenomenon, so remarkable as the zodiacal light, should only first have attracted the attention of natural philosophers and astronomers about the middle of the 17th century, and how it could have escaped the observant Arabians in Ancient Bactria, on the Euphrates, and in the south of Spain. The tardy observation of the nebulæ in Andromeda and Orion, first described by Simon Marius and Huygens, excites almost equal astonishment. The first distinct description of the zodiacal light is contained in Childrey's Britannia Baconica(61), of the year 1661 ; the first observation upon it may have been made two or three years earlier ; but Dominic Cassini has the indisputable merit of having, in the spring of 1683, investigated the phenomenon in all its relations in space. The luminous appearance which he observed in 1668, at Bologna, and which was seen at the same time in Persia by the celebrated traveller, Chardin, (the court-astrologers of Ispahan called this light, which they had never seen before, nyzek, or little lance,) was not, as has been frequently said(62), the zodiacal light, but the monstrous tail of a comet, whose head was hidden amidst the vapours of the horizon, and which, in point of length and appearance, presented many points of resemblance to the great comet of 1843. It might be maintained, with no slight show of probability, that the remarkable light, rising pyramidally from the earth, which was seen in the eastern sky for forty nights in succession, on the lofty plateau of Mexico in 1509, was the zodiacal light. I find this phenomenon mentioned in an ancient Aztekan manuscript (Codex Telleriano-Remensis) of the Royal Library at Paris(63).

The Zodiacal Light, of primeval antiquity, doubtless, though first discovered in Europe by Childery and Cassini, is not the luminous atmosphere of the sun itself ; for this, from mechanical laws, cannot be more oblate than in the ratio of two to three, and not more dilated than 9-20ths of Mercury's distance. The same laws determine that, in the case of a revolving planetary body, the height or distance of the extreme limits of its atmosphere—the point, namely, where gravity and the centrifugal force are in equilibrium—is that alone in which a satellite can revolve around this in the same time as the primary rotates upon its axis(64). Such a limitation of the sun's atmosphere in its present concentrated state, comes to be more particularly remarkable when we compare the central body of our system with the nucleus of other nebulous stars. Herschel discovered many in which the semidiameter of the burr which surrounds the star appears under an angle of 150″. Assuming a parallax which does not quite reach 1″, we find the outermost nebulous layer of such a star 150 times farther from its centre than the earth is distant from the sun. Were the nebulous star in the place of our sun, consequently, its atmosphere would not merely include the orbit of Uranus, but would extend 8 times beyond it(65).

With the narrow limits of the sun's atmosphere now indicated, there is great probability in the hypothesis which assumes the existence of an extremely oblate ring of nebulous or vaporous matter revolving freely in space between the orbits of Venus and Mars, as the material cause of the zodiacal light[66]. Meantime, of its proper material dimensions, of its increment by emanations from the tails of myriads of comets which approach near to the sun[67], of the singular variability of its extent —for it seems at times not to extend beyond the orbit of the earth, and lastly, of its very probable close connection with the denser world-ether in the vicinity of the sun—nothing certain can be concluded. The vaporiform particles of which the ring consists, and which circulate about the sun in conformity with planetary laws, may either be self-luminous, or lighted by the sun. Even a terrestrial haze or fog (and the fact is very remarkable) appeared at the time of the new moon (1743), which at midnight was so phosphorescent that objects at the distance of 600 feet could be plainly distinguished by its light[68]. In the tropical climate of South America, the variable strength of light of the zodiacal gleam struck me at times with amazement. As I there passed the beautiful nights in the open air, on the banks of rivers and in the grassy plains (Llanos) for several months together, I had opportunities of observing the phenomenon with care. When the zodiacal light was at its very brightest, it sometimes happened that but a few minutes afterwards it became notably weakened, and then it suddenly gleamed up again with its former brilliancy. In particular instances, I believed that I remarked—not any thing of a ruddy tinge, or an inferior arched obscuration, or an emission of sparks, such as Mairan describes, but a kind of unsteadiness and flickering of the light. Is it that there are any processes going on in the vaporous ring itself? or is it not more likely that, though I could detect no change, by the meteorological instruments, in the temperature and moistness of the regions of the atmosphere immediately above the ground, and though small stars of the fifth and sixth magnitudes appeared to shine with undiminished strength of light, that in the superior strata of the atmosphere condensations were proceeding which modified the transparency, or rather the reflection of the light, in a peculiar and, to us, unknown manner? For the assumption of such meteorological processes on the limits of our atmosphere, the "explosions and pulsations" observed by the acute Olbers[69], "which, in the course of a few seconds, went trembling through the whole of a comet's tail, with the effect now of lengthening, now of abridging it by several degrees," appear to vouch. "As the several parts of the millions-of-miles-long tail are at very different distances from the earth, the laws of the velocity and propagation of light do not permit us to suppose that actual alterations in a body filling an extent of space so vast, could be perceived by us in such short intervals of time." These considerations by no means exclude the reality of varying emanations around the condensed nuclear envelopes of a comet, the reality of suddenly supervening brightenings of the zodiacal light, through internal molecular movements, through alternately augmented or diminished reflections of light by the matter of the luminous ring; they should only make us careful to distinguish between them and all that belongs to the celestial ether—to universal space itself, or to the aërial strata composing the atmosphere through which we see. What in other respects takes place in the outer limits of our atmosphere—the subject of great diversity of opinion—is, as well-observed facts indicate, by no means to be completely or satisfactorily explained. The wonderful lightness of many whole nights of the year 1831, in which small print could be read at midnight in Italy and the north of Germany, is in obvious contradiction with all that the latest and ablest observations on the crepuscular theory, and the height of our atmosphere, make known[70]. Luminous phenomena are dependent on conditions that are yet unexplored, the unstableness of which, within the limits of the twilight, as well as in connection with the zodiacal light, strike us with astonishment.

Thus far we have considered what belongs to our sun, and the world of formations that is ruled by him—the primary and secondary planets, comets of shorter and longer periods of revolution, meteoric asteroids which move singly in closed rings, or in multitudes like a stream; finally, a luminous nebulous ring which circles round the sun near to the orbit of the earth, and which from its position may remain with its name of zodiacal light. Every where the Law of Return prevails in the motions, how different soever the measure of the projectile velocity and the quantity of conglobated material parts; the asteroids alone, which fall from space into our atmosphere, are interrupted in their planetary round, and united to a larger planet. In the solar system, whose limits the attractive force of the central body determines, comets, at the distance of forty-four times the distance of Uranus from the sun, are compelled to return in their elliptical orbits; in these comets themselves, indeed, whose nuclei, from the smallness of the masses they comprise, present themselves to us in the guise of flitting cosmic clouds, these nuclei, nevertheless, bind, by their attractive force, the very outermost particles of the tail that is streaming away at the distance of millions of miles from them. The central forces, therefore, are the forming, the fashioning, and even the preserving forces of a system.

Our sun, in its relations to all the returning or circulating, greater or smaller, denser or almost vaporiform bodies that belong to it, may be regarded as at rest; yet does it revolve around the common centre of gravity of the whole system, which, however, still falls within itself; which, in other words, despite the variable position of the planets, still remains attached to its material bounds. Altogether different from this phenomenon, is the motion of translation of the sun—the progressive motion of the centre of gravity of the entire solar system in Universal space. This goes on with such velocity, that, according to Bessel, the relative motions of the sun and of the 61st star in Cygnus do not amount to less than 834,000 geographical miles in a day[71]. This change of place of the whole solar system would remain unknown to us, were it not that the wonderful perfection of modern astronomical instruments for taking measurements, and the advances of the astronomy of observation, render our progress obvious towards distant stars as towards objects on a coast apparently in motion. The proper motion of the 61st star in the constellation of the Swan, for example, is so considerable, that in the course of 700 years it will have amounted to a whole degree.

The measure or quantity of alteration in the heaven of the fixed stars—of alteration in the relative positions of the self-luminous stars to one another—can be determined with more of certainty than the phenomenon itself can be genetically explained. Even after we have allowed for all that belongs to the precession of the equinoxes and the nutation of the earth's axis, as consequences of the influence of the sun and moon upon the spheroidal figure of our planet, to the propagation or aberration of light, and to the parallax produced by diametrically opposite positions of the earth in its orbit round the sun—when a correction has been made for each and all of these particulars, there is always a quantity in the remaining annual motion of the fixed stars, which is the consequence of the translation of the whole solar system in space, and which is the consequence of the proper and actual motion of the stars themselves. The difficult numerical separation of these two elements, of the proper from the apparent motion, has been made possible by the careful specification of the directions in which the motions of the several stars take place, and by the reflection that, were all the other stars absolutely at rest, they would appear to recede perspectively from the point towards which the sun was moving in his course. The final result of the investigation, which the calculus of probabilities confirms, is this: that both the stars and our sun change their place in the Universe. From the admirable researches of Argelander(72), who in Abo extended and materially improved upon the labours begun by the elder Herschel and Prevost, it appears that the sun is in motion towards the constellation of Hercules, very probably towards a point in this constellation, which lies in a combination of 537 stars (for the equinox of 1792·5) in 257° 49′ Right Ascension; + 28° 49′·7 Declination. In this class of investigations it is always matter of great difficulty to separate the absolute from the relative motion, and to determine what belongs to the solar system in particular and alone.

If the non-perspective proper motions of the stars be considered, many of them appear groupwise opposed in their directions; and the data hitherto collected make it at least not necessary to suppose that all the parts of our astral system, or the whole of the star-islands which fill the universe, are in motion about any great, unknown, luminous, or non-luminous central mass. The longing to reach the last or highest fundamental cause, indeed, renders the reflecting faculty of man as well as his fancy disposed to adopt such a supposition. The Stagirite himself has said—"All that is in motion refers us to a Mover, and it were but an endless adjournment of causes were there not a primary immoveable Mover"(73).

The manifold changes of place exhibited by the fixed stars in groups, not parallactic motions, dependent on changes in the position of the observer, but actual and ceaseless motions in universal space, reveal to us in the most incontrovertible manner, through a particular class of phenomena, namely the motions of the double stars, and the measure of their slower or more rapid motions in different parts of their elliptical orbits, the empire of the laws of gravitation beyond the limits of our solar system, in the remotest regions of creation. The curiosity that is inherent in the nature of man needs not any longer to seek satisfaction upon this field of inquiry in gratuitous assumptions, in the limitless ideal-world of analogies. By the progress of the astronomy of observation and calculation, it stands at length even here upon stable ground. It is not so much the numbers of the double and multiple stars that have been discovered (2,800 to the year 1837!) circulating about a centre of gravity lying beyond the confines of either or any of them, that excites our amazement; it is the extension of our knowledge of the fundamental force of the whole material world, the indications of the universal dominion of mass-attraction, that arrest us, and that belong to the most brilliant discoveries of our age. The time of revolution of double stars of different colours presents the greatest imaginable diversity; it extends from a period of 43 years, as in η Coronæ, to one of several thousands, as in 66 Ceti, 38 Geminorum, and 100 Piscis. Since Herschel's measurements in 1782, the nearest leader in the triple system of ζ Cancri, has now accomplished more than a complete revolution. By a skilful combination of observations of altered distances and angles of position(74), the elements of the orbits of more than one of the double stars have been discovered—nay, conclusions as to the absolute distance of double stars from the earth, and comparisons of their masses with the mass of the sun, have even been made. But whether here, and in our solar system, the quantity of matter is the sole measure of the force of attraction, or whether specific attractions, not in proportion to the mass, are at the same time efficient, as Bessel first showed, is a question the solution of which it remains with late posterity to accomplish(75).

If we compare our sun, with the other so-called fixed stars in the Astral system to which we belong, with other self-luminous suns, therefore, we discover, in connection with several of them at least, ways opened up, which enable us to approximate, within certain extreme limits, to a knowledge of their distance, of their volume, of their mass, and of the rapidity with which they change their places. If we assume the distance of Uranus from the sun, at 19 of the distances of the earth from the sun then is the central body of our planetary system 11,900 Uranus distances from the star α Centauri, almost 31,300 of these distances from 61 Cygni, and 41,600 of the same measures from α Lyræ. The comparison of the volume of the sun with the volume of fixed stars of the first magnitude, depends on an extremely uncertain optical element; viz., the apparent diameter of the fixed stars. If, with Herschel, we assume the apparent diameter of

Arcturus at but one-tenth part of a second, the actual diameter of this star would still come out eleven times greater than that of our sun([76]). The distance of the star 61 Cygni, for the discovery of which we are indebted to Bessel, has led us approximatively to a knowledge of the quantity of material particles, which, as a double star, it contains. Although the portion of the apparent path which has been passed through since Bradley's observations, is not yet sufficiently great to enable us to conclude with perfect certainty upon the true path, and the semi-axis major of the same, it has still become matter of probability to the great astronomer of Konigsberg, "that the mass of the double star in question is not materially either less or more than half the mass of our sun([77])." This is the conclusion from actual measurement. Analogies which are derived from the greater masses of the moon-attended planets of our solar system, and from the fact that Struve finds six times as many double stars among the brighter fixed stars as among the telescopic ones, have led other astronomers to conjecture that the mass of the greater number of the twin-stars is in the mean greater than that of the sun([78]). General results, however, cannot be looked for in this direction for long years to come. With reference to proper motion in space, our sun, according to Argelander, belongs to the class of fixed stars which are in rapid motion.

The view of the heavens inlaid with stars, the relative position of the stars and nebulous spots, as also the distribution of their luminous masses, the charms of the landscape, if I may here make use of the expression, presented by the firmament at large, will depend, in the course of millenniums, relatively on the proper actual motions of the stars and nebulæ, on the translation of our solar system in space, on the bursting out of new stars, and on the disappearance, or sudden diminution in the intensity of light in old stars; finally, and especially, on the alterations which the axis of the earth experiences through the attraction of the sun and moon. The beautiful stars of the Centaur and the southern Cross will one day become visible in these northern latitudes, whilst other stars and constellations, Sirius and Orion's belt, will have sunk. The stationary north pole will be indicated in succession by stars in Cepheus (β and α), and the Swan (δ), until, after the lapse of 12,000 years, Vega in Lyra will appear as the most brilliant of all the possible polar stars. These statements serve to bring sensibly before us the vastness of the motions which in infinitely small divisions of time go on incessantly like an eternal clock—the timepiece of the Universe. If we imagine, as in a vision of the fancy, the acuteness of our senses preternaturally sharpened, even to the extreme limit of telescopic vision, and incidents compressed into a day or an hour, which are separated by vast intervals of time, everything like rest in spacial existence will forthwith disappear. We shall find the innumerable host of the fixed stars commoved in groups in different directions; nebulæ drawing hither and thither, like cosmic clouds; the milky way breaking up in particular parts, and its veil rent; motion in every point of the vault of heaven, as on the surface of the earth, in the germinating, leaf-pushing, flower-unfolding organisms of its vegetable covering. The celebrated Spanish botanist, Cavanilles, first conceived the thought of "seeing grass grow," by setting the horizontal threads of a micrometer attached to a powerful telescope, at one time upon the tip of the shoot of a Bambusa, at another upon that of the fast-growing flowering stem of an American aloe (Agave Americana), precisely as the astronomer brings a culminating star upon the cross wires of his instrument. In the aggregate life of nature, organic as well as sidereal, Being, Maintaining, and Becoming, are alike associated with motion.

The disruption of the milky way, to which I have alluded above, seems to require a more particular explanation in this place. William Herschel, our safe and admirable guide in these regions of space, discovered, by means of his star-gaugings, that the telescopic breadth of the milky way is six or seven degrées greater than it appears upon our maps of the heavens, and than the star-glimmer indicates it to the unassisted eye([79]). The two brilliant nodes in which both branches of the milky zone unite, in the regions of Cepheus and Cassiopea, as in those of Scorpio and Sagittarius, appear to exercise a powerful attraction upon the neighbouring stars; betwixt β and γ Cygni, however, in the most brilliant region, of 333,000 stars that lie in 5° of latitude, one-half draw towards one side, the other half towards the opposite side. Here Herschel suspects that the stratum breaks up([80]). The number of the distinguishable telescopic stars of the milky way—stars that are broken by no nebulæ—has been estimated at eighteen millions. In order, I will not say to give any idea of the magnitude of this number, but to contrast it with something analogous, I will remind the reader, that of stars between the 1st and 6th magnitude, that are visible to the naked eye, there are but some 8,000 scattered over the whole face of the heavens. In the barren astonishment, excited by vastness of number and of space, without reference to the spiritual nature or the faculty of perception inherent in man, extremes in respect of dimensions of the things that exist in space, likewise meet and contrast—the heavenly bodies with the smallest forms of animal life: a cubic inch of the tripoli of Bilin, contains, according to Ehrenberg, 40,000 millions of the siliceous coverings of the Galionellæ!

To the milky way of stars, to which, according to Argelander's acute observation, many of the bright stars of the firmament appear remarkably to approximate, there is a milky way of nebulæ opposed almost at right angles. The former, according to Sir John Herschel's views, forms a ring, a detached and somewhat remote girdle, from the lenticular star-island similar to the ring of Saturn. Our planetary system lies excentrically, nearer to the region of the Cross than to the diametrically opposite point of Cassiopea([81]). The form of our astral stratum, and the parted ring of our milky way, present themselves reflected with wonderful similarity in a nebula discovered by Messier, in 1774, but imperfectly seen by him([83]). The milky way of the nebulæ does not properly belong to our astral system; it surrounds this, without having any physical connection with it, at a vast distance, and passes nearly in the form of a great circle through the thick nebulosity of Virgo (particularly in the northern wing), through the Coma Berenices, the Great Bear, the girdle of Andromeda, and the Northern Fish. It probably intersects the starry milky way in Cassiopea, and connects its poles, which are poor in stars, made desolate by cluster-forming forces, at the place where the stratum of stars is of least thickness in space(83).

It follows, from these considerations, that whilst our cluster of stars bears traces, in its diverging branches, of greater transformations effected in the lapse of time, and strives, through secondary points of attraction, to resolve and decompose itself, it is surrounded by two rings, one vastly remote, made up of nebulæ, and one nearer, consisting of stars. The latter ring, which forms our milky way, is a mixture of unnebulous stars, on an average from the 10th to the 11th magnitude(84), but, severally observed, of very dissimilar magnitudes, whilst isolated clusters of stars have almost always the character of sameness.

Wherever the vault of heaven is searched with powerful space-penetrating telescopes, stars, though perchance telescopic only, and from the twentieth to the twenty-fourth in order, or luminous nebulæ, are discovered. Numbers of these nebulæ will probably resolve themselves into stars, when they come to be examined with yet more powerful instruments. Our retina receives the impression of single or of thickly aggregated luminous points; whence, as Arago has lately shown, totally different photometrical relations of the sensibility to light result(85). The cosmic nebulosity, formless or fashioned, generally diffused, producing heat by condensation, probably modifies the transparency of space, and lessens the equal intensity of luminousness which, according to Halley and others, must result, were every point of the vault of heaven beset with an endless succession of stars in the direction of its depth(86). The assumption of any such continuous inlaying of stars contradicts observation; which, in fact, shows us vast starless regions—OPENINGS IN HEAVEN, as William Herschel calls them—one in Scorpio, four degrees in breadth, and another in the loin of Ophiucus; in the vicinity of both of which, and close to their edges, we discover resolvable nebulæ. That which is situated on the western edge of the opening in Scorpio, is one of the richest and most thickly set clusters of small stars that ornament the heavens. Herschel himself ascribes the openings, the starless regions in the sky, to the attraction and cluster-forming force of these marginal groups(87). "They are portions of our star-stratum," says he, in the fine liveliness of his style, "which have suffered great desolations from time." If we picture to ourselves the telescopic stars that lie one behind another, as forming a starry canopy investing the whole of the visible vault of heaven, then, I believe, are those starless regions of the Scorpion and Serpent-bearer, to be regarded as tubes, through which we see into the farthest regions of space. The layers of the canopy are interrupted; other stars, indeed, may lie within the gaps, but they are unattainable to our instruments. The sight of fiery meteors had already led the ancients to the idea of clefts and chasms in the canopy of heaven; but these were regarded as passing or temporary only. Instead of being dark, they were luminous and fiery, by reason of the translucent igneous ether that lay behind them(88). Derham, and even Huyghens, appear not indisposed to explain the mild light of nebulæ on some such grounds(89).

When we compare the brilliant, and on an average certainly nearer, stars of the first magnitude, with the telescopic or resolvable nebulæ, and contrast the nebulous stars with the wholly unresolvable nebulæ (with the one in Andromeda, for example), or even with the so-called planetary nebulæ, in the contemplation of distances so different, plunged, as it were, in the boundlessness of space, we have a fact revealed to us by the world of phenomena, and the reality, which, in causal connection with it, always forms its substrate—the fact of THE PROPAGATION OF LIGHT. The rate of this propagation, according to Struve's latest researches, is 41,518 geographical [166,072 English] miles in a second; nearly a million times greater, therefore, than the rate of sound. From what we know through the measurements of Maclear, Bessel, and Struve, of the parallaxes and distances of three fixed stars of very unequal magnitudes—α Centauri, 61 Cygni, and α Lyræ—a ray of light requires 3 years, 9¼ years, and 12 years, to reach us from these celestial bodies severally. In the short but remarkable period from 1572 to 1604, from Cornelius Gemma and Tycho to Kepler, three new stars blazed suddenly forth in Cassiopea, in Cygnus, and in the foot of Ophiucus. The same phenomenon showed itself in 1570 in the constellation of the Fox; but here it recurred several times. In the very latest times, since 1837, Sir John Herschel during his sojourn at the Cape of Good Hope observed the star η of the constellation Argo increase in brilliancy from a star of the second magnitude to one of the first(90). Such incidents in the universe belong, however, in their historical reality, to other times than those in which the phenomena of light notify their commencement to the inhabitants of the earth; they are the voices of the past which reach us. It has been well said, that with our mighty telescopes we penetrate at once into space and into time. We measure the former by the latter, the latter by the former; an hour of travel for the ray of light is one hundred and forty-eight millions of geographical miles passed through. Whilst the dimensions of the universe are expressed in the theogony of Hesiod by the fall of heavy bodies—"the brazen anvil falls in no more than nine days and nine nights from heaven to earth"—Herschel, the Father(91), believed "that the light of the farthest nebulæ, which his forty-feet reflector showed him, took about two millions of years to reach the earth." Much, therefore, has long disappeared, much has already been otherwise arranged, before it becomes visible to us. The aspect of the starry heavens presents us with evidences of diversity in point of time; and diminish as we will the millions or even thousands of years which serve us as measures for the distance of the unresolvable nebulæ with their soft lustre, and of the resolvable nebulæ with their twilight gleamings, bring them as close to us as we may, it still remains more than probable, from the knowledge we have of the velocity of light, that the light of the remote celestial bodies offers the oldest sensible evidence of the existence of matter. So rises reflecting man, from his stance on simple premises, to solemn and noble views of natural formations to the deep fields of space, where flooded with everlasting light—

"Myriads of worlds spring up like the grass of night."(92)

From the region of celestial formations, from the children of Uranos, we now descend to the narrower domain of terrestrial forces, to the children of Gæa. A mysterious band surrounds and binds together both classes of phenomena. In the import of the old Titanian Mythus (93), all the powers of the universal life, the whole mighty order of nature, is connected with the co-operation of the heavens and the earth. And, indeed, if the terrestrial ball, like all the other planets, belongs, in virtue of its origin, to the central body, the sun, and to its atmosphere, once parted into nebulous rings, an intercourse is still kept up, by means of light and radiant heat, with this neighbouring sun, as with all the farther suns that sparkle in the firmament. The diversity of the mass of these influences must not restrain the physical astronomer from referring in a natural picture to the connection and the dominion of common and similar forces. A small fraction of the terrestrial heat belongs to that of the universal space through which our planetary system pursues its way, and which, the product of all the light-radiant stars, is nearly of the mean temperature of our icy circumpolar regions, according to Fourier. But what it is that excites the light of the sun more powerfully in the atmosphere and upper strata of the earth—how, producing heat, it gives rise to electrical and magnetical currents—how it magically kindles and beneficially feeds the flame of life in the organic forms that people the earth—all this will form the subject of our considerations by and by.

Whilst we here apply ourselves exclusively to the telluric sphere of nature, then, we shall first take a glance at the relative proportions of the Solid and the Fluid, at the figure of the earth, its mean density, and the partial distribution of this density in the interior of the planet; at the contained heat, and the magnetic charge of the earth. These relations in respect of space, and these forces inherent in matter, lead to the reaction of the interior upon the exterior of our earth; they lead through the special consideration of an universally diffused natural force—sub-terrestrial heat—to the not always merely dynamic phenomena of earthquakes in circles of concussion of various extent, to the outbreak of hot springs, and the mightier operations of volcanic processes. The crust of the earth shaken from below, now in pulses, suddenly and violently, now smoothly and continuously, and therefore scarcely perceptibly, alters in the course of centuries the relations in point of elevation between the Dry and the surface-level of the Fluid; nay, the form of the bed of the ocean itself. There are, at the same time, either temporary cracks, or more permanent openings formed, through which the interior of the earth comes into relationship with the atmosphere. Welling up from unknown depths, molten masses flow in narrow streams along the slopes of the mountains, here precipitously, there slowly, gently, until the fiery spring runs dry, and the lava, emitting vapours, solidifies beneath a crust which it has formed for itself. New rocky masses then arise before our eyes, whilst older ones, already formed by Plutonic forces, suffer change, rarely through immediate contact, more frequently from their vicinity to heat-radiating centres or masses. In situations where there is no eruption, crystalline particles are still displaced, and then combined into denser textures.

The waters present us with formations of a totally different nature: aggregations of the remains of plants and animals; earthy, cretaceous, and clayey deposits; conglomerates of finely pulverized mountain species, overlaid by layers of siliceous-shelled infusoria, and bone-containing drift, the resting place of the remains of animals that peopled a former world. All that we see engendered in such variety of ways beneath our eyes, and arranged in layers, all that we observe so variously cast down, and bent, and raised again, under the influence of opposing pressure and volcanic force, leads the reflective observer, who yields himself to the guidance of simple analogies, to the comparison of the Present with times that have long gone by. Through combination of actual phenomena, through ideal amplification in reference to the extent as well as to the mass of the forces in operation, we reach at length the long-desired, the dimly-imagined, but first, in the course of the last century, firmly-founded domain of geognosy.

It has been acutely observed, that, "with all our looking through powerful telescopes, we actually know more of the interior of other planets than of their exterior—the moon, perhaps, excepted." They have been weighed, and their volumes have been measured; their masses and their densities are known, in either case—thanks to the progress of the astronomy of observation and calculation—with still increasing numerical certainty. Over their physical constitution there hangs a deep obscurity. It is only in our own earth that immediate vicinity brings us into contact with the various elements of organic and inorganic creation. Here the garner of matter, in its multifarious diversity, in its endlessness of admixture and modification and change, in the ever-varying play of forces evoked, presents the spirit with its proper food: the joys of investigation, the unbounded field of observation, which, cultivating and strengthening the faculty of thought, gives to the intellectual sphere of man's existence a portion of its grandeur, of its sublimity. The world of sensible phenomena reflects itself in the deeps of the ideal world: the abundance of nature, the mass of things discernible, passes gradually into the domain of knowledge approved by reason.

And here, again, I touch upon an advantage to which I have already alluded several times—the advantage of that knowledge which has a home origin, and of which the possibility is most intimately connected with our earthly existence. The description of the heavens, from the far-gleaming nebulous stars (with their suns) down to the central body of our own system, we found limited to such general conceptions as volume and quantity of matter. No vital movement is there revealed to our senses. It is only after resemblances, often after fanciful combinations, that we arrive at conjectures as to the specific nature of matters of different kinds, as to its [presence or] absence in this or in that planetary body. The heterogeneousness of matter, its chemical diversity, and the regular forms into which its particles arrange themselves, as crystals and granules; its relations to the penetrating deflected or decompounded waves of light, to radiating, transmitted, or polarized heat, to the brilliant, or invisible, but not on that account less powerful, phenomena of electro-magnetism—all this vast treasury of physical knowledge, which so exalts our views of nature, we owe to the surface of the planet we inhabit, and to the solid rather than the fluid element in its constitution. How this knowledge of natural things and natural forces, how the measureless variety of objective perceptions, calls forth the intellectual activity of our kind, and hastens our progress in improvement, has been already observed upon above. These relations as little require farther development in this place, as the enchainment of the causes of that material force which the control of a portion of the elements has given to particular nations.

If it was imperative on me to direct attention to the difference which exists betwixt the nature of our telluric knowledge, and our knowledge of heavenly space and its contents, so is it also necessary for me to indicate the narrowness of the field from which the whole of our knowledge of the heterogeneousness of matter is derived. This field is somewhat inappropriately called THE CRUST OF THE EARTH; it is the thickness of the strata that lie nearest the surface of our planet, and that are exposed in deep chasm-like valleys, or by the labour of man in his boring and mining operations. These works scarcely attain a perpendicular depth of more than two thousand feet (less than $\frac{1}{11}$th of a German mile) below the level of the sea; consequently only $\frac{1}{9800}$th of the semidiameter of the earth([94]). The crystalline masses which are ejected by active volcanoes, and which are mostly of the same nature as the rocky matters of the surface, come from unknown, certainly sixty times greater absolute depths than those which the labours of man have reached. In situations where seams of coal dip to rise again at distances determinable by accurate measurements, it is easy to ascertain the depth of the basin in which the strata lie. In this way we learn, that in some places (Belgium, for example) the coal measures, together with the organic remains of a former world, which they contain, frequently lie more than five, or even six, thousand feet below the present level of the sea([95]): aye, that the mountain limestone and Devonian basin-shaped bent strata, descend even to twice that depth. If we now contrast these subterraneous basins with the mountain summits which have hitherto been held as the highest portions of the uplifted crust of the earth, we obtain a distance of 37,000 feet, or nearly $\frac{1}{524}$th of the earth's semidiameter betwixt the point of extreme descent and that of highest elevation. This, in the perpendicular dimension and space-filling superposition of rocky strata, would still be the only theatre of geognostic investigation, even did the general surface of the earth reach the height of Dhawalagiri, in the Himalaya chain, or of Sorata, in Bolivia. All that lies under the sea level deeper than the basins referred to above, than the works of man, than the bottom of the ocean, attained in various places with the plumb-line (Sir James Ross sounded with 25,400 feet of line, without reaching the bottom), is even as much unknown to us as is the interior of the other planets belonging to our system. We also know but the mass of the whole earth and its mean density, compared with the superior and to us solely accessible strata. Where all knowledge of the chemical and mineralogical natural constitution of the interior of the earth fails us, we are again thrown upon conjecture, just as we are with reference to the farthest bodies that revolve about the sun. We can determine nothing with certainty upon the depth at which the rocky strata of the crust of the globe should be regarded as existing in a tenacious softened state, or as a molten liquid; upon the cavities filled with elastic vapours; upon the condition of liquids when they are heated red-hot under enormous pressures; or upon the law of the increment of density from the surface of the earth down to its centre.

The consideration of the increment of temperature of the interior of our planet with increasing depths, and of the reaction of the interior upon the surface, has led us to the extensive series of volcanic phenomena. These manifest themselves as earthquakes, effusions of gaseous fluids, hot springs, mud-volcanoes, and lava-streams, from craters; the influence of elastic force is also shown in unquestionable alterations in the level of the general surface. Extensive levels, variously-partitioned continents, are upheaved or sunk; the solid is parted from the fluid; but the ocean itself, traversed by hot and cold currents that flow through it like rivers, congeals at either pole, and sets into solid rocky masses, here stratified and immoveable, there broken into moveable packs and islets. The boundaries of the sea and land, of the fluid and the solid, are variously and frequently changed. Plains, too, oscillate upwards and downwards. After the elevation of continents, long clefts or chasms took place, mostly parallel to one another, and then, in all probability, at similar epochs in time, and through them, were mountain-chains upheaved: salt pools and great inland seas, which were long inhabited by the same creatures, were forcibly separated. The fossil remains of shells and zoophytes bear witness to their original connection. And so we come, following the relative dependence of phenomena, from the consideration of the fashioning forces, working deep in the interior of the earth, to that which shakes and shatters its upper crust, and which, through the force of elastic vapours, flows out as a molten stream of earth (lava) from open fissures.

The same forces that uplifted the Andes and

Himalaya chains, even to the regions of eternal snow, produced new admixtures and new textures in the rocky masses, and altered the strata which had been thrown down at earlier periods, from waters teeming with life and organized matters. We recognize here the succession of formations, separated according to their age and superposed, in their dependence upon the alterations in form of the surface, upon the dynamical relations of the upheaving forces, upon the chemical actions of outbreaking vapours upon the fissures.

The form and distribution of continents—in other words, of the dry land—of that portion of the crust of the earth which is susceptible of the vigorous evolution of vegetable life, stands in intimate relationship, and potential reciprocity of action, with the all-surrounding sea. In this the organizing force is almost wholly expended upon the animal world. The liquid element, again, is invested by the gaseous atmosphere, an aerial ocean, into which the mountain chains and lofty plateaus of the dry land rise like reefs and shoals, induce a vast variety of currents and changes of temperature, collect moisture from the region of the clouds, and by the running streams that furrow their sides, spread motion and life over all.

If the Geography of Plants and Animals depends on these intricate contrasts in the distribution of sea and shore, in the formation of the surface, and the direction of isothermal lines (or zones of mean annual temperature), so, on the other hand, are characteristic differences in the races of men and their relative numerical distribution over the face of the earth—the last and noblest object of a physical description of the globe—influenced not by these natural relations alone, but at the same time, and especially by progress in civilization, in mental improvement, in political superiority grounded upon national cultivation. Some races, clinging to the soil, are supplanted and annihilated by the dangerous vicinity of more politic communities: a faint historical trace is soon all that remains of them ; other races, in numbers not the strongest, put forth upon the liquid element ; and almost omnipresent by means of this, have they alone, though late, attained to a general graphical knowledge of the surface, of all the seaboards at least, of our planet from pole to pole.

Here, then, and before I have touched upon the individual, in our NATURAL PICTURE OF THE TELLURIC SPHERE OF PHENOMENA, I have shown in General, how from considerations on the form of the globe, and on the ceaseless manifestations of force in its electro-magnetism and subterranean heat, the relations of the earth's surface in horizontal extension and elevation, the geognostic type of mineral formations, the realm of the ocean, and of the atmosphere with its meteorological processes, the geographical distribution of plants and animals, and, finally, the physical gradations of the human race, alone, but in all circumstances susceptible of spiritual culture, may be comprised in one and the same contemplative survey. This unity of contemplation presupposes an enchainment of phenomena according to their intimate connections. A mere tabular arrangement of phenomena would not accomplish the purpose I prescribed myself; it does not satisfy the want of that COSMICAL REPRESENTATION which the aspect of nature by sea and land, the diligent study of formations and forces, and the lively impression of a natural whole, which has been made upon my mind in the course of my travels in various and dissimilar climates of the globe. Much that in this essay is so exceedingly defective, with the accelerated rate at which knowledge of all the departments of physical science advances, will probably ere long be corrected and filled up. It lies, indeed, in the path of development which every science pursues, that that which long stood isolated, becomes connected by degrees and subjected to higher laws. I but point out the EMPIRICAL way, along which I, and many minded like myself, advance, full of expectation that "Nature," as Plato tells us Socrates once desired, "shall have interpretation according to reason"(96).

Our account of terrestrial phenomena, in their principal features, must begin with the form and relations in space of our planet. And here, too, it may be said, that not merely does the mineral constitution, the crystalline, the granular, the dense masses filled with petrefactions, but also the geometrical figure of the earth itself, bear witness to the mode of its origin ; its figure is its history. An elliptical spheroid of rotation indicates a once soft or semi-fluid mass. To the oldest geognostic incidents, writ down, and clearly legible to the understanding eye, in the book of nature, belongs the flattening [of the poles of the earth], and to adduce another and nearly related instance, the perpetual direction of the greater axis of the moon's spheroid towards the earth ; *i. e.* the accumulation of matter upon that half of the moon which we see, and which determines the relation between the period of rotation and that of revolution. And the same law extends to the oldest formative epochs of all the satellites. "The mathematical figure of the earth is that which it would have were its surface covered with water in a state of repose ;" to this are referred all geodetic measurements of degrees reduced to the sea-level. From this mathematical surface of the earth, the physical one, with all its accidents and inequalities of the solid, differs(97). The whole figure of the earth is determined when the quantity of oblateness and the magnitude of the equatorial diameter are known. To obtain a complete picture of the figure, however, it were necessary to have measurements in two directions perpendicular to each other.

Eleven measurements of degrees, or determinations of the curvature of the earth's surface in different countries, of which nine belong exclusively to the present century, have given us accurate information on the dimensions of the earth, which Pliny long ago designated as "a point in the infinity of space"(98). If these measurements do not agree in the curvature of different meridians under the same degrees of latitude, this very circumstance vouches for the sufficiency of the instruments and of the methods employed, for the accuracy of partial results true to nature. The inference from the increase of attractive force proceeding from the equator towards the pole, in reference to the figure of a planet, depends on the distribution of density in its exterior. If Newton, upon theoretical grounds, and also excited to the inquiry by Cassini's discovery of the flattening of Jupiter's poles in 1666[99], determines the flattening of the earth as a homogeneous mass at $\frac{1}{230}$th, in his immortal work, the Principia, actual admeasurements, under the influence of the new and more perfect analysis, have shown that the oblateness of the earth's spheroid, the density of the strata being assumed to go on increasing towards the centre, amounts to $\frac{1}{306}$th very nearly.

Three methods have been employed to determine fundamentally the curvature of the earth's surface: measurements of degrees, pendulum experiments, and certain inequalities of the moon's orbit. The first of these methods is an immediate geometro-astronomical one; in the other two, conclusions are drawn from carefully observed motions, in regard to the forces which occasion these motions, and, from these forces, in regard to their causes, viz. the oblateness of the earth in its polar axis. I have here, in the general picture of nature, referred exclusively to the application of these methods, because their certainty reminds us forcibly of the intimate concatenation of natural phenomena in their forms and forces, because this application has itself become the happy occasion of improving all our instruments, whether optical or those that are employed in the measurement of space or of time—the very foundation of astronomy and mechanics in reference to the moon's motions, and the determination of the resistance which the oscillation of the pendulum experiences—and because it has even served to open up peculiar and untrodden paths to analysis. After the researches on the parallax of the fixed stars, which led to the discovery of aberration and nutation, the history of the sciences presents us with no problem second in importance to that in which the result sought is a knowledge of the mean oblateness of the earth, and the certainty that the figure of our planet is not a regular one. In none of the long and laborious ways by which the goal is attained in scientific investigations, is higher general cultivation, or more perfect knowledge of mathematical and astronomical science required than in this. The comparison of eleven measurements of degrees, among which three extra European—the old Peruvian one, and two East-Indian—are included, calculated in conformity with the severe theoretical requirements of Bessel, has given $\frac{1}{299}$th as the measure of oblateness of the polar diameter of the earth[100]. From this it appears that the polar semidiameter is 10,938 toises, about $2\frac{7}{8}$ geographical miles, shorter than the equatorial semi-diameter of the elliptical spheroid of rotation. The bulging under the equator, therefore, in consequence of the curvature of the surface of the spheroid in the direction of gravity, comes to something more than $4\frac{3}{7}$ times the height of Mont Blanc, only $2\frac{1}{2}$ times the probable height of Dhawalagiri, in the Himalaya range. The moon's equation, in other words the perturbation in longitude and latitude of the moon, from the latest researches of Laplace, give nearly a similar degree of oblateness as the measurement of degrees of the meridian—viz. $\frac{1}{299}$th. Experiments with the pendulum indicate a much more considerable amount of flattening—viz. $\frac{1}{288}$th[101].

Galileo, when a boy, during divine service, and somewhat inattentive to the matter in hand, as it would seem, perceived that the whole height of a roof might be ascertained from the dissimilar times in which chandeliers, suspended at different elevations, oscillated; but he certainly did not imagine that the pendulum would one day be carried from pole to pole, with a view to determine the figure of the earth; or rather to afford evidence of the length of the seconds-pendulum being affected by strata of the earth of unequal density. These local attractions are complex, undoubtedly; but over extensive districts of country they show themselves almost identical in point of amount. These geognostic relations of an instrument for the measurement of time; this peculiar property of the pendulum to act as a plumb-line, and give us intelligence of the unseen deep, even in volcanic islands[102], and on the acclivities of uplifted continental mountain chains[103], to indicate dense masses of basalt and melamphyx instead of caverns, combine to render difficult, despite the wonderful simplicity of the method, the attainment of any general result as to the figure of the earth from observations on the oscillation of the pendulum. Even in the astronomical part of the measurement of a degree of latitude, the occurrence of mountain masses, or of denser strata in the ground, have a disturbing and prejudicial influence, although not to the same extent as in pendulum experiments.

As the figure of the earth exerts a powerful influence on the motion of other planetary bodies, especially on that of her immediate satellite, so, on the other hand, does the very perfect knowledge we possess of the motion of the moon enable us to draw counter-conclusions in regard to the figure of the earth. From this, as Laplace[104] has significantly observed, might an astronomer, "without leaving his observatory, by a comparison of the lunar theory with positive observations, determine, not only the figure and magnitude of the earth, but farther, its distance from the sun and from the moon; results which have only been obtained by long and toilsome journeys undertaken to the remotest countries of either hemisphere." The oblateness which has been deduced from the inequalities of the moon has this advantage, possessed neither by single measurements of degrees nor pendulum observations, that it is a MEAN applicable to the whole planet. Contrasted with the velocity of rotation, it informs us, moreover, of the increase of density of the earth's strata from the surface towards the centre; an increase which the comparison of the relation of the axes of Jupiter and Saturn with their periods of rotation also reveals in both of these great planets. In this way does knowledge of mere external configuration lead to conclusions in regard to the internal constitution of the heavenly bodies.

The northern and southern hemispheres appear to have nearly like curvatures under equal parallels of latitude[105]; but pendulum experiments, and measurements of degrees of the meridian, give such different results in reference to particular portions of the surface, that nothing like a regular figure can be inferred which would accord with the whole of the results hitherto obtained in these ways. The true figure of the earth stands in the same relation to a regular figure, "as the uneven surface of ruffled stands to the even surface of unruffled water."

After the earth has been MEASURED, it must be WEIGHED. Pendulum vibrations and the plumb-line have alike served to determine the mean density of the earth—whether the relative density was investigated by a combination of astronomical and geodetical operations, through the deflection of a plumb-line from the perpendicular in the vicinity of a mountain, or by contrasting the length of the pendulum beating seconds on a plain and on the summit of a neighbouring height, or, finally, by the application of the torsion-balance, which may be regarded as a delicate horizontally swinging pendulum. Of these three methods(106), the last is the safest, inasmuch as it is independent of the difficult determination of the density of the minerals composing the spherical segment of a mountain in the neighbourhood of which the observations are made. The latest researches, which are those of Reich, give 5·44 as the mean density of the whole earth; that is to say, the earth is nearly 5½ times more dense than pure water. But as the mineral species which constitute the dry land have a mean density of no more than about 2·7, and the dry land and the ocean together a density of but 1·6, it follows from this assumption how much the elliptical unequally oblated strata of the interior must increase in density through pressure, or through heterogeneousness of material towards the centre. And here we see, again, with what propriety the pendulum, both that which swings perpendicularly and that which swings horizontally, has been designated a geognostical instrument.

But the conclusions to which the use of such an instrument leads, have induced distinguished natural philosophers to take entirely opposite views of the constitution of the earth's interior. It has been calculated at what depth liquid, and even aëriform bodies, would come to surpass platinum, and even iridium, in density, through the proper pressure of their own superimposed strata; and in order to bring the oblateness of the earth's spheroid, known within a very small quantity, into harmony with the assumption of a single and infinitely compressible substance, the acute Leslie has gone so far as to have described the nucleus of the earth as a hollow sphere, filled with "imponderable matter of enormous repulsive powers." These daring and arbitrary conjectures have given rise to still more fantastical dreams in non-scientific circles. The hollow sphere has, by degrees, been peopled with plants and animals, and furnished, moreover, with a couple of small subterranean planets—Pluto and Proserpine, which there dispense their gentle light. An unvarying temperature reigns in this internal space, and the air, self-luminous by compression, might well make the presence of the subterraneous planets, Pluto and Proserpine, unnecessary. Near the north-pole, under the 82d parallel of latitude, where the aurora borealis streams up into the sky, there is an enormous opening, through which it were easy to descend into the hollow sphere. To such a subterranean expedition the late Sir Humphry Davy and I were repeatedly and publicly invited by Captain Symmes. So strongly is the morbid disposition of man inclined, unencumbered with the contradictory testimony of well-established facts or generally admitted natural laws, to fill unseen space with marvellous forms! But the celebrated Halley himself, at the end of the 17th century, had hollowed out the earth in the course of his magnetical speculations: a subterraneous freely rotating nucleus, by its varying position, occasions the diurnal and annual variations of the magnetical declination! What was a mere lively fiction with the clever Holberg, has, in our days, with tedious solemnity, been attempted to be decked out in a scientific garb.

The figure of the earth, and the degree of solidity or density which it possesses, stand in intimate connection with the forces which animate our globe, in so far, namely, as these forces are not excited or awakened from without by our planetary position opposite to a self-luminous central body. The oblateness, a consequence of the operation of the centrifugal force upon a rotating mass, reveals the pristine or former state of fluidity of our planet. On the setting or solidification of this fluid, which we are accustomed to conjecture as existing in the shape of a vaporiform matter, originally heated to a very high temperature, an enormous amount of latent caloric became free. If the process of consolidation began in the way Fourier will have it, by radiation from the surface into celestial space, the parts of the earth which are situated towards the centre must still be hot and molten. While, after long radiation of the heat of the central parts towards the surface, a state of stability in the temperature of the earth is finally attained, it is at the same time assumed that, with an increase in depth, there will also be a regular progressive increase of temperature. The temperature of the water which flows from bores of great depth into the bowels of the earth (Artesian wells), immediate experiments on the temperature of the rocks in mines, above all, however, the volcanic activity of the earth, in other words, the discharge of molten mineral streams through fissures in the surface, bear testimony in the most incontestable manner to this increase of temperature in the upper strata of the earth at considerable depths. From conclusions which, it is true, are only founded on analogy, it is more than probable that the temperature goes on increasing in a still greater degree towards the centre.

The conclusions which have been presented to us by an ingenious, and, for this class of inquiries, singularly perfect analytical calculus, on the motion of heat in homogeneous metallic spheroids(107), can only be applied, with many precautions, to the actual constitution of our planet, in consequence of our ignorance of the matter of which the earth is composed, of the various capacities for heat and powers of conduction inherent in the superimposed masses, and of the chemical transformations which solid and fluid bodies undergo under enormous pressures. Most difficult of all, for our powers of comprehension, is the conception of the boundary line betwixt the fluid masses of the interior and the concrete mineral species of the outer crust of the earth, of the gradual increase of solidity in the strata, and the state of tenacious semi-fluidity of earthy matters, to which the known laws of hydraulics can only apply under considerable modifications. The sun and moon, which keep the ocean in a state of alternate ebb and flow, act in all likelihood even down to these depths. Beneath a vault of already consolidated mineral strata, periodical rises and falls of a molten mass may, indeed, be readily enough conceived as taking place, and occasioning inequalities in the pressure exerted against the vault. The amount and the influence of such oscillations can, however, be but small; and if the relative position of the attracting heavenly bodies must here also produce spring-tides, it is still certain that the concussions of the earth's surface which take place, are not to be ascribed to these, but to other more powerful internal forces. There are groups of phenomena, the existence of which it is still useful to adduce in illustration of the universality of the attractive influences of the sun and moon upon the external and internal life of the globe, however little we may feel ourselves in a condition to determine numerically their amount.

From experiments on Artesian wells, which agree pretty closely, the temperature of the upper crust of the earth appears, on an average, to increase 1° of the centigrade thermometer for each 92 Paris feet in perpendicular depth. Did this increase go on in arithmetical progression, then, as I have already had occasion to observe(108), would a granitic stratum at the depth of $5\frac{2}{10}$ geographical miles (from four to five times the depth of the highest peak in the Himalaya range) be in a molten state.

In the body of the earth there are three kinds of motion of heat to be distinguished: the first is periodical, and, according to the position of the sun and the season of the year, alters the temperature of the earth's strata according as the heat penetrates from above downwards, or as it passes in the same way from below upwards. The second kind of motion is likewise an effect of the sun, and is of extraordinary slowness: part of the heat which has penetrated the equatorial regions is propagated along the interior of the crust of the earth towards the poles, and there escapes into the atmosphere and distant space. The third kind of motion is the slowest of all: it consists in the secular cooling of the body of the earth, in the dissipation of the small amount of the primitive heat of the planet which at the present time is still given off from its surface. This loss which the central heat suffers was very considerable at the epochs of the oldest revolutions of the globe; since the commencement of the historical period, however, it is scarcely mensurable by our instruments. The surface of the earth, from the foregoing view, is intermediate between the red heat of the interior strata, and the temperature of space, which is probably below the congealing point of mercury.

The periodical variations of temperature which the altitude of the sun and the meteorological processes of the atmosphere occasion, are propagated in the interior of the earth, but only to very small depths. This slow conduction of heat by the ground, however, lessens the loss of warmth in the winter, and is favourable to deeply-rooted trees. Points which lie at different depths in a vertical line come to the maximum and minimum of the communicated temperature in very different times. The more distant they are from the surface, the smaller are the differences of these extremes. On the continent of Europe, between the parallels of 48° and 52°, the stratum of invariable temperature occurs at from 55 to 60 feet deep; even at half this depth the oscillations of the thermometer, in consequence of the influence of the seasons, scarcely amount to half a degree. In tropical climates, on the contrary, the stratum of invariable temperature is met with at no more than a foot below the surface; and this fact has been used by Boussingault, in an able manner, as a convenient and, in his opinion, accurate way of determining the mean temperature of the air of a place(109). This mean temperature of the air at a determinate point, or in a group of points of the surface lying near to one another, is, in a certain measure, the fundamental element of the climatic relations, and also of the relations in reference to civilization of a country; but the mean temperature of the whole surface is very different from that of the earth itself. The oft-repeated questions, whether, in the course of centuries, this has suffered any considerable change? whether the climate of a country has become deteriorated? whether the winters have not become milder, and the summers in the same proportion colder? can only be decided by the thermometer; and the discovery of this instrument scarcely dates three half-centuries back; its rational application no more than about 120 years. The nature and novelty of the means, therefore, prescribe very narrow bounds to inquiries into the temperature of the air. It is quite otherwise with the solution of the great problem of the internal heat of the whole globe. In the same way as from the unaltered rate of a pendulum we can conclude on the unchanged preservation of its temperature, so does the unaltered velocity of rotation of the earth on its axis inform us of the degree of stability of its mean temperature. This perception of the relations between the length of the day and the earth's temperature, is one of the most brilliant applications of a long knowledge of the heavenly motions to the thermal condition of our planet. The velocity of rotation of the earth, to wit, depends on its volume: precisely as the axis of rotation of the mass that was cooling gradually by radiation would become shorter, so through diminution in temperature must the velocity of rotation be increased, and the length of the day be abridged. Now by a comparison of the secular inequalities of the moon's motions with the eclipses that have been observed in the more ancient times, it appears that since the age of Hipparchus, for full 2000 years therefore, the length of the day has not varied by the one-hundredth part of a second. From this, again, and, within the utmost limits of the decrease[110], the mean temperature of the body of the earth is discovered not to have altered, in the course of 2000 years, by the $\frac{1}{170}$th part of a thermometrical degree.

This invariableness of form farther implies great invariability in the distribution of density in the interior of the earth. The translatory movements effected by the eruptions of our present volcanoes, the outbursts of ferruginous lavas, and the filling up of empty chasms and hollows with dense masses of rock, are therefore to be regarded as mere superficial phenomena, as peculiarities of parts of the earth's crust, which, in point of magnitude, when contrasted with the semidiameter of the earth, are utterly insignificant.

The internal heat of the planet, in its course and distribution, I have described almost exclusively from the results and beautiful experiments of Fourier. Poisson, however, doubts the uninterrupted increase of the terrestrial heat from the surface to the centre. He believes that all the heat has penetrated from without inwards, and that the temperature of the interior of the earth depends on the very high or very low temperature of the universal space through which the solar system has moved. This hypothesis, devised by one of the most profound mathematicians of the age, has satisfied himself only; it has met with little countenance from other natural philosophers and geologists.

But whatever be the cause of the internal temperature of our planet, and of its limited or unlimited increase in the deeper strata, it still leads in this Essay to present a general picture of nature, through the intimate connection of all the primary phenomena of matter, and through the common bond which surrounds the molecular forces, into the obscure domain of MAGNETISM. Changes of temperature elicit magnetical and electrical currents. Terrestrial magnetism, whose principal character in the threefold manifestation of its force is an uninterrupted periodic changeableness, is ascribed either to the unequally heated mass of the earth itself[111], or to those galvanic currents which we consider as electricity in motion, as electricity in a circuit returning into itself[112]. The mysterious march of the magnetic needle is equally influenced by the course of the sun, and change of place upon the earth's surface. The hour of the day can be told between the tropics by the motion of the needle, as well as by the oscillations of the mercury in the barometer. It is suddenly, though only passingly, affected by the remote Aurora, by the glow of heaven, which emanates in colours at one of the poles. When the tranquil hourly motion of the needle is disturbed by a magnetical storm, the perturbation frequently proclaims itself over hundreds and thousands of miles, in the strictest sense of the word simultaneously, or it is propagated gradually, in brief intervals of time, in every direction over the surface of the earth[113]. In the first case the simultaneousness of the storm might serve, like the eclipses of Jupiter's satellites, fire signals, and well-observed shooting stars, within certain limits, for the determination of geographical longitudes. It is seen with amazement, that the tremblings of two small magnetic needles, were they suspended deep in subterraneous space, measure the distance that intervenes between them; that they tell us how far Kasan lies east from Göttingen, or from the banks of the river Seine. There are regions of the earth where the seaman, enveloped for days in fog, without sight of the sun or stars, without all other means of ascertaining the time, can still accurately determine the hour by the variation of the dip of the needle, and know whether he be to the north or south of the port towards which he would steer his course[114].

If the sudden perturbation of the needle in its hourly course makes known the occurrence of a magnetic storm, the seat of the perturbing cause—whether it be to seek in the crust of the earth itself, or in the upper regions of the air—remains, to our extreme regret, as yet undetermined. If we regard the earth as an actual magnet, then are we compelled, according to the decision of the deep-thinking founder of a general theory of terrestrial magnetism, Frederick Gauss, to admit that every eighth of a cubic metre, or $\frac{37}{16}$ths of a cubic foot of the earth, possesses, on an average, at least as much magnetism as a one-pound magnetic bar[115]. If iron and nickel, and probably cobalt also — not chrome, as was long supposed[116], be the only substances which become permanently magnetic, and retain polarity by a certain coercive force, the phenomena of Arago's rotative magnetism, and Faraday's induced currents, assure us, on the other hand, that probably all terrestrial substances may passingly comport themselves magnetically. From the experiments of the first of the great natural philosophers just mentioned, water, ice[117], glass, and charcoal, affect the oscillations of the needle precisely as quicksilver does in the rotatory experiments. Almost all substances show themselves in a certain degree magnetic when they are conductors; that is to say, when they are traversed by a current of electricity.

How ancient the knowledge of the attractive power of natural magnetic iron appears to have been among the western nations (and this historically well-authenticated fact is remarkable enough), the knowledge of the polarity or directive force of the magnetic needle, and its connection with terrestrial magnetism, was, nevertheless, confined to the extreme east of Asia, to the Chinese. A thousand years and more before the commencement of our era, in the dark epoch of Codru and the return of the Heraclidæ to the Peloponnesus, the Chinese had already magnetic cars, upon which the moveable arm of a human figure pointed invariably to the south, as a means of finding the way through the boundless grassy plains of Tartary; in the third century, indeed, of the Christian era, at least seven hundred years, therefore, before the introduction of the ship's compass upon European seas, Chinese craft were sailing the Indian ocean under the guidance of MAGNETIC SOUTHERN INDICATION[118]. I have shown in another work[119], what advantages this method of determining topographical position, this early knowledge and application of the magnetic needle, wholly unknown in the west, gave the Chinese geographers over those of Ancient Greece and Rome, to whom, for example, the true course of the Apennines and Pyrenees was never known.

The magnetic force of our planet reveals itself on its surface in three classes of phenomena, one of which shows the variable *intensity* of the force, the two others indicate the variable direction in the *inclination* or *dip*, and in the horizontal departure, or *declination*, from the terrestrial meridian of the place, the aggregate outward effect of which may be graphically exhibited by means of three systems of lines, one isodynamical, another isoclinial, a third isogonial; or lines of equal force, of equal dip, and of equal variation. The distance and relative position of these ever-moved, oscillatingly-progressive curves, do not always remain the same. The total variation or declination of the magnetic needle has not, however, changed appreciably, or at all in certain parts of the earth(120), in the Western Antilles and in Spitzbergen, for example, in the course of a whole century. Even so, the isogonial curves, when, in the course of their secular movement, they have passed from the surface of the sea to a continent or island of considerable magnitude, are seen to linger long upon it, and then they curve off again in their farther progress.

These gradual transformations which accompany the translation, and in the course of time extend the empire of the Eastern and Western variations so unequally, render it difficult, in the graphic representations that belong to different centuries, to discover the transitions and analogies of the forms. Every branch of a curve has its own history; but this history, among the Western nations, nowhere mounts higher than to the remarkable epoch, the 13th of September, 1493, when the rediscoverer of the New World recognized a line of no variation, three degrees west from the meridian of Flores, one of the Azores(121). The whole of Europe, a small portion of Russia alone excepted, has, at the present time, western variation; whilst, at the end of the 17th century, first in London (1657), and then in Paris (1669), with a difference of twelve years, consequently, despite the short distance between them, the needle pointed directly to the north pole. In East Russia, to the east of the mouth of the Wolga, of Saratow, Nijni-Novogorod and Archangel, the Eastern variation presses in upon us from Asia. Two excellent observers, Hansteen and Ad. Erman, have given us intelligence of the remarkable double curvature of the variation-lines in the wide-spread realms of Northern Asia; convex towards the pole betwixt Obdorsk and Obi and Turuchansk, concave betwixt lake Baikal and the bay of Ochotsk. In this last part of the earth, in the north-east of Asia, betwixt the Werchojansk mountains, Jakutsk and Northern Corea, the isogonial lines form a remarkable system enclosed within itself. This ovoidal formation(122) is more regularly repeated, and on a larger scale, in the South Sea, nearly in the meridian of Pitcairn island and the Marquesas group, betwixt the parallels of 20° N. and 45° S. latitude. One might feel disposed to regard so singular a configuration of self-included, almost concentric lines of variation, as the effect of a peculiar local constitution of the body of the earth; but should these apparently isolated systems move on in the course of centuries, then, as in all grand natural forces, must some more general cause of the phenomenon be presumed.

The hourly changes in the variation, dependent on the true time, and apparently determined by the sun so long as it is above the horizon of a place, decrease in their angular amount with the magnetic latitude. Near the Equator, in Rawak Island, for example, they are scarcely more than from 3 to 4 minutes, whilst in the middle of Europe they amount to from 13 to 14 minutes. Now, as the north end of the needle, in the whole of the northern hemisphere, travels, on an average, between half-past 8 A.M. and half-past 1 P.M. from east to west, and in the southern hemisphere the same north end traverses from west to east during the same period of time, it has been recently, and with reason, remarked(123), that there must be a region of the earth situated, probably, between the terrestrial and the magnetic equator, in which no horary changes of the variation will be observed. But this fourth curve, that of no-movement, or rather of no change in horary variation, has not yet been discovered.

As the points of the earth's surface where the horizontal force disappears, are called magnetic poles, and a greater degree of importance has been attached to these points than belongs to them of right(124), in the same way is that curve called the magnetic equator upon which the dip of the needle is nothing. The position of this line, and its secular variations of form, have been made objects of particular investigation in recent times. From the admirable work of Duperrey(125), who, between the years 1822 and 1825, crossed the magnetic equator six times, it appears that the two points in which the line of no dip cuts the terrestrial equator, and so passes from one hemisphere into another, are so unequally divided, that, in the year 1825, the node by the island of St. Thomas, on the west coast of Africa, lay in a direct line 188½° from the node in the South Sea by the little Gilbert's Island (nearly in the meridian of the Viti group), in the Southern Pacific. In the beginning of the present century, at an elevation of 11,200 feet above the level of the sea, in 70° 1′ S. lat. and 48° 40′ W. long., I was enabled astronomically to determine the point at which the Andes betwixt Quito and Lima, in the interior of the New Continent, are crossed by the magnetic equator. From this point, proceeding westward, it lingers in the southern hemisphere, through almost the whole of the South Sea, slowly approaching the terrestrial equator. It first crosses over into the northern hemisphere shortly before it reaches the Indian Archipelago; it then just touches the south point of Asia, and enters the African continent westward from Socotora, close to the straits of Babelmandel, where it is at its greatest elongation from the terrestrial equator. Traversing the unknown regions of central Africa in a south-western direction, the magnetic equator returns, in the gulph of Guinea, into the southern tropic, and in its course across the Atlantic separates so far from the terrestrial equator, that it meets the coast of Brazil at Os Ilheos, to the north of Porto Seguro, in 15° S. latitude. From thence to the lofty plains of the Cordilleras, betwixt the silver mines of Micuipampa and the old seat of the Incas, Caxamarca, where I had an opportunity of observing the inclination, it traverses the whole of South America, which, in these southern latitudes, like the interior of Africa, remains a magnetic terra incognita up to the present time.

Late observations collected by Colonel Sabine(126), inform us that the node of the Island of St. Thomas has travelled four degrees, from east to west, between 1825 and 1837. It would be of the highest importance to know whether the opposite node of Gilbert's Island, in the South Pacific, had not travelled as far westward, towards the meridian of the Carolinas. The general survey now given must suffice to connect the different systems of not perfectly parallel isoclinal lines with the great phenomenon of equilibrium which manifests itself in the magnetic equator. It is no small advantage for the establishment of the laws of terrestrial magnetism, that the magnetic equator, whose fluctuating alterations of form, and whose nodal motion in the midst of the various magnetic latitudes, exert an influence(127) upon the dip of the needle in the remotest countries of the world, is, with the exception of one-fifth, wholly oceanic ; it is therefore, through the remarkable relations betwixt the sea and the land, by so much the more accessible, as we are now in possession of a means of determining both variation and dip, with great accuracy, on shipboard, whilst the vessel is holding her course.

We have now portrayed the distribution of magnetism upon the surface of our planet, according to the two forms of variation and dip. The third form, that of intensity of the force, still remains, and this is graphically expressed by isodynamic curves (lines of equal intensity). The investigation and measurement of this force, in its terrestrial relations, by the oscillations of a vertical or horizontal needle, have only excited general and lively interest since the beginning of the nineteenth century. The measurement of the horizontal force has been made capable of a degree of accuracy, particularly by the application of delicate optical and chronometrical instruments, which far exceeds that of all the other magnetical determinations. If, with reference to the immediate application to navigation and steering, the isogonal lines be the more important, the isodynamic, especially those that indicate the horizontal force, present themselves, according to the most recent views, as those which promise the richest harvest for the theory of terrestrial magnetism(128). One of the earliest facts discovered by observation, was this : that the intensity of the sum of the force increases from the equator towards the pole(129).

For a knowledge of the measure of this increase, and the establishment of all numerical relations of the law of intensity, embracing the whole earth, we are especially indebted to the ceaseless activity of Colonel Sabine, who, ever since the year 1819, after he had made observations on the same needle oscillating at the American north pole, in Greenland, in Spitzbergen, on the coast of Guinea, and in the Brazils, has been incessantly engaged in collecting and arranging whatever may serve to illustrate the direction of the isodynamic lines. I have myself given the first plan of an isodynamical system, divided into zones, for a small part of South America. These isodynamic lines are not parallel to the lines of equal dip ; the intensity of the force is not, as was at first believed, weakest at the magnetic equator ; it is not once equal at any part of the same. If Erman's observations in the southern portion of the Atlantic, where a zone of declining intensity runs from Angola, over the island of St. Helena, to the coast of Brazil (0·706), be compared with the very latest observations of that distinguished navigator Sir James Clark Ross, it is found that the force upon the surface of our planet increases nearly in the ratio of one to three towards the magnetic south pole, and where Victoria Land stretches away from Cape Crozier towards Mount Erebus, that volcano which rises from everlasting ice to the height of 11,600 feet above the level of the sea(130). If the intensity in the vicinity of the magnetic south pole be expressed by 2·052 (—the intensity which I found on the magnetic equator in North Peru is still assumed as unity, or 1·000), Sabine found it, in Melville Island, 24° 27′ N. lat., near the magnetic north pole, only 1·624 ; whilst, in the United States, near New-York—nearly under the same parallel of latitude as Naples, consequently—it was 1·803.

Through the brilliant discoveries of Oersted, Arago, and Faraday, the electrical charge of the atmosphere has been brought to approximate more closely to the magnetical charge of the earth. If Oersted found that electricity induced magnetism in the vicinity of the body which was conducting it, so, on the other hand, it was shown in Faraday's experiments that free magnetism gave rise to electricity. Magnetism is one of the numerous forms in which electricity manifests itself. The ancient suspicion of the identity of electrical and magnetical attraction has been demonstrated in the present age. "If electrum" (amber), says Pliny (131), in the sense of the Ionic natural philosophy of Thales, "becomes inspired by friction and warmth, it attracts bark and dried leaves, exactly like the magnetic iron stone." The same words occur in the literature of a people inhabiting the easternmost parts of Asia, in the discourse, laudatory of the magnet, of the Chinese natural philosopher, Kuopho(132). It was not without surprise that I myself observed, among the children at play on the woody banks of the Orinoco, the offspring of native tribes in the lowest grade of civilization, that the excitement of electricity by friction was known. The boys rubbed the dry, flat, and shining seeds of a creeping leguminous plant (probably a negretia), until they attracted fibres of cotton wool and chips of the bamboo. This amusement of these coppery children is calculated to leave a deep and solemn impression behind it. What a chasm lies between the electrical play of these savages, and the discovery of the lightning conductor, of the chemically decompounding pile, of the light-evolving magnetical apparatus ! In such gulphs, millenniums in the history of the intellectual progress of mankind lie buried !

The ceaseless change, the fluctuating movements which are observed in all magnetical phenomena—those of the dip, variation, and intensity, according to the hour of the day and even of the night, according to the season and the lapse of whole years, permit us to suspect the existence of very dissimilar partial systems of electrical currents in the crust of the earth. Are these currents, as in Seebeck's experiments, thermo-magnetical, and immediately excited by unequal distribution of heat? Or shall we not rather regard them as induced by the position of the sun, and through the influence of his heat?[133] Has the rotation of our planet and the accident of the different velocities impressed upon the several zones, according to their distance from the equator, any influence upon the distribution of magnetism? Shall the seat of the currents, in other words, of the electricity in motion, be sought for in the atmosphere, in the interplanetary spaces, or in the polarity of the sun and moon? Galileo, in his celebrated Dialogo, is disposed to ascribe the parallel direction of the earth's axis to a magnetic point of attraction in space.

When the interior of the earth is regarded as molten and subjected to an enormous pressure, as raised to a degree of temperature such as we have no means of estimating, then must the idea of a magnetical nucleus of the earth be abandoned. All magnetism is certainly lost at a white heat[134]; it is still manifested when iron is raised to a dull red; and however different the modifications undergone by the molecular condition, and the coercive force of matter dependent on it, may be in experiments, there still remains a considerable thickness of the crust of the earth which might be assumed as the seat of magnetic currents. In what regards the old explanation of the horary variations of the deflection, by the progressive heating of the earth in the apparent course of the sun from east to west, it must be owned that we are here limited to the very outermost surface; inasmuch as the thermometers now sunk in the ground in so many places, and so carefully observed, show us how slowly the sun's heat penetrates even to the moderate depth of a few feet. And then the thermal state of the surface of the ocean, covering two-thirds of the globe, is little favourable to such an explanation, when the question is one of immediate mean influence, not of induction from the aërial and vaporous covering of our planet.

To all questions as to the ultimate physical cause of phenomena so complicated, there is no satisfactory answer to be given in the present state of our knowledge. It is only in reference to the three-fold manifestations of the earth-force, to that which meets us as mensurable relations of Space and of Time, as the Normal or conformable to laws in the Variable, that brilliant advances have lately been made, through the determination of numerical mean values. Since the year 1828, from Toronto, in Upper Canada, to the Cape of Good Hope and Van Dieman's Land, from Paris to Pekin, the earth has been covered with magnetical observatories[135], in which uninterrupted and simultaneous observations are made of every regular and irregular excitement of the earth-force. A decrease of the magnetic intensity amounting to the $\frac{1}{40000}$th part is measured; at certain epochs, observations are noted every 2½ minutes through an entire period of 24 hours. An illustrious English astronomer and natural philosopher[136] has calculated that the mass of observations accumulated in the course of three years, which remain for discussion, amounts to 1,958,000! Never has there been so grand, so delightful an effort made to get at the root of the Quantitative in the laws of a natural phenomenon. We may therefore be permitted to entertain a well-grounded hope, that these laws, compared with those which prevail in the atmosphere, and still more distant spaces, will gradually bring us nearer and nearer to the Genetical in magnetic phenomena. Until now we can only boast that a greater number of ways which might possibly lead to information have been opened up. In the physical doctrine of terrestrial magnetism, which must not be confounded with the purely mathematical one, as in the doctrine of the meteorological processes of the atmosphere, some completely satisfy themselves by conveniently denying as realities all the phenomena which cannot be explained in conformity with their views.

Terrestrial magnetism, the electro-dynamic forces which have been calculated by the able Ampère[137], stands at the same time in intimate relationship with the Earth- or Northern-Lights [Aurora borealis], as with the internal and external temperature of our globe, whose magnetic poles must be regarded as poles of cold[138]. If Halley[139], some 128 years ago, gave it out as a mere bold conjecture that the northern light was a magnetic phenomenon, Faraday's brilliant discovery of the evolution of light through magnetic power has raised that conjecture to the rank of an empirical certainty. There are heralds or harbingers of the northern lights. In the course of the day on which the lights are to appear, irregular horary movements of the magnetic needle usually indicate an interruption of equilibrium in the distribution of the terrestrial magnetism. When this disturbance has attained a great intensity, the equilibrium of the distribution is restored by a discharge, accompanied with an evolution of light. "The northern light itself is not, therefore, to be regarded as an external cause of the disturbance, but rather as a terrestrial activity raised to the pitch of a luminous phenomenon, one of the sides of which is the light, the other the oscillations of the needle"[140]. The splendid phenomenon of coloured northern lights is the act of discharge, the conclusion of a magnetic storm; in the same way as, in the electrical storm, an evolution of light—lightning—indicates the restoration of the disturbed equilibrium in the distribution of electricity. The electrical storm is usually limited to a small space, beyond which the state of the electricity remains unchanged. The magnetic storm, on the contrary, reveals its influence on the march of the needle over large portions of continents, as Arago first observed, and far from the place where the development of light is visible. It is not improbable that, as in the case of heavily charged and threatening clouds, and of frequent transitions of the atmospheric electricity into opposite states, it does not always come to discharges by lightning, so also may magnetic storms produce great disturbances in the horary motions of the needle over extensive circles, without there being any necessity for explosions, for luminous effusions from the pole to the equator, or from one pole to another, in order to restore the equilibrium.

He who would have all the particulars of the phenomenon embraced in one picture, should have the origin and course of a complete appearance of the northern lights set before him. Deep on the horizon, nearly in the situation where it is intersected by the magnetic meridian, the heaven, up to this moment clear, grows black. There is a kind of hazy bank or screen produced, which rises gradually, and attains to an altitude of from 8 to 10 degrees. The colour of the dusky segment passes over into brown or violet. Stars are visible in it, but they are seen as in a portion of the sky obscured with dense smoke. A broad bright luminous arc or seam, first white, then yellow, bounds the dusky segment; but as the brilliant bow arises later than the smoky-grey segment, it is impossible, according to Argelander([141]), to ascribe the latter to the effect of mere contrast with the bright luminous border. The highest point of the luminous arc, when it has been carefully measured([142]), has usually been found to be not exactly in the magnetic meridian, but to vary between 5 and 18 degrees from it, towards the side on which the magnetic declination of the place of observation lies. In high northern latitudes, very near the north pole, the smoky-looking spherical segment appears less dark; sometimes it is even entirely absent. In the situation, too, where the horizontal force is least, the middle of the luminous arc is seen to depart farthest from the magnetic meridian.

The luminous bow, in constant motion, flickering and changing its form incessantly, sometimes remains visible for hours before anything like rays and pencils of rays shoot from it, and rise to the zenith. The more intense the discharges of the northern lights, the more vividly do the colours play from violet and bluish-white, through every shade and gradation, to green and purplish-red. In our ordinary electricity produced by friction, in the same way, the spark first becomes coloured when the tension is high, and the explosion is violent. The magnetic fiery columns shoot up at one time singly from the luminous arch, even mingled with black rays, like thick smoke; at another, many columns arise simultaneously from several and opposite points of the horizon, and unite in a flickering sea of flame, to the splendour of which no description can do justice, and whose luminous waves assume another and a different shape at every instant. The intensity of the northern light is at times so great, that Lowenörn perceived its oscillations, in bright sunshine, on the 29th of January, 1786. The motion increases the brilliancy of the phenomenon. Around the point of the vault of heaven which corresponds with the direction of the dipping needle, the rays at length collect together, and form the corona or crown of the northern lights. This surrounds the summit, as it were, of a vast canopy, the dome of heaven, with the mild radiance of its streaming but not flickering rays. It is only in rare instances that the phenomenon proceeds the length of forming the corona completely. With its appearance, however, the whole is at an end. The rays now become rarer, shorter, less intensely coloured. The crown and the luminous arches break up. By and by nothing but broad, motionless, and almost ashy-grey, pale gleaming fleecy masses, appear irregularly dispersed over the whole vault of heaven; these vanish, in their turn, and before the last trace of the murky fuliginous segment, which still shows itself deeply on the horizon, has disappeared. Of the whole brilliant spectacle, nothing at length remains but a white delicate cloud, feathered at the edges, or broken up, as a cirro-cumulus, into small rounded masses or heaps, at equal distances.

This connection of the polar light with the most delicate cirrus-clouds, deserves to be particularly mentioned; inasmuch as it shows us the electro-magnetic evolution of light as part of a meteorological process The terrestrial magnetism here manifests itself in its effects upon the atmosphere, in a condensation of the watery vapour which it holds dissolved. The observations, made in Iceland by Thienemann, who regards the cirro-cumulus, or divided fleecy cloud, as the substrate of the northern lights, have been confirmed in later times by Franklin and Richardson, near the North American magnetic pole, and by Admiral Wrangel, on the Siberian coasts of the icy sea. All observed "that the northern lights sent forth the most brilliant rays when masses of cirro-stratus floated in the upper regions of the atmosphere; and when these were so thin, that their presence was only known by the formation of a halo about the moon." These light clouds occasionally arranged themselves, by day, in the same manner as the rays of the Aurora, and had the same effect as these in disturbing the magnetic needle. After a grand nocturnal display of the northern lights, the same streaks of clouds that had been luminous over night, were discovered in the morning arranged in the same manner([143]). The apparently converging polar zones of clouds (streaks of clouds, in the direction of the magnetic meridian), which constantly attracted my attention in the course of my travels on the lofty platforms of Mexico, as well as in Northern Asia, belong apparently to the same group of diurnal phenomena([144]).

Southern lights have been frequently seen in England by that able and diligent observer, Dalton; northern lights in the southern hemisphere, as low as 45° of latitude (Jan. 14, 1831). In instances that are not very rare, the magnetic equilibrium is disturbed at both poles simultaneously. I have distinctly stated that northern polar lights are seen within the tropics, even as far south as Mexico and Peru. It is necessary to distinguish, however, between the sphere of a simultaneous apparition of the phenomenon, and the zone of the earth in which the phenomenon is displayed almost every night of the year. As each observer sees his own rainbow, so also, doubtless, does he see his own polar light. A great portion of the earth engenders the radiating Light-phenomenon at the same time. Many nights can be mentioned in which it was observed simultaneously in England, in Pennsylvania, in Rome, and in Pekin. When it is maintained that the northern lights decline with the decrease of latitude, this must be understood as referring to magnetic latitude, measured from the magnetic pole. In Iceland, Greenland, and Newfoundland, on the banks of the Slave lake, and at Fort Enterprise (in North Canada), the Aurora is lighted up, at certain seasons, almost every night, and with its shifting, shivering rays, performs its "merry dance" through the sky, as the natives of the Shetland Islands term it([145]). Whilst in Italy the northern light is a great rarity, it is seen with extreme frequency in the latitude of Philadelphia (39° 57′ N. L.), in consequence of the southern position of the American magnetic pole. But in the districts of the new continent, and also of the shores of Siberia, which are remarkable for the frequency of the phenomenon, there occur what may be called especial regions of the northern lights—longitudinal zones in which they are peculiarly splendid([146].) Local influences are, consequently, not to be overlooked. Wrangel observed their brilliancy decline as he left the shores of the icy sea, about Nijne-Kolymsk, behind him. The experience of the Northern Polar Expedition seems to indicate that the evolution of light is not greater in the immediate vicinity of the magnetic pole than it is at some distance from this spot.

What we know of the altitude of the northern light is based on measurements, which, by reason of the incessant oscillations of the luminous rays, and the consequent uncertainty of the parallactic angle, cannot be greatly depended on. The conclusions come to (not to speak of older estimates) vary between several miles and three or four thousand feet([147]). It is not improbable that the northern light is at very different distances at different times. The latest observers are disposed to connect the phenomenon, not with the outer limits of the atmosphere, but with the region of the clouds itself; they even believe that the northern streamers may be moved by winds and currents of air, if the luminous phenomenon, by which alone the existence of electro-magnetic emanations becomes obvious to us, be actually connected with material collections of vesicular vapour, or, to speak more correctly, penetrates these collections, darting over from one vesicle to another. Captain Franklin saw a streaming Aurora on Bear lake, which he believed illuminated the under side of the stratum of cloud; whilst Kendal, who had the watch through the whole of the night, and never lost the heavens for a minute from his sight, at the distance of but 4½ geographical miles, observed no luminous phenomenon whatsoever. The statement, repeated several times of late, to the effect that streamers of the northern light have been observed close to the ground, and between the observer and a neighbouring height, is one of those points, which, like lightning and the fall of fire-balls, is exposed to the manifold dangers of optical deception.

Whether or not the magnetic storm, of which we have just quoted a remarkable example of local circumscription within very narrow bounds, have the noise, besides the light, in common with the electrical storm, is now rendered extremely doubtful, since the testimony of the Greenland sledgers, and the Siberian fox-hunters, is no longer taken unconditionally. The northern lights have become more silent since they have been examined more carefully with the eye and the ear. Parry, Franklin and Richardson, near the north pole; Thienemann, in Iceland; Gieseke, in Greenland; Lottin and Bravais, at the North Cape; Wrangel and Anjou, on the shores of the icy sea, have, altogether, looked at thousands of northern lights, yet never heard any noises. If this negative testimony be not admitted against two positive witnesses, Hearne, at the mouth of the Coppermine river, and Henderson, in Iceland, it must still be remembered that Hood heard the same noises—as of musket balls shaken rapidly together, and slight cracklings, during the occurrence of the northern lights, indeed, but also on the following day, when there was no Aurora in the heavens; and then it must not be forgotten, that Wrangel and Gieseke were firmly convinced that the noises heard were owing to contractions of the ice and crust of snow, in consequence of a sudden cooling of the air. The belief in a crackling noise did not take its origin among the people, but with learned travellers, and in this way: the flashing of electricity in attenuated atmospheres having been known from an early period, the northern light was forthwith declared to be an effect of atmospheric electricity, and then the noises were heard that ought to have been heard. Recent experiments with the most delicate electrometers, however, contrary to all expectation, have hitherto given merely negative results; the state of the aërial electricity has not been found altered during the prevalence of the most brilliant Auroras.

All the three manifestations of force of the terrestrial magnetism — Declination, Inclination, and Intensity, on the contrary, are affected at once by the northern lights. In one and the same night, and from hour to hour, the Aurora affects the same end of the needle differently, now attracting it, now repelling it. The assertion that the facts collected by Parry at Melville Island, near the magnetic pole, lead to the conclusion that the northern lights do not disturb the needle, but rather have a "calming effect" upon it, is completely contradicted by a more careful perusal of Parry's own journal([148]), by the beautiful observations of Richardson, Hood, and Franklin in North Canada, and more lately still, by Bravais and Lotten in Lapland. The process in the northern lights is, as we have above observed, the act of restoration of an equilibrium disturbed. The effect upon the needle varies according to the measure of force in the explosion. It was only unobservable at the nocturnal winter station at Bosekop,* when the luminous phenomenon showed itself very feebly and deep on the horizon. The upshooting radiate cylinders of the northern light have been aptly compared to the flame which, in the closed circuit of the Voltaic pile, arises between two charcoal points at a distance from one another, or, according to Fizeau, between a silver and a charcoal point, and to that which is drawn or thrown off from the magnet. This analogy at all events renders superfluous the assumption of those metallic vapours in the atmosphere which some natural philosophers

* [Vide Kaemtz's Complete Course of Meteorology, by C. V. Walker, (Plates, 8vo. Lond. 1845), for a full account of the Aurora.—TR.]

have imagined as the substrate of the northern lights.

If the luminous phenomenon which we ascribe to a galvanic current, *i. e.* a motion of electricity in a circuit returning into itself, be designated by the indefinite name of the Northern light, or the Polar light, nothing more is thereby implied than the local direction in which the beginning of a certain luminous phenomenon is most generally, but by no means invariably, seen. What gives this phenomenon its greatest importance is the fact which it reveals, *viz.* that the EARTH IS LUMINOUS; that our planet, beside the light which it receives from the central body, the sun, shows itself capable of a proper luminous act or process. The intensity of the Earth-light, or rather the degree of luminosity which it diffuses, exceeds by a little, in the case of the brightest coloured rays that shoot up to the zenith, the light of the moon in her first quarter. Occasionally, as on the 7th of January, 1831, a printed page can be read without straining the sight. This light-process of the earth, which the Polar regions exhibit almost incessantly, leads us by analogy to the remarkable phenomenon which the planet Venus presents. The portion of this planet which is not illuminated by the sun, glows occasionally with a proper phosphorescent gleam. It is not improbable that the Moon, Jupiter, and Comets, besides the reflected sun-light recognizable by the polariscope, also emit light produced by themselves. Without insisting on the problematical but very common phenomenon of *sheet-lightning*, in which the whole of a deep massy cloud is flickeringly illuminated for several minutes at a time, we find other examples of terrestrial evolutions of light. To this head belong the celebrated *dry-fogs* of 1783 and 1831, which were luminous by night; the steady luminousness of large clouds, perfectly free from all flickering, observed by Rosier and Beccaria; and even the pale, diffused light, as Arago has well observed[149], which serves to guide us in the open air, in thickly clouded autumn and wintry nights, when there is neither moon nor star in the firmament, nor snow upon the ground. As in the phenomenon of the Polar lights occurring in high northern latitudes, in other words, in electro-magnetic storms, floods of flickering and often parti-coloured light stream through the air, so, in the hotter zones of the earth, between the tropics, are there many thousand square miles of ocean which are similarly light-engendering. Here, however, the magic of the light belongs to the organic forces of nature. Light-foaming flashes the bursting wave, the wide level glows with lustrous sparks, and every spark is the vital motion of an invisible animal world. So manifold is the source of terrestrial light. And shall we conceive it latent, not yet set free in vapours, as a means of explaining Moser's *pictures*—a discovery in which reality still presents itself to us as a vision shrouded in mystery?

As the internal heat of our planet is connected on one hand with the excitement of electro-magnetic currents and the light-producing process of the earth (a consequence of the bursting of a magnetic storm), so on the other hand does it also manifest itself as a principal source of geognostic phenomena. These we shall consider in their connection, and in their transition from a merely dynamic concussion, and from the upheaving of continents and mountain masses, to the production and effusion of gases and liquids, of boiling mud, and of red hot and molten earths, which harden into crystalline rocks. It is no trifling advance in the newer geognosy (the mineralogical portion of the physics of the globe), that it has firmly founded the concatenation of phenomena here indicated. The views of modern geognosy lead off from mere hypothesis, which trifles or plays with its subject, and seeks to explain, severally and apart, every manifestation of force of the old globe; they shew the connection of the various matters ejected with what appertains only to change in reference to space—concussion, elevation, depression; they arrange side by side groups of phenomena which at first sight present themselves as extremely heterogeneous—thermal springs, effusions of carbonic acid gas, escapes of sulphureous vapours, harmless eruptions of mud, and the awful devastations of burning mountains. In a grand picture of nature all this becomes fused in the single conception of the reaction of the interior of a planet upon its crust and surface. So do we recognize in the depths of the earth, in its temperature increasing with the distance from the surface, at once the germs of concussive movements, of the gradual elevation of entire continents, or of mountain chains through lengthened chasms, of volcanic eruptions, and of the varied production of mineral species and rocky masses. But it is not inorganic nature alone that has felt the force of this reaction of the interior upon the exterior. It is extremely probable that in the primitive world immense discharges of carbonic acid gas mingled with the atmosphere, excited the faculty possessed by vegetables of separating carbon from the air, and that thus, in revolutions which destroyed extensive forests, inexhaustible supplies of combustible matter—lignites and coals of different kinds—have been buried beneath the upper strata of the earth. The destiny of man we even recognize as in part dependent on the fashion of the outer crust of the globe, on the partitioning of continents, on the direction of their mountain chains and high lands. To the inquiring spirit is it given to mount from link to link in the chain of phenomena, till the point is gained at which in the incipient consolidation of our planet, in the first transition of the conglobated matter from the vaporous form, the internal heat of the earth, that heat which does not belong to the action of the sun, was developed.

In our survey of the causal connection of geognostical phenomena, we shall begin with those which, in their principal features, are dynamical, which consist in motion and a change in space. Earthquakes of every kind and degree are distinguished by a series of perpendicular, or horizontal, or rotatory vibrations following each other in rapid succession. In the course of the considerable number of earthquakes which I have felt in both hemispheres of the globe, on shore and at sea, the two first kinds of motion have appeared to me very frequently to take place together. The explosive movement such as is produced by the firing of a mine—the perpendicular action, from below upwards—was displayed most conspicuously on the occasion when the town of Riobamba was destroyed (1797), when the bodies of many of the inhabitants were thrown upon the hill of La Culla, which is several hundred feet high, and rises on the other side of the Lican rivulet. The propagation of the motion generally takes place in a linear direction, in waves, and with a velocity of from five to seven G. geographical miles in a minute. Sometimes it is in circles, or in great ellipses, from the centre of which the vibrations are propagated with decreasing force towards the circumference. There are districts which belong to or fall within two mutually intersecting circles of concussion. In North Asia, which the father of history(150), and, after him, Simocatta(151) characterize as "the Scythian territories free from earthquakes," I found the southern part of the Altai Mountains, so rich in mineral treasures, subject to the influence of the concussive foci both of Lake Baikal and the volcanoes of Thian-Schan, or the Celestial Mountain(152). When thé circles of concussion intersect each other—when, for instance, a lofty plain lies between two simultaneously active volcanoes — then may several systems of waves exist at once, and not interfere with each other, just as in the case of fluids. Interference, however, can be conceived here, as in mutually intersecting waves of sound. The magnitude of the transmitted wave of succession is increased at the surface, in conformity with the general laws of mechanics, according to which, when motion is communicated in elastic bodies, the outermost free-lying stratum tends to detach itself from the others.

The waves of succession can be pretty accurately measured in their direction and total strength, by the pendulum and the *sismometer* bowl, but in no way investigated in the intimate nature of their alternations and periodical intumescences. In the city of Quito, which stands at the foot of an active volcanic mountain—the Rucu-Pichincha, 8,950 feet above the level of the sea, and boasts of beautiful cupolas, lofty fanes, and massive houses several stories high, I have frequently been surprised at the violence of the earthquakes by night, which nevertheless very rarely occasion rents in the walls; whilst in the plains of Peru, apparently much weaker oscillations injure lowly houses built of cane. Natives who have stood the shocks of many hundred earthquakes, believe that the difference of effect is less connected with the length or shortness of the waves, with the slowness or rapidity of the horizontal oscillation(153), than with the equality of the motion in opposite directions. Circular or rotatory concussions are the rarest, but they are the most dangerous of all. Twistings round of walls without throwing them down; plantations of trees, which had previously stood in parallel rows, deflected; the direction of the ridges of fields covered with various kinds of grain altered, were observed on occasion of the great earthquake of Riobamba, in the province of Quito (February 4th, 1797), as well as of those of Calabria (February 5th and March 28th, 1783). With the latter phenomenon of rotation, or the transposition of fields and cultivated plots of ground, of which one has occasionally taken the place of another, there is connected a translatory motion, or mutual penetration of several strata. When taking the plan of the ruined city of Riobamba, I was shown a place where the whole of the furniture of one dwelling-house had been found under the ruins of another. The loose earth of the surface had run in streams like a fluid, of which it must be conceived that it was first directed downwards, then horizontally, and finally upwards. Disputes about the property, in those instances where things were carried many hundred toises from their original stances, were adjusted by the Audiencia, or Court of Justice.

In countries where earthquakes are comparatively much rarer, in the south of Europe for example, a very general belief, grounded upon an imperfect induction, prevails(154); viz. that calms, oppressive heats, and a misty state of the horizon, are always preludes to an earthquake. The erroneousness of this popular belief is not, however, shown by my own experience only; it is farther gainsaid by the observations of all who have lived long in countries where earthquakes are frequent and violent, as in Cumana, Quito, Peru, and Chili. I have experienced earthquakes when the air was clear and a fresh east wind was blowing, as well as during rain and thunder storms. Even the regularity in the horary variations in the declination of the magnetic needle, and in the pressure of the air(155), remained unaffected within the tropics on the day of the earthquakes. The observations which Adolphus Erman made in the temperate zone on the occasion of an earthquake at Irkutsk, near lake Baikal, on the 18th of March, 1829, agree perfectly with my experience. During the violent earthquake of Cumana which happened on the 4th of November, 1799, I found the declination of the needle and the magnetic intensity unaffected; but to my astonishment the dip was diminished by 48′(156). I had no suspicion of any error; yet in all the other earthquakes which I have experienced in the high lands of Quito and in Lima, the dip of the needle remained equally unaffected with the other elements of the terrestrial magnetism. If in a general way the acts that proceed deep in the interior of the earth are announced beforehand by no special meteorological phenomenon, by no peculiar aspect of the heavens, it is on the contrary not improbable, as we shall see immediately, that in certain very violent earthquakes the atmosphere has sympathized or partaken in some measure, and that these, therefore, do not always act in a purely dynamical manner. During the prolonged tremblings of the ground in the Piedemontese valleys of Pelis and Clusson, extreme changes in the electrical tension of the atmosphere were observed, whilst the heavens were free from storm.

The strength of the dull noise which generally accompanies an earthquake does not by any means increase in the same measure as the strength of the vibrations. I have satisfactorily made out that the grand concussion in the earthquake of Riobamba (Feb. 4th, 1797), one of the most awful catastrophes in the physical history of our earth, was accompanied by no noise whatever. The great noise (*el gran ruido*) which was heard under the cities of Quito and Ibarra, but not nearer the centre of the motion in Tucunga and Hambato, occurred from eighteen to twenty minutes after the proper catastrophe. In the celebrated earthquake of Lima and Callao (28th Oct. 1746), the sound was first heard like a subterraneous peal of thunder in Truxillo a quarter of an hour later, and without any trembling of the ground. In like manner, long after the earthquake of New Granada (Nov. 16th, 1827), which has been described by Boussingault, subterraneous detonations were heard in the whole of the valley of Cauca, with great regularity at intervals of thirty seconds. The nature of the noises heard on such occasions is very various: rolling, rattling, clanking like chains, occasionally in the town of Quito like thunder close at hand; or it is clear and ringing, as if masses of obsidian or other vitrified matters were struck in caverns underground. As solid bodies are excellent conductors of sound, as sound, for example, is transmitted with ten or twelve times the velocity in burnt clay that it is in air, the subterraneous noise, it may be easily imagined, will be apt to be heard at great distances from the place where it is occasioned. In Caraccas, in the grassy plains of Calabozo, and on the banks of the Rio Apure, which falls into the Orinoco, in the whole of a region of 2300 square miles in superficial extent, there was heard an extraordinary thundering noise, without any shock of an earthquake, on the 30th of April, 1812, at the very time that the volcano of the Island of St. Vincent, lying 158 geographical miles off, was pouring an immense stream of lava from its crater. This, in respect of distance, was as if an eruption of Vesuvius were to be heard in the north of France. In 1744, on the occasion of the great eruption of Cotopaxi, subterraneous cannonadings were heard at Honda on the Rio Magdalena. The crater of Cotopaxi, however, is not only 17,000 feet above the level of Honda, but the two points are separated by the colossal mountain masses of Quito, Pasto, and Popayan, as well as by valleys and precipices innumerable, besides lying 109 geographical miles apart. The sound was certainly transmitted not through the air, but through the earth from a great depth. In the violent earthquake of New Granada (February, 1835), subterraneous thunder was heard at the same time in Popayan, Bogota, Santa Martha, and Caraccas (in the latter for a period of seven hours without any shock), in Haiti, Jamaica, and round the lake of Nicaragua in Mexico.

These sonorous phenomena, when they are accompanied by no perceptible shocks, leave a remarkably deep impression even with those who have long dwelt in districts subject to repeated earthquakes. All seem to expect with alarm what is to follow the subterraneous rumbling. The most remarkable example of uninterrupted subterraneous noises, without any trace of earthquake, and comparable with nothing else, was presented by the phenomenon which was known in the high lands of Mexico under the name of the subterraneous bellowings and thunderings (*bramidos y truenos subterraneos*) of Guanaxuato([157]). This celebrated and flourishing mining town lies far remote from any active volcano. The noise continued from midnight of the 9th of January, 1784, for more than a month. I have been able to give a particular account of it from the report of many witnesses, and from the documents of the municipality which I was permitted to use. It was (January 13—16th) as if heavy thunder-clouds lay under the feet of the inhabitants, in which slowly rolling thunder alternated with sharper claps. The sound drew off as it had come on with decreasing loudness. It was confined to a limited space; at the distance of a few miles off, in a district abounding in basalt, it was not heard at all. Almost all the inhabitants fled the town in alarm, although great piles of silver bars were contained in it; the more courageous becoming accustomed to the subterraneous noise, by and by returned and disputed possession with the bands of robbers who had seized on the treasure. Neither on the surface of the ground, nor in the workings at the distance of 1500 feet below it, was there the slightest movement of the earth perceived. Over the whole of the Mexican highlands no noise of the same kind had ever been heard before, neither has the alarming incident recurred. Thus do chasms in the interior of the earth open and close; and the sonorous waves either reach us or are interrupted in their progress.

The influence of a volcanic mountain in action, however terrific or picturesquely grand as an object of sense, is still always limited to a very narrow space. It is very different with the shocks of earthquakes, which are scarcely appreciable to the eye, but their undulations occasionally extend simultaneously to the distance of thousands of miles. The great earthquake which desolated Lisbon on the 1st of November, 1755, and whose influences have been so admirably investigated by the great philosopher Emanuel Kant, was felt among the Alps, on the coast of Sweden, in the West Indian islands, Antigua, Barbadoes, and Martinique, and on the great Canadian lakes, as well as in the small inland lakes of the basaltic plains of Thuringia and the northern flats of Germany. Distant springs were interrupted in their course, an incident in earthquakes to which Demetrius the Galatian directed attention in ancient times. The hot springs at Tepliz ran dry, and then returned deeply tinged with a ferruginous ochre, flooding every thing. At Cadiz the sea rose sixty feet high; in the lesser Antilles it became of an inky black colour, and the tide, which generally rises but about twenty-six or twenty-eight inches, mounted twenty feet above its usual level. It has been calculated that a territory more than four times the superficial extent of Europe was shaken by the earthquake of November 1st, 1755. There is, therefore, no other outward manifestation of force known —the murderous inventions of our race included—through which, in the brief period of a few seconds or minutes, a larger number of human beings have been destroyed: in 1793, sixty thousand perished in Sicily; from thirty to forty thousand fell victims in the catastrophe of Riobamba of 1797, and perhaps five times as many in Lesser Asia and Syria under Tiberius and Justin the Elder, about the years 19 and 526 of the Christian era.

There are instances among the Andes of

South America of the earth having quaked incessantly for several days together; but I only know of shocks that were felt almost every hour for several months, having occurred *far from any volcano*, on the eastern slopes of the Alps of Mont-Cenis, about Fenestrella and Pignerolo, from April, 1808; in the United States of America, betwixt New Madrid and Little Prairie([158]), to the north of Cincinnati, after December, 1811; in the Paschalic of Aleppo, in the months of August and September, 1822. As the vulgar mind can never rise to general views, and therefore always ascribes great phenomena to local processes of the earth or of the air, wherever succussions continue for any length of time, fears for the appearance of new volcanoes take their rise. In single, rare instances, this fear has indeed shown itself well-founded, as in the case of the sudden rise of volcanic islands, and in the production of the volcano of Jorullo, a new mountain, rising 1580 feet above the old neighbouring level, on the 29th of September, 1759, after ninety days of earthquakes and subterraneous thunderings.

Could we have daily news of the state of the whole of the earth's surface, we should, in all probability, become convinced that some point or another of this surface is ceaselessly shaken; that there is uninterrupted reaction of the interior upon the exterior going on. This constancy and general diffusion of a phenomenon, which is probably connected with the high temperature of the deepest strata of the earth, explains its independence of the nature of the rocky masses among which it is manifested. Shocks of an earthquake have been experienced even in the loosest alluvial deposits of Holland, around Middleburg and Flushing. Granite and mica slate are shaken in the same way as mountain limestone and sandstone, as trachytic and amygdaloidal formations. It is not the chemical nature of the constituents, but the mechanical structure of the mineral species, that modifies the propagation of the motion (the wave of succussion). Where the wave proceeds regularly along a coast, or by the foot, and in the direction of a mountain-chain, it is occasionally observed that there is an interruption suffered at certain points. This has been noticed for centuries. The undulation advances along the depths, but at the points in question it is never felt at the surface. The Peruvians say of these unshaken superior strata, that "they form a bridge"([159]). As mountain-chains appear upheaved through fissures, the walls of these cavities may very well favour or influence the course of the undulations that run parallel with the chain; occasionally, however, the waves of succussion cut across several chains, almost at right angles. We thus see them break through the littoral chains of Venezuela and the Sierra Parime in South America. In Asia, the earthquakes of Lahore and the foot of the Himalayas (Jan. 22d, 1832) were propagated transversely through the chain of Hindoo-Cusch to Badakhchan, to the Upper Oxus, and even to Bokhara([160]). Unfortunately, too, the circles of concussion enlarge, in consequence of a single extremely violent shock. It is only since the destruction of Cumana (14th Dec. 1797) that every shock of the southern coast is felt in the mica-slate strata of the peninsula of Maniguarez, which lies opposite the limestone or chalk-hills of the fortress. In the almost incessant undulations of the ground of the valleys of the Mississippi, Arkansas, and Ohio, which occurred from 1811 to 1813, the progress of the motion from south to north was very striking. It was as if subterranean impediments had been gradually overcome, and the wave of commotion then advanced upon each occasion along the way which had been opened up.

If an earthquake appear, at first sight, to be a phenomenon of motion wholly dynamical, having reference to space only, it is still recognized, on the grounds of the most careful experience, that it is not only competent to raise whole districts above their old level (Ulla-Bund, eastward from the delta of the Indus, for example, after the earthquake of Cutch, in June, 1809, and the coast of Chili, in November, 1822), but farther, that during the shock, hot water (Catania, 1818), hot steam (valley of the Mississippi, near New Madrid, 1812), mephitic or irrespirable gases, which are injurious to the pasturing herds and flocks of the Andes, mud, black smoke, and even flames (Messina, 1782, Cumana, 14th Nov. 1797), have been discharged. During the great earthquake of Lisbon, Nov. 1, 1755, flames and a column of smoke were seen to rise from a newly-formed fissure in the rock of Alvidras, near the city. The smoke became on each occasion where it appeared, by so much the more dense as the subterraneous noise increased in loudness([161]). When the town of Riobamba was destroyed in 1797, the earthquake was not accompanied by any eruption of the volcano which is so close at hand; but *Moya*, a singular mass, compounded of carbon, crystals of augite, and the siliceous coats of infusory animalcules, was pushed out of the ground in numerous small and progressive cones. The escape of carbonic acid gas during the earthquake of New Granada (16th Nov. 1827), from fissures in the Magdalena valley, caused the suffocation of many snakes, rats, and other creatures that live in holes. Sudden changes in the weather, too, the setting in of the rainy season at unusual periods in the tropics, have occasionally followed great earthquakes in Quito and Peru. Do gaseous fluids, escaping from the interior of the earth, then become mingled with the atmosphere? or, are these meteorological processes the effect of a disturbance of the atmospherical electricity by the earthquake? In the countries of tropical America, where sometimes not a drop of rain falls for ten months, the inhabitants look upon repeated shocks of earthquakes, which cause no danger to their low cane huts, as a happy indication of plenty of rain, and consequently of fertility.

The intimate connection of all the phenomena now described is still buried in obscurity. Elastic fluids are undoubtedly the cause, as well of the slight and uninjurious tremblings of the earth, which continue for many days (as in 1816 at Scaccia, in Sicily, previous to the elevation of the new island called Julia), as of the frightful explosions which are announced by noises. The focus of the mischief, the seat of the moving power, lies deep beneath the crust of the earth; how deep, we know even as little as we do what the chemical nature of the vapour of such high tension may be. Encamped on the edge of two craters, on Vesuvius, and on the castellated rock which overlooks the vast gorge of Pichincha, near Quito, I experienced periodical and very regular shocks, and, each time, from 20 to 30 seconds before red-hot ashes or vapours were ejected. The shocks were by so much the stronger as the explosions were later of occurring, and the vapour consequently had been longer accumulating. In this simple fact, confirmed by the experience of so many travellers, lies the general solution of the phenomenon. Active volcanoes are to be regarded as safety-valves for surrounding districts. The danger of the earthquake increases when the opening of the volcano is stopped up, and there is no longer a free communication with the atmosphere; but the destruction of Lisbon, of Caraccas, Lima, Cashmir (1554)[162], and of so many towns of Calabria, Syria, and Asia Minor, teaches us, that on the whole the force of earthquakes is by no means greatest in the vicinity of still active volcanoes.

As the pent-up force of a volcano acts in shaking the ground, so does the concussion react, in its turn, upon the volcanic phenomenon. The occurrence of fissures favours the rise of the cones through which eruptions take place, and the processes which go on within these cones in free contact with the atmosphere. A column of smoke, which had been seen for months rising from the volcano of Pasto, in South America, disappeared suddenly on the occurrence of the great earthquake of Riobamba, in the province of Quito, at the distance of 48 geographical miles to the south (Feb. 4, 1797). After the earth had long continued to tremble in the whole of Syria, in the Cyclades, and in Cubœa, the convulsions ceased suddenly upon the eruption of a stream of "red-hot mud" (lava from a crack) in the Lelantine plain, near Chalcis([163]). The admirable geographer of Amasia, who has preserved the record of this fact, adds: "Since the mouths of Etna have been opened, through which the fire belches forth, and since, in this way, heated masses and water can be ejected, the lands by the sea-shore are no longer so frequently shaken as they were in times before the separation of Sicily from Lower Italy, when there was no communication with the surface."

In earthquakes, therefore, we have evidence of a volcano-producing force; but such a force, as universally diffused as the internal heat of the globe, and proclaiming itself everywhere, rarely gets the length of actual eruptive phenomena; and when it does so, it is only in isolated and particular places. The formation of extensive veins or dykes, in other words, the filling up of fissures with crystalline matter ejected from the interior, such as basalt, melaphyre, and greenstone, interferes by degrees with the free escape of vapours; which, confined, become operative, through their tension, in three ways: *concussively; explosively*, or suddenly up and down; and, as first observed in a large portion of Sweden, *liftingly* or continuously, and only in long periods of time perceptibly altering the relative level of the sea and land.

Before we quit this great phenomenon, which has been here considered not so much in its individual as in its general physical and geognostical relations, we must advert to the cause of the indescribable, deep, and quite peculiar impression which the first earthquake we experience makes upon us, even when it is accompanied by no subterranean noises. The impression here is not, I believe, the consequence of any recollection of destructive catastrophes presented to our imagination by narratives of historical events: what seizes upon us so wonderfully is the disabuse of that innate faith in the fixity of the solid and sure-set foundations of the earth. From early childhood we are habituated to the contrast between the mobile element, water, and the immobility of the soil on which we stand. All the evidences of our senses have confirmed this belief. But when suddenly the ground begins to rock beneath us, the feeling of an unknown mysterious power in nature coming into action, and shaking the solid globe, arises in the mind. The illusion of the whole of our earlier life is annihilated in an instant. We are undeceived as to the repose of nature, we feel ourselves transported to the realm, and made subject to the empire, of destructive unknown powers. Every sound—the slightest rustle in the air—sets attention on the stretch. We no longer trust the earth upon which we stand. The unusual in the phenomenon throws the same anxious unrest and alarm over the lower animals. Swine and dogs are particularly affected by it; and the very crocodiles of the Orinoco, otherwise as dumb as our little lizards, leave the shaken bed of the stream and run bellowing into the woods.

To man the earthquake presents itself as an all-pervading unlimited something. We can remove from an active crater; from the stream of lava that is pouring down upon our dwelling we can escape; with the earthquake we feel that whithersoever we fly we are still over the hearth of destruction. Such a mental condition, though evoked in our very innermost nature, is not, however, of long duration. When a series of slighter shocks occur in a district one after another, every trace of alarm soon vanishes among the inhabitants. On the rainless coasts of Peru nothing is known of hail, nor of explosions of lightning and rolling thunder in the bosom of the atmosphere. The subterraneous noise that accompanies the earthquake there comes in lieu of the thunder of the clouds. Use and wont for a series of years, and the very prevalent opinion that dangerous earthquakes are only to be apprehended two or three times in the course of a century, lead the inhabitants of Lima scarcely to think more of a slight shock of an earthquake than is thought of a hail-storm in the temperate zone.

Having now taken a general survey of the activity, and likewise of the internal life of the globe: in its contained heat, in its electro-magnetic tension, in its luminous emanations at the poles, in its irregularly-recurring phenomenon of motion, we come to elementary MATERIAL PRODUCTION, to chemical changes in the crust of the earth, and in the composition of the atmosphere, which are in like manner the consequence of planetary vital activity. From the ground we see effusions: of watery vapour and of gaseous carbonic acid, mostly free from all admixture of azote([164]); of carburetted hydrogen gas (in the Chinese province of Sse-Tschuan([165]) for thousands of years, and in the State of New York, where, in the village of Fredonia, it has lately been employed for economical purposes in heating and lighting; of sulphuretted hydrogen gas; of sulphur fumes, and more rarely of sulphurous and hydrochloric acid vapours([166]). Such emanations from fissures in the ground do not only indicate the dominion of volcanoes long extinct or still burning; they are farther observed exceptionally in districts in which neither trachyte nor any other volcanic rock meets the eye exposed upon the surface. In the Andes of Quindiu I have seen sulphur precipitated from hot sulphureous vapours issuing out of mica slate, at a height of 6410 feet above the level of the sea([167]); whilst the same, and, as it used to be regarded, primitive rock, in the Cerra Cuelo, near Ticsan, south of Quito, exhibits an enormous bed of sulphur in pure quartz.

Of all the air-springs which the earth pours forth, those of carbonic acid gas are still at the present time the most important both in number and extent. Germany, in her deeply-cut valleys of the Eifel, in the neighbourhood of Lake Lach, in the Kesselthal of Wehr, and in Western Bohemia, as also in the burning hearths of the primeval world, or their vicinity, shews us these effusions of carbonic acid as a kind of last effort of volcanic activity. In former epochs, where, with a higher temperature of the earth, and the frequency of fissures yet unfilled, the processes which we are here describing proceeded more actively, where carbonic acid gas and watery vapours were mingled with the atmosphere in larger quantities than at present, the youthful vegetable world, as Adolph Brongniart([168]) has acutely observed, must have attained almost everywhere, and independently of geographical position, to the most rank luxuriance and evolution of its organs. In the ever hot, ever moist atmosphere, surcharged with carbonic acid, vegetables must have found such vital excitement, such superfluity of nourishment, as enabled them to supply the material of those beds of coal and lignite, the exhaustion of which it is difficult to conceive, and which now serve as foundations for the physical strength and the welfare of nations. Such beds are principally contained in basins, and are peculiar to certain parts of Europe. They are abundant in the British Isles, in Belgium, in France, on the Nether Rhine, and in Upper Silesia. In the same primeval times of all-pervading volcanic action, too, must those enormous quantities of carbonaceous matter have issued from the bowels of the earth which all the limestone rocks contain, and which, separated from oxygen, and represented in the solid form, composes about an eighth part of the absolute bulk of these great mountain masses ([169]). The carbonic acid which the atmosphere still contained, and which was not absorbed by the alkaline earths, was gradually consumed by the vegetation of the primeval world, so that the atmosphere, purified by the processes of vegetable life, by and by contained no more of the gas than was uninjurious to the organization of such animals as people the earth at the present time. Sulphurous or sulphuric acid vapours, too, occurring more frequently and much more abundantly then than now, occasioned the destruction of the inhabitants of the inland waters—mollusca and numerous genera of fishes, as well as the formation of the strangely-twisted beds of gypsum which have often apparently been shaken by earthquakes.

Under precisely similar physical relations there were further thrown out from the bosom of the ground various gases and liquids, mud, and, from the eruption-cones of volcanoes, which are but a species of intermitting springs, streams of molten earths([170]). All these matters owe their temperature and the nature of their chemical constitution to the place of their origin. The mean temperature of ordinary springs is lower than that of the atmosphere of the place where they appear, when the water is derived from high levels; their temperature increases with the depth of the strata with which they come in contact at their origin. The numerical law of this increase has been stated above. The mixture of the waters which come from the mountain elevations and from the depths of the earth, renders the position of the isogeothermal lines([171]), or lines of like internal heat of the earth, difficult of determination, when the conclusion has to be come to from the temperature of springs as they rise. So, at least, did I and my friends find it in some experiments which we made in Northern Asia. The temperature of springs, which has been so constant an object of physical investigation for the last half century, depends, like the height of the line of perpetual snow, on numerous and highly complex causes. It is a function of the temperature of the stratum in which they have their origin, of the capacity for heat of the ground, and of the quantity and temperature of the atmospheric or meteoric water that falls([172]), which last, again, according to the mode of its origin, differs in its temperature from that of the lower strata of the atmosphere([173]).

Cold springs, as they are called, can only give the mean temperature of the air if unmixed with water that is rising from great depths, or that is descending from considerable heights, and when they have flowed for a very long way under the surface of the earth—in our latitudes from 40 to 60 feet, in the equinoctial zone, according to Boussingault, one foot([174]). These depths are those, in fact, of the stratum of earth in which, in the temperate and torrid zone respectively, the point of invariable temperature begins, in which the hourly, diurnal, or monthly variations in temperature of the air are no longer perceived.

Hot springs burst out of the most diversified mineral strata; the hottest of all the permanent springs which have yet been observed, and which I myself discovered, flow remote from all volcanoes. I here refer to the Aguas calientes de las Trincheras, between Puerto Cabello and New Valencia, and to the Aguas de Comangillas, near Guanaxuato in Mexico. The first spring, issuing from granite, indicated 90·3° C.; the second, which breaks from basalt, shewed 96·4° C. The depth of the source of water of these temperatures, from what we know of the law of increase of temperature in the interior of the earth, must probably be about 6700 feet (more than half a geographical mile). If the cause of the heat of thermal springs, as well as of active volcanoes, be the universally diffused heat of the earth, then would mineral species produce an effect only through their capacity for, and their power of conducting heat. The hottest of all the permanent springs, those, namely, from 95° to 97° C. (204° to 207·6° F.), it is remarkable, are the purest, are those that contain the smallest quantity of mineral matter in solution. Their temperature appears on the whole to be less permanent than that of springs between 50° and 74° C., the invariableness of which, both in regard to temperature and mineral impregnation, has been maintained so wonderfully, within the confines of Europe at least, during the last fifty or sixty years, *i. e.* since accurate thermometrical observations and chemical analyses were made. Boussingault found that the thermal springs of las Trincheras had risen in temperature in the course of twenty-three years (from 1800, when my journey was performed, to 1823), from 93·3° to 97° C.(175). This very smoothly flowing spring is consequently at this time 7° C. higher in temperature than the intermitting Geyser and Strokr, the temperature of which has been lately very carefully ascertained by Krug of Nidda. One of the most remarkable proofs of the origin of these hot springs being due to the percolation of cold meteoric water into the interior of the earth, and its contact there with a volcanic focus, was presented in the preceding century in connection with the volcano of Jorullo in Mexico, which was unknown to geography till after my South American journey. When this mountain suddenly made its appearance in September, 1759, rising to a height of 1580 feet above the surrounding level, the two small streams, Rios de Cuitimba y de San Pedro disappeared; but by and by they made their appearance again, under the dreadful shocks of an earthquake, as hot springs. In 1803 I found their temperature 65·8° C.

The springs of Greece still flow apparently in the same places as they did in the times of Hellenic antiquity. The source of Erasinos, two leagues south of Argos, on the declivity of Chaon, is even mentioned by Herodotus. At Delphi, the Cassotis, under its name of Stream, of St. Nicholas, still rises to the south of the Lesche, and flows under the Temple of Apollo; the Castalia, too, at the foot of the Phædriadæ, and the Pirene at Acrocorinth, are there, as well as the hot baths of Ædepsum in Cubœa, in which Sulla bathed at the time of the Mithridatic war(176). I gladly adduce these particulars, because they forcibly remind us how, in a country exposed to earthquakes so frequent and so violent, the interior of our planet has been able to preserve its fashion for 2000 years at least; the small, branching, and open fissures that convey the water of these springs have not altered. The Fontaine jaillissante of Lillers in the department of the Pas de Calais, was bored in the year 1126, and ever since then has the water flowed uninterruptedly to the same height, and in the same quantity; the excellent geographer of the Caramanian coasts, Captain Beaufort, moreover, observed the same flame, fed by a stream of inflammable gas, which escapes in the district of Phaselis, which Pliny(177) describes as the flame of Chimæra in Lycia.

The observation made by Arago in 1821, that the deeper Artesian wells are the warmer(178), was the first means of throwing a great light upon the origin of thermal springs, and led to the discovery of the law of the increase of the temperature of the earth according to the depth. It is remarkable, and only noticed in very recent times, that St. Patricius(179), probably bishop of Pertusa, was led to a very correct view of the phenomenon which presented itself in the appearance of the hot springs near Carthage at the end of the third century. When questioned as to the cause of the boiling hot water which poured out from the earth, he answered: "Fire is nourished in the clouds and in the interior of the earth, as Etna, and another mountain in the neighbourhood of Naples, inform you. The subterranean waters rise as through syphons; and the cause of the heat of hot springs is this: the waters that are more remote from the subterraneous fire show themselves colder; those that flow in closer proximity to the fire, warmed by it, bring an insupportable heat to the surface which we inhabit."

As earthquakes are frequently accompanied by eruptions of water and watery vapour, so do we perceive in the volcanoes that pour out mud a transition from the alternating phenomena presented by jets of vapour and thermal springs to the grand and destructive activity of the mountains that vomit lava. If these, as springs of melted earths, produce volcanic rocks, so do the thermal springs that are charged with carbonic acid and sulphurous gas [and earthy matters], produce by incessant precipitation either horizontal beds of limestone (travertin), or they form conical hillocks, as in the north of Africa (Algeria), and the Baños of Caxamarca, on the western declivity of the Peruvian Andes. In the travertin of Van Dieman's Land, not far from Hobart Town, they are contained, according to Mr. Charles Darwin, the remains of an extinct flora. By lava and travertin, two species of rock the production of which goes on under our eyes, we here indicate the grand antitheses in geognostical relations.

Mud volcanoes (*Salsen*) deserve a greater share of attention than geologists have hitherto bestowed upon them. The extent of the phenomenon has been overlooked, because in the two states in which it presents itself to us, the one of repose is that which has been principally dwelt upon, and in this state of repose mud volcanoes often continue for centuries. The production of mud volcanoes is accompanied by earthquakes, subterranean thunder, the elevation of a whole district of country, and the eruption of flames, which rise high, but last only for a short time. When the mud volcano of Iskmali made its appearance in the peninsula of Abscheron, eastward from Baku, on the Caspian Sea (on the 27th of November, 1827), flames burst forth, and blazed up to an extraordinary height for a period of three hours; for the next succeeding twenty hours they scarcely rose three feet above the surface of the crater that discharged the mud. The column of flame mounted to such a height near the village of Baklichi, westward from Baku, that it was seen at the distance of six [German] miles. Great blocks of stone, torn from their foundations beneath, were scattered widely around. Similar blocks are observed about the now slumbering mud volcanoes of Monte Zibio, near Sassuolo, in the north of Italy. The second state, or that of activity, has continued for 1500 years in the mud volcano of Girgenti (Macalubi), in Sicily, which is described by the ancients. Many conical hillocks of 8, 10, and even 30 feet high, though the height as well as the form of these varies at different times, are there seen arranged near one another. From the superior very small basin, which is full of water, along with periodic escapes of gas, there are periodic streams of clayey mud discharged. The mud of these volcanoes is generally cold, but occasionally, as at Damak, in the province of Samarang, island of Java, it is of high temperature. The gases, which escape with a rushing noise, are also of different kinds—hydrogen gas, mixed with naphtha, carbonic acid, and, as Parrot and I ascertained (in the peninsula of Taman and the South American Volcancitos de Turbaco), almost pure nitrogen gas(180).

Mud volcanoes, after the first forcible outburst of flame, which perhaps is not common to all in the same measure, present the observer with a picture of an activity of the interior of the earth that proceeds incessantly but feebly. The communication with the deep strata in which a high temperature prevails is speedily interrupted again; and the cold discharges of mud volcanoes seem to indicate that the seat of the phenomenon, in its state of continuance, cannot be very remote from the surface. The reaction of the interior of the earth upon its outer crust is exhibited in a very different degree of force in the proper volcanoes, or burning mountains; in other words, in those points of the earth where a permanent communication, or, at all events, a communication that is renewed from time to time, is established between the surface and the deep focus of ignition. We must carefully distinguish between more or less exaggerated volcanic phenomena, such as these: Earthquakes, hot springs and jets of steam, mud volcanoes, the appearance of unopened dome-shaped trachytic mountains, the opening of these mountains, or the upheaval of basaltic beds as craters of elevation, the final rise of a permanent volcano within the upheavement crater itself, or amongst the fragments of its previous constitution. At different times, along with different degrees of activity and force, permanent volcanoes throw out jets of aqueous vapour, acids, glowing ashes and scoriæ, and, when the resistance can be overcome, fiery streams of melted earthy matters.

As a consequence of a great but local manifestation of force in the interior of our planet, elastic vapours raise either single parts of the crust of the earth into dome-shaped, unopened masses of felspathic trachyte and dolerite (Puy de Dôme and Chimborazo); or the upheaved strata are broken through, and inclined outwards in such wise that upon the opposite inner aspect a steep rocky edge is produced. This edge then becomes the boundary of an upheavement crater. When this has risen from the bottom of the sea, which does not by any means happen in every case, it then presents the whole of the characteristic physiognomy of the upheaved island. This is the origin of the circular form of Palma, which Leopold von Buch has described so carefully and so ably, as well as of Nisyros, in the Ægean Sea(181). Occasionally, one half the ring-like edge is destroyed, and in the bay which the sea that has flowed in then forms, the social coral insects establish themselves, and produce their cellular dwellings. Craters of elevation on continents are also frequently found filled with water, when they contribute to beautify the landscape in an extraordinary and quite peculiar manner.

Their origin is not connected with any special mountain formations; they break out in basalt, trachyte, and leucitic porphyry (Somma), or in doleritic mixtures of augite and labrador. Hence the very dissimilar natures and external forms of this kind of crater edge. "No eruptive phenomena take place from such boundaries; through them there is no permanent channel of communication established with the interior, and it is only very rarely that traces of still active volcanic power are discovered in the precincts or within the circuit of such craters. The force competent to bring about such important effects must long have gathered itself together, and gained strength in the interior, before it could overcome the resistance of the superincumbent masses. On the formation of new islands, it throws up granular rocky masses, and conglomerates (layers of tufa full of marine plants) above the level of the sea. Compressed gases escape through the crater of elevation; but a mass of such magnitude thus upheaved sinks down again, and closes forthwith the openings which are only formed for such manifestations of force. No volcano is produced"(182).

A proper volcano only arises where a permanent connection is established between the interior of the earth and the atmosphere. Here the reaction of the interior upon the exterior proceeds for lengthened periods. It may, as in the case of Vesuvius (Fisove)[183], be interrupted for centuries, and exhibit itself anew with renovated vigour. In the time of Nero it was already customary, in Rome, to rank Ætna among the number of the gradually-expiring volcanic mountains(184); Ælian, indeed, at a later period, maintained that the seamen began to see the sinking summit at a less distance on the high seas than formerly(185). Where the evidence of the eruption, I might say the old scaffolding, has been perfectly preserved, the volcano shows itself rising from a crater of elevation; there a high rocky wall, a rampart of greatly-inclined strata, surrounds the isolated cone in the manner of a circus. Sometimes there is not a trace of this circus-like inclosure visible, and the volcano, not always conical in figure, then arises as an elongated ridge immediately from the elevated platform. This is the case with Pichincha, at the foot of which stands the city of Quito.

As the nature of mountain masses, in other words, the combination or grouping of simple minerals into granite, gneiss, and mica-slate, into trachyte, basalt, and dolerite, independently of present climates, and under the most dissimilar zones, is still the same; so do we everywhere observe the same laws of formation proclaiming themselves in the realm of inorganic nature, laws according to which the strata of the crust of the earth stand in a certain relationship to one another, under the influence of elastic forces, and break through one another as dykes. This recurrence of the same phenomena is particularly striking in volcanoes. When the navigator, among the islands of distant seas, finds himself surrounded by palms and strange forms of vegetation, and no longer sees the same stars, in the individualities of the landscape he still traces Vesuvius, the dome-shaped summit of Auvergne, the craters of elevation of the Canaries and Azores, the fissures of eruption of Iceland repeated and reflected; a glance at the attendant of our planet, the moon, generalizes still farther the analogy of formation here adverted to. In the maps of the moon, drawn from the image reflected in powerful telescopes, in our satellite, without atmosphere and without water, we can distinguish vast craters of elevation, which surround conical mountains, or support them on their circular walls: unquestionable effects of the reaction of the interior of the moon upon her exterior, aided by the influence of diminished gravity.

If in many languages volcanoes are very properly designated Burning Mountains, it would still be a great mistake to suppose that they were produced by any gradual accumulation of the streams of lava that have flowed from them; their origin appears to be much more generally the consequence of a sudden upheaval of tenacious masses of trachyte or augitic rock, including Labrador spar. The measure of the upheaving force reveals itself in the height of the volcano; and this is so different, that in one case it is a mere hillock (as in Cosima, one of the Japanese Kuriles), in another it is a cone that rises to a clear elevation of 18,000 feet. It has seemed to me as if the relative height had a great influence upon the frequency of the eruptions; as if these were much more common in the lower than in the loftier volcanoes. I will call attention to the following series: Stromboli (2,175 feet high), Guacamayo, in the province of Quiros, which thunders almost every day (I have frequently heard it in Chillo, near Quito, at a distance of 22 miles), Vesuvius (3,637 feet high), Ætna (10,200 feet high), the Peake of Teneriffe (11,424 feet high), and Cotopaxi (17,892 feet high). If the focus of these several volcanoes be at the same depth below the surface, a greater force will be required to raise the molten masses to a 6 or 8 times higher level. Whilst the lowly Stromboli (Strongyle) has laboured restlessly, at least since the times of the Homeric traditions, and serves as a light-house to the Tyrrhenian Sea, guiding the seaman with its fiery signal on his course, the more lofty volcanoes are characterized by lengthened periods of repose. The eruptions of the greater number of the colossal volcanoes that crown the Andes, occur at intervals almost of a century apart: where exceptions to this rule have been observed—and I long ago directed attention to them—they may probably be connected with the circumstance, that the communication between the volcanic focus and the crater of eruption is not, cannot be conceived to be, equally or permanently free in every volcano at all times. In the less elevated volcanoes the channel of communication may be closed for a season; so that their eruptions become rarer, without their being, on this account, any nearer to extinction.

With the consideration of the relation between the absolute heights of volcanoes and the frequency of their activity, in so far as this is externally visible, the place at which the lava flows out is closely connected. Eruptions from the crater are extremely rare in the case of many volcanoes; they generally occur from lateral fissures (as noticed by the celebrated historian, Bembo, in the 16th century, whilst yet a youth), at places where the flanks of the uplifted mountain, in consequence of their formation and inclination, offer the least amount of resistance(186). Upon these fissures cones of eruption are occasionally raised. The larger of these are of such dimensions that they are often erroneously designated by the title of new volcanoes; ranked side by side, they show the direction of a fissure which has again become closed; the smaller ones frequently occur in groups, thickly set together, and cover whole districts, as it were with bell-shaped, or beehive like, elevations. To the latter class belong the *hornitos* of Jorullo(187), and the cone of eruption of Vesuvius of October, 1822, of the volcano of Awatscha, according to Postels, and of the lava field near the Baidare mountains, in the peninsula of Kamtschatka, according to Erman.

When volcanoes do not rise free and isolated from a plain, when, on the contrary, they are surrounded by table-lands from 9 to 12,000 feet high, as in the double chain of the Andes of Quito, this circumstance may very well give rise to the fact, that the most violent eruptions, when red-hot ashes and scoriæ are thrown out with detonations that are heard for hundreds of miles around, are never accompanied with streams of lava(188). This is the case with the volcanoes of Popayan, of the lofty plains of Los Pastos, and of the Andes of Quito; the single volcano of Antesana, among the latter, perhaps excepted.

The height of the cone of ashes, and the dimensions and form of the crater, are the elements in the figure of volcanoes which more particularly impress upon each of them an individual character; but both of these elements, both the cone and the crater, are perfectly independent of the magnitude of the whole mountain. Vesuvius is not one-third of the height of the Peake of Teneriffe, yet its cone of ashes forms one-third of the whole height of the mountain, whilst the cone of the Peake is only one twenty-second of the entire elevation. In the case of another volcano of much greater height than the Peake, that of Rucu-Pichincha, namely, the relations come nearer to those of Vesuvius. Of all the volcanoes which I have seen in either hemisphere, Cotopaxi is that of which the conical form is the most regular and beautiful. A sudden melting of the snow of its ashy cone indicates the proximity of an eruption. Before there is even any smoke visible in the thin strata of the atmosphere that surround the summit and the crater's mouth, the walls of the ashy cone are sometimes heated through, when the entire mountain presents the most threatening and ill-omened aspect of inky black.

The crater which, except in very rare cases, occupies the summit of the volcano, forms a cauldron-like, and often accessible valley, whose bottom is subject to incessant changes. The greater or less depth of the crater is, in many volcanoes, an indication of the proximity or remoteness of an eruption. In the cauldron-like crater extensive fissures open and close again alternately; through these vapours of various kinds find vent, or small rounded fiery throats, filled with molten matters, are formed upon them. The ground rises and falls, and on it are piled hillocks of ashes and cones of eruption, which occasionally rise high above the edges of the crater, and give the volcano its characteristic physiognomy for years; but on the occurrence of fresh eruptions, they sink suddenly down, and disappear. The openings of these cones of eruption, which rise from the floor of the crater, must not, as is too frequently done, be confounded with the crater itself, which encircles them. When the crater is inaccessible, from its vast depth, and the perpendicular inward slope of its sides, as in the case of Rucu-Pichincha (14,946 feet high), one can still look down from the edges, upon the summits of the monticules which rise within the cauldron-like crater, partially filled with sulphureous vapours. A more wonderful or grander natural prospect I have never enjoyed. In the interval between two eruptions, the crater of a volcano may exhibit no luminous phenomenon, but merely open fissures and jets of watery vapour; or hillocks of ashes that can be approached without danger, are found upon its scarcely heated bottom. These often gratify the wandering geologist, without making him run any risk, by casting out glowing masses, which fall on the edges of the cone of scoriæ, their appearance being regularly announced by slight, and entirely local shocks—earthquakes on a small scale. Lava occasionally flows from open fissures, or small fiery gorges, into the crater itself, without bursting through its walls, or overflowing its edges. But if it does break through, the molten spring generally flows smoothly, and in such a determinate direction, that the great cauldron-like valley, called the crater, can still be visited during the period of the eruption. Without a particular description of the conformation, and also of the normal structure of burning mountains, phenomena cannot be rightly comprehended which have been distorted by fantastical descriptions, and the various significations attached to the words crater, volcano, and cone; or, rather, to the indefinite and indeterminate use of these words. The edges of the crater sometimes show themselves much less liable to change than might be expected. A comparison of De Saussure's measurements with my own, yields the remarkable result, in connection with Vesuvius at least, that the north-west edge of the volcano, the Rocca del Palo, may be regarded as having remained for forty-nine years (1773–1822) almost without change in its elevation above the level of the sea. Any difference that appears may be looked on as with in the possible errors of measurement(189).

Volcanoes which lift their summits far above the limits of eternal snow, like those of the Andes, present a variety of peculiar features. The sudden melting of the snow in the course of an eruption, not only occasions destructive floods, torrents in which heaps of smoking ashes are floated away on blocks of ice; but the accumulation of ice and snow goes on producing its influence uninterruptedly, and by filtration into the trachytic rocks, even whilst the volcano is perfectly quiescent. Caverns are thus gradually produced on the declivities or at the foot of the burning mountain, and these become subterraneous reservoirs of water, which communicate in various ways, and by narrow mouths, with the Alpine rivulets of Quito. The fishes of these Alpine streams multiply greatly, particularly in the gloom of the caverns; and then, when the earthquakes come, which precede all eruptions of volcanoes in the Andes, and the whole mass of the mountain is shaken, the subterraneous caverns at once give way, and pour out a deluge of water, fishes, and tufaceous mud. This is the singular phenomenon which the presence of the Pimelodes Cyclopum(190), the Preñadilla of the inhabitants of the lofty plains of Quito, attests. When, in the night between the 19th and 20th of June, 1698, the summit of Carguairazo, a burning mountain 18,000 feet high, crumbled together, so that no more than two enormous rocky horns of the crater's edge remained, the country for nearly two square miles was desolated with liquid tuff and argillaceous mud (lodazales) inclosing dead fishes. So also was the putrid fever of the mountain town, Ibarra, to the north of Quito, which occurred seven years before, ascribed to an eruption of fish from the volcano Imbaburu.

Water and mud which, in the volcanoes of the Andes, do not pour down from the crater itself, but from cavities in the trachytic mass of the mountain, ought not, consequently, in the strict sense of the phrase, to be reckoned among the number of proper volcanic phenomena. They are only mediately connected with the activity of volcanoes, nearly in the same measure as the irregular meteorological process, which, in my earlier writings, I have spoken of under the title of the Volcanic storm. The hot, watery vapour which rises from the crater, and mingles with the atmosphere during the eruption, forms a cloud as it cools, with which the column of ashes and fire, many thousand feet in height, is surrounded. So sudden a condensation of vapour, and the production of a cloud of enormous superficial dimensions, increase the electrical tension, as Gay Lussac has shown. Forked lightnings dart from the column of ashes, and the rolling thunder of the volcanic storm is then plainly distinguishable from the rumbling in the interior of the mountain. This was well observed towards the end of the eruption of Vesuvius in the month of October, 1822. The lightning, which proceeded from the volcanic steam-cloud of the Katlagia burning-mountain in the Island of Iceland, according to Olaffen's account, upon one occasion (17th October, 1755), killed eleven horses and two men.

Having now, in our physical delineation, portrayed the general structure and dynamic activity of volcanoes, we have still to cast a glance at the material diversity of their products. The subterraneous forces separate old combinations of elements, in order to bring about new combinations; they farther, and at the same time, put the matters transformed or changed into motion, so long as they are dissolved by heat and moveable. The solidification of the tenaciously or more limpidly fluid and moveable mass, under different degrees of pressure, appears to be the principal cause determining differences in the structure of Plutonic and volcanic rocks or mineral species. The mineral mass which has flowed in a liquid state from a volcanic opening—a molten mineral spring—is called lava. Where several streams of lava have encountered and severally restrained each other in their course, they spread out and fill extensive basins, where they cool into stratified beds. These few points comprise the whole of the general features in the productive activity of volcanoes.

Minerals which merely break through a volcano often remain enclosed in the products of its igneous activity. I have, for instance, seen angular masses of syenite, rich in felspar, contained in the black augitic lava of the Mexican volcano, Jorullo; but the masses of dolomite and granular limestone, which contain beautiful druses or cavities lined with crystallized minerals—vesuviane and garnets, mejonite, nepheline, and sodalite—are not ejections of Vesuvius: "they rather belong to a very extensive formation, tuff-strata, older than the upheaval of Somma and Vesuvius, and are probably products of submarine volcanic influences, at great depths below the surface"(191). Among the products of our present volcanoes there are five metals: iron, copper, lead, arsenic, and selenium, discovered by Stromeyer in the crater of Volcano. Through the smoking fumaroles, the chlorides of iron, copper, lead, and ammonium, are sublimed; iron-glance(192), and common salt (the latter often in large quantities), are seen filling veins in recent streams of lava, or covering fresh fissures of the crater's edges.

The mineral composition of lavas differs according to the nature of the crystalline rock of which the volcano consists; according to the height of the point at which the eruption takes place—as it is near the foot of the mountain, or in the vicinity of the crater; and according to the temperature of the interior. Vitreous volcanic products, obsidian, pearlstone, or pumice, are entirely wanting in some volcanoes, and in others are only ejected from the crater itself, or from some considerably elevated point. These important and complex relations can only be ascertained by careful crystallographic and chemical researches. My companion in my Siberian journey, Gustavus Rose, and after him, Herman Abich, have begun, with much acumen and success, to throw clear light upon the compact texture of such a variety of volcanic minerals.

The greater portion of the vapour that rises is pure steam or watery vapour. Condensed and flowing away as a rivulet, it is used by the goatherds of the island of Pantellaria. The stream which was seen flowing from a lateral fissure in the crater of Vesuvius on the morning of the 26th of October, 1822, and was long regarded as hot water, was found by Monticelli to be dry ashes, which poured forth like drift-sand; it was lava ground to dust by attrition. The appearance of ashes, however, which darken the air for hours, and even for days, and which, by adhering to the leaves, become so destructive to vineyards and olive trees, in their columnar ascent, borne up by vapours, indicate the termination of every great eruption. This is the magnificent spectacle which the younger Pliny describes in the celebrated letter to Cornelius Tacitus, and which he compares, in point of shape, to a lofty-branched and shady pine tree. What has been described as flame in the eruption of ashes, is certainly not, any more than the light of the glowing red cloud that floats above the crater, to be ascribed to hydrogen gas on fire. It is rather the reflection of light from the upheaved molten masses; sometimes, too, it may be the light from the depths of the fiery gorge cast upon the ascending vapours and reflected by them. But as to what those flames may be, which have been occasionally seen ever since Strabo's time during the activity of volcanoes on the coast, and that have risen from the bosom of the sea immediately before the upheaval of a volcanic island, I do not pretend to decide.

If we are asked what it is that burns in volcanoes, what it is that produces the heat which melts and mixes the earths and metals, and even imparts an elevated temperature, for many years, to streams of lava of great thickness?(193) there is always the presumption that, as in the case of the coal fields which catch fire and go on burning, volcanoes must necessarily be connected with the presence of certain substances calculated to support combustion. According to the various phases of chemical science, we have had bitumen, iron pyrites, the moist contact of finely divided sulphur and iron, pyrophoric substances, and the metals of the alkalies and earths, assigned as the cause of volcanic phenomena in their highest intensity. The great chemist to whom we are indebted for our knowledge of the most combustible of the metallic substances, Sir Humphrey Davy, has himself renounced his bold chemical hypothesis in the last volume he published—"Consolations in Travel, and the Last Days of a Philosopher"—a work that excites painful feelings of regret in the mind of the reader. The great mean density of the earth (5·44), compared with the specific gravity of potassium (0·865), of sodium (0·972), and of the metals of the earths (1·2), the absence of hydrogen in the gaseous emanations of the fissures of volcanoes, and the streams of lava that have not yet cooled, many chemical considerations, in a word, rise up in opposition to the earlier conjectures of Davy and Ampère(194). Were hydrogen evolved during eruptions of lava, how enormous must its quantity prove in cases where, from the low level of the point whence the eruption flows, the outpouring mass spreads over many square miles of surface, and, dammed up in its course, acquires a thickness of several hundred feet; as happened in the remarkable eruption at the foot of the Skaptar-Jokul in Iceland (11th of June to 3d of August, 1783), which has been described by Mackenzie and Soemund Magnussen. The same difficulties present themselves in connection with the small quantities of azote that are evolved, when the atmospheric air is conceived as penetrating by the crater, or, as such an act has been figuratively expressed, when the earth is imagined as inspiring. So general, so deeply effective, and, in reference to the interior of the earth, so extensive an action as that of volcanoes, cannot well have its source in the chemical affinities, in the contact of individual and only locally distributed substances. Modern geognosy prefers seeking for this source in the temperature increasing with the depth under every parallel of latitude, in the great internal heat of the globe, which is due to its original consolidation, to its formation in space, to the spherical contraction of vaporous matter circulating in an elliptical orbit. Beside positive knowledge, stand Conjecture and Opinion. A philosophical science of nature strives to rise beyond the limited requirements of a bare description of nature. It consists not, as we have several times reminded the reader, in the barren accumulation of isolated facts. The curious, the inquiring spirit of man, must be suffered to make excursions from the present into the past, still to surmise what cannot be positively known, and to revel in the old, and, under various shapes, ever recurring *myths* of geognosy. If we consider volcanoes as irregular intermitting springs, which pour out a liquefied mixture of oxidized metals, alkalies and earths, that flow smoothly, silently enough, once the mixture, uplifted by the vast force of compressed vapour, finds a vent, we are involuntarily reminded of Plato's geognostical phantasies, according to which hot springs, as well as all the varieties of volcanic fiery streams, are effusions of Periphlegethon, a cause universally present in the interior of the globe([195]).

Volcanoes, in their mode of distribution over the surface of the earth, independently of all climatic differences, are well and characteristically referred to two classes, viz.: Central volcanoes and Linear volcanoes, "according as a central and common point of eruption for many foci all around is established, or as several vents extending in one direction, and at no great distance from each other, are formed, along the course apparently of a lengthened fissure. Linear volcanoes, again, are of two kinds: they either rise as insulated cones from the bottom of the sea, and are accompanied usually on one side by a primitive mountain mass running in the same direction, the foot of which they seem to indicate; or they stand upon the crest of the mountain chain, and form its loftiest summits"([196]). The Peake of Teneriffe, for example, is a central volcano, the middle point of the volcanic group to which the outbreaks of Palma and Lancerote belong. The lengthened chain of the Andes, which runs like a wall from Southern Chili to the northwest coasts of America, here singly, there in two and three parallel lines, connected at intervals by narrow transverse yolks, presents an instance upon the grandest scale of the occurrence of linear volcanoes on dry land. The vicinity of active volcanoes in the line of the Andes is proclaimed by the sudden appearance of certain species of rocks, such as dolerite, melaphyre, trachyte, andesite, and dioritic porphyry, which separate the so-called primitive rocks, as well as the schistose and sandstone transition strata and the tertiary or flœtz formations. A phenomenon of this kind constantly recurring, begot a persuasion in my mind at an early period, that these sporadic rocks had been the seat of volcanic phenomena, and had been determined by volcanic eruptions. At the foot of the great Tunguragua, near Pe nipe, on the banks of the Rio Puela, I for the first time, and distinctly, saw a mica schist, which rested upon granite, broken through by volcanic rocks.

The linear volcanoes of the New World, where they lie near to one another, are partially in a state of reciprocal dependance; it is even obvious that the volcanic activity has been gradually advancing for centuries in particular determinate directions—in the Province of Quito, for example, from north to south([197]). The hearth or focus itself lies under the whole of the elevated lands of this province; the particular openings by which communications are established with the atmosphere constitute the mountains which we designate by special names, such as Pichincha, Cotopaxi or Tunguragua, and which, by their grouping, as well as by their height and form, present the grandest and most picturesque prospect that is anywhere to be seen within a small compass in a volcanic country. As the outermost members of such groups of linear volcanoes are connected with one another by subterraneous communications, as multiplied experience shows, this fact reminds us of Seneca's old and truthful sentence([198]), "that the burning mountain is but the passage to deeper-lying volcanic forces." In the Mexican highlands, likewise, the volcanoes (Orizaba, Popocatepetl, Jorullo, Colima), which I have shown([199]) all to lie in one direction, between 18° 59′ and 19° 12′ N. latitude, appear to indicate a transverse fissure extending from sea to sea, and to be mutually dependant. The volcano of Jorullo broke out on the 29th of September, 1759, exactly in this direction, upon the same transverse fissure, and rose to a height of 1580 feet above the surrounding level. This mountain never threw out but one stream of lava; precisely like Epomeo in Ischia, in the year 1302.

But if Jorullo, distant as it is some German miles from every active volcano, be, in the strictest sense of the word, a *new* mountain, nevertheless it must not be confounded with the appearance of the Monte Nuovo near Pozzuolo (19th September, 1538), which is to be reckoned among the number of upheavement craters. I have already said that it were more in conformity with nature to assimilate the eruption of the newly produced Mexican volcano with the upheaval of the hill of Methone (now Methana), upon the peninsula of Træzene. This upliftment, described by Strabo and Pausanias, has led one of the most imaginative of the Roman poets to propound views which agree in a very remarkable manner with those of modern geognosy: "A tumulus is seen at Træzene, rugged, and without wood; once a level, now a mountain; the vapours pent up in dark caverns sought in vain for a crevice of escape. They swelled the expanding soil under the force of the compressed vapour, like a bladder filled with air; it swelled like the skin of a two-horned goat. The upheavement remains upon the spot; the high, uplifted hill became hardened in the course of time into a naked rocky mass." So picturesquely, and, also, as analogous appearances lead us to believe, so truly, does Ovid describe the grand natural incident which occurred between Træzene and Epidaurus, 282 years before the commencement of our era, and therefore 45 years before the volcanic separation of the island of Thera (Santorin) from Therasia(200).

Of all the islands belonging to the series of linear volcanoes, Santorin is the most important. "It comprises in itself the entire history of upheaved islands. For full two thousand years, so long as history and tradition extend, Nature has not ceased from her attempts to form a volcano within the circuit of the crater of elevation"(201). Similar insular upheavements, at almost regularly recurring intervals of 80 or 90 years, are exhibited in the island of St. Michael, one of the group of the Azores(202), though here the bottom of the sea has not been uplifted quite at corresponding points. The island named Sabrina by Captain Tillard unfortunately appeared at a time when the political state of the maritime nations of the west of Europe was little favourable to scientific investigations (30th Jan., 1811); so that this great event did not attract the same degree of attention as was bestowed upon the island of Ferdinandea,* which appeared on the 2d of July, 1831, but soon fell to pieces again, between the limestone coast of Sciacca and the purely volcanic Pantellaria in the Sicilian Sea(203).

The geographical distribution of the volcanoes which have continued active since the historical epoch, their frequent situation by the sea-shore, and on islands, to say nothing of the recurrence, from time to time, of temporary eruptions from the bottom of the sea, appears at an early period to have begotten the belief, that volcanic activity was connected with the vicinity of the sea, and could not continue without it. "Ætna and the Œolian isles," says Justin(204), or, rather, Trogus Pompeius, whom he copies, "have already been burning for many centuries; and how were this long continuance possible, did not the neighbouring sea supply food for the fire?" To explain the necessity for the neighbourhood of the sea, the hypothesis of the penetration of sea-water to the hearth of the volcano, *i. e.*, to the deep-lying strata of the earth, has in recent times been again proposed. If I embrace all that occurs to me, derived either from personal observation or from carefully collected facts, it seems to me that everything in this difficult inquiry depends upon the way in which the following questions are answered: Whether the undeniably large quantities of watery vapour, which volcanoes emit, even in their state of repose, be derived from sea-water loaded with salts, or from sweet atmospheric water? Whether, with different depths of the volcanic hearth (a depth, for example, of 88,000 feet, at which the expansive force of the vapour of water would be exerted under a pressure of 2,800 atmospheres), the expansive force of the vapour engendered would be competent to counterbalance the hydrostatic pressure of the sea, and admit the access of its water to the volcanic hearth, under certain conditions?(205) Whether the many metallic chlorides, the appearance, indeed, of common salt in the fissures of craters, and the admixture of hydrochloric acid vapours with the watery vapour emitted, lead necessarily to the conclusion, that the sea must have access to the volcano? Whether the repose of the volcano, be this temporary only, or final and complete, depends on the stoppage of the channels which previously conducted the sea, or the meteoric water, to the volcanic hearth? Whether the absence of flame and of hydrogen gas—for sulphuretted hydrogen belongs to the *solfataras* rather than to the active volcanoes—is not rather in open contradiction with the assumption of any extensive decomposition of water?

The discussion of physical questions of such importance does not fall within the scope of a Picture of Nature. Here we attach ourselves to the narration of phenomena; to facts in the geographical distribution of yet active volcanoes. Now facts inform us, that in the New World, three of these—Jorullo, Popocatepetl, and La Fragua—are 20, 33, and 39 geographical miles distant from the sea-shores, and that in central Asia (and M. Abel-Remusat(206) first directed the attention of geologists to the fact), there is a great volcanic mountain chain, Thian-schan, or the Celestial Mountains, with the lava-emitting Pe-schan, the solfatara of Urumtsi, and the burning mountain of Turfan (Ho-tscheu), the several members of which are at nearly equal distances—370 to 382 geographical miles—from the shores of the Icy Sea and of the Indian ocean. The distance of Pe-schan from the Caspian Sea is also full 340 geographical miles; and from the great lakes, Issikul and Balkasch, it is 43 and 52 miles(207). It is farther remarkable, that of the four great parallel mountain chains—the Altai, the Thian-schan, the Kuen-luen, and the Himalaya, which cross the continent of Asia from east to west—it is not the Himalaya, or the chain that is nearest the ocean, but the two minor chains, the Thian-schan and the Kuen-luen, at the distance respectively of 400 and 180 geographical miles from the sea, that are found vomiting fire like Ætna and Vesuvius, and producing ammonia, like the volcanoes of Guatimala. The Chinese writers describe, in unmistakable terms, streams of lava, 10 Li long, as occurring in the eruptions of flame and smoke which took place from Pe-schan, and spread far and wide, in the 1st and 7th centuries of our era. "Burning masses of rock," say they, "flowed as thin as melted fat." These few compressed facts, which have not been sufficiently attended to, make it probable that the vicinity of the sea, and the access of sea-water to the burning focus, are not indispensably necessary to the breaking out of subterranean fires, and that coasts are only favourable to volcanic eruptions, because they form the sides or edges of the deep sea-basin, which, covered with strata of water, offers less resist-

* The Graham Island of English geologists; *vide* Lyell's *admirable* Principles of Geology, vol. ii., p. 266. Sixth edit., Lond., 1840.—Tr.

K

ance, and lies many thousand feet lower, than inland and more lofty countries.

The volcanoes that are active at the present time, and that communicate permanently by craters with the interior of the earth and the atmosphere, became open at so late an epoch, that the superior cretaceous deposits, and the whole of the tertiary formations, were already in existence when they arose. This is proclaimed by the trachytes, and also by the basalts, which frequently form the walls of the upheavement craters. Melaphyres extend to the middle tertiary strata ; but have already begun to show themselves under the Jura formations, when they appear breaking through the variegated sandstone(208). The active volcanoes of the present time, communicating with the air by craters, must not be confounded with those older eruptions of granite, quartzose porphyry, and euphotide, through open, but speedily-closed fissures (forming veins), which occur in the old transition strata.

The extinction of volcanic activity is either partial only, so that the subterranean fire finds another vent in the same mountain chain ; or it is total, as in Auvergne ; later examples are supplied, in perfectly historical times, by the volcano Mosychlos(209), on the island dedicated to Hephœstos, whose "upward flickering fiery glow" was known to Sophocles, and by the volcano of Medina, which, according to Burckhardt, threw out a stream of lava on the 2d of November, 1276. Each stage of the volcanic activity, from its first excitement to its extinction, is characterized by peculiar products : first, by fiery scoriæ, by trachytic, pyroxenic, and vitreous lavas in streams, by scoriæ and tuff ashes, accompanied by the evolution of large quantities of generally pure watery vapour ; at a later period as solfataras, when there is an evolution of watery vapour mixed with sulphuretted hydrogen and carbonic acid gases ; lastly, when all has cooled, by exhalations of carbonic acid gas alone. Whether that singular class of burning mountains which discharge no lava, but dreadful devastating streams of hot water(210), loaded with burning sulphur, and rocks ground down to powder—such, for instance, as Galunggung, in the island of Java —present us with what may be called a normal condition, or only a certain transitory modification of the volcanic process, will remain a question undecided, until they have been visited by geologists possessed at the same time of a knowledge of modern chemistry.

Such is the very general view of volcanoes, so important an element in the life of the earth, which I have here endeavoured to throw together. It is based, in part, upon my own observations ; in the generality and comprehensiveness of its outlines, however, upon the labours of my friend of many years, Leopold von Buch, the greatest geologist of our age, who was the first to recognize the intimate connection of volcanic phenomena, and their mutual interdependence in regard to their actions and their relations in space.

The reaction of the interior of a planet upon its outer crust and surface, as manifested in the phenomena of volcanoes, was long considered as a mere isolated phenomenon, and peculiar only with reference to the destructive agency of its dark and subterraneous forces ; it is but very lately, and greatly to the advantage of that geology which is founded on physical analogies, that the volcanic forces have begun to be regarded as formative of new species of rocks, and as transformative of older mineral masses. Here, indeed, is the point already alluded to, where a more deeply-grounded doctrine of volcanoes in a state of activity, and either casting out fire or vapour, leads us, in our general Picture of Nature, by a double way, the one to the mineralogical portion of geognosy, or the doctrine of the structure and succession of the strata composing the crust of the earth ; the other to the form and fashion of the continents and groups of islands raised above the level of the sea, or the doctrine of the geographical forms and outlines of the several portions of the earth. Enlarged views of such an enchainment of phenomena is a consequence of the philosophical direction which the serious study of geognosy has now so generally taken. Greater perfection of the sciences leads, as in the political improvement of mankind, to connection and agreement, where there had formerly been separation and distinction.

When we class rocks or mineral masses not according to differences in the form and arrangement of their constituent particles, into stratified and unstratified, schistose and massy, normal and abnormal rocks, but look at the phenomena of formation and transformation which are still going forward under our eyes, we discover a four-fold process of production in connection with rocks : 1st. Eruptive rocks, rocks thrown out from the interior of the earth, in a liquefied, or softened and more or less tenacious state (volcanic and Plutonic rocks). 2d. Sedimentary rocks, rocks deposited from fluids in which the particles had been either dissolved or suspended, but from which they were precipitated and deposited upon the surface of the crust of the earth. The greater number of the floetz and tertiary groups. 3d. Metamorphic rocks, rocks altered in their intimate structure and stratification, either through the contact and vicinity of a Plutonic or volcanic (endogenous) (211) ejected rock, or—and this is more commonly the case—altered by the penetration of the vaporiform sublimed matters(212), which accompany the escape of certain molten ejected masses. 4th. Conglomerates —coarse or fine-grained sandstones, breccias— rocks made up of mechanically divided masses of the three former species.

These four-fold rock-formations, which still go on at the present day, through the effusion of volcanic masses in the shape of streams of lava, through the influence of these masses upon rocks consolidated at a former period, through mechanical separation or chemical precipitation from liquids charged with carbonic acid, finally, through the cementation of fragments often of totally different kinds of rocks, are phenomena and formative processes which can, however, only be regarded as weak reflections of what went on under the higher intensity of action in the life of the earth during the chaotic state of the primitive world, and under totally different conditions of pressure and high temperature, not only of the whole crust of the earth, but of the atmosphere, surcharged with moisture and of much greater extent than it is at the present day. If at the present time, on surfaces as extensive as Europe, we scarcely find four openings (volcanoes) through which eruptions of fire and molten matters can take place, the firm crust of the earth was traversed in former periods by vast open fissures, through which mountain chains were upheaved, or into which streams of molten rock—granite, porphyry, basalt, and melaphyre—were injected, and by which they were variously stopped and filled up. At former epochs, in the much and variously fissured, thinner, and upwardly and downwardly fluctuating crust of the earth, there were almost everywhere passages of communication between the molten interior and the atmosphere. Gaseous emanations arising from very dissimilar depths, and therefore bringing chemically different substances, then animated the Plutonic formative and transformative processes. The sedimentary formations, too, the precipitations from liquids, which we designate travertin, and which we see proceeding in the neighbourhood of Rome as well as of Hobart Town in Australia, from cold and hot springs and river waters, give but a very poor idea of the origination of the floetz formations. Our seas, in virtue of processes which have not yet been examined generally enough, or with sufficient care, gradually form by precipitation, by overflowing and by cementation, small calcareous banks, which, at some points, almost approach Carrara marble in hardness(213). This process goes on upon the Sicilian coasts, the Island of Ascension, and King George's Sound in Australia. On the coasts of some of the West India islands these formations of the present ocean now enclose earthenware vessels and other products of human manufacturing industry; and in the Island of Guadaloupe, even skeletons of the Carib race of men. The negroes of the French colonies characterize this formation as the "Masonry of God" (Maçonne-bon-Dieu) (214). In the Island of Lancerote, one of the Canaries, there is a small oolitic stratum, admitted to be a product of the sea and of storms, but which, despite its newness, reminds us of the Jurassic limestone(215).

The compound rocks are determinate associations of certain simple minerals—felspar, mica, solid silicic acid, augite, and nepheline. Very similar rocks, *i. e.* rocks made up of the same elements but otherwise grouped, are produced by volcanic processes under our eyes, at the present time, just as they were in former epochs of the world's history. The independence of rocks in respect of geographical position or relationship, is so great, that, as we have already observed(216), the geologist sees with amazement, to the north and south of the equator, in the farthest zones of the earth, the same familiar appearances in the rocks, the repetition of the minutest details in the periodic series of the Silurian strata, and in the effects of contact with augitic masses, the products of eruptions.

If we now take a closer view of the four fundamental forms of rock (the four phases in the formative process) in which the stratified and unstratified portions of the crust of the earth present themselves to us, we may designate among the endogenous or eruptive rocks, (the massive and abnormal rocks of some modern geologists), the following principal groups, as immediate evidences of subterraneous activity, viz.:

Granite and Syenite—of very different relative ages, but frequently penetrating both granite and syenite of more recent formation in veins(217). Along with these it is also proper to consider the forcing or upheaving power. "Where granite protrudes in evenly vaulted ellipsoids, in great masses, like islands, whether this be in the Harzforest, or in Mysore, or in Lower Peru, it is always covered with layers that have become fissured into blocks. Such a rocky sea probably owes its origin to a contraction of the upper surface of the granitic vault, which, on its protrusion, and originally, must have been very much expanded"(218). In Northern Asia also(219), in the charming, the romantic neighbourhood of Lake Kolyvan, on the north-western declivity of the Altai range, as also on the slopes of the maritime chain of Caraccas, near Las Trincheras(220), I observed the granite subdivided into blocks or piles, in consequence, possibly, of such contractions, but which in these cases appear to have extended deeply into the interior. Farther to the south of Lake Kolyvan, towards the confines of the Chinese province Ili, between Buchtarminsk and the river Narym, the characters of the entire mass of ejected rock, which is here unaccompanied by gneiss, are more striking than I have observed them in any other part of the globe. The granite, always scaling and crumbling on the surface, and splitting up into tabular masses, rises in the steppes here in low semi-globular hillocks, not more than six or eight feet high, there in basalt-like knolls, which run out at opposite sides, as it were, into thin wall-like effusions(221). By the cataracts of the Orinoco, as well as in the Fichtelgebirge (Seissen), in Gallicia, and betwixt the Southern Ocean and the lofty platforms of Mexico (at Papagallo), I have seen granite in great depressed globular masses, which, like basalt, split or scaled off in concentric layers. In the valley of the Irtisch, between Buchtarminsk and Ustkamenogorsk, the granite covers the clay-slate for a mile in length(222), and penetrates the same strata from above in slender veins, which are numerously branched, and wedge-shaped at their extremities. I have adduced these particulars by way of examples, only that I may illustrate the individual characters of an eruptive rock in one of the most widely diffused of the mineral masses. In the same way as the granite overlies the schists in Siberia, and in the Department of Finisterre (Isle de Michau), so does it cover the Jurassic limestone in the mountains of Oisons (Fermonts), and syenite, and chalk with syenite interposed, near Weinböhla, in Saxony(223). In the Ural mountains near Mursinsk, the granite shows drusy cavities, and the druses here, like the fissures and druses of newer volcanic productions, are the Plutonic seat of numerous beautiful cystals, particularly of beryl and topaz.

Quartzose Porphyry, from its relations of stratification, having frequently the character of veins. The base is generally a finely granular mixture of the same elements which present themselves to us as large embedded crystals. In granitic porphyry, which is very poor in quartz, the felspathic base is at once granular and foliaceous(224).

GREENSTONE or DIORITE—granular mixtures of white albite and blackish-green hornblende, constituting dioritic porphyry, when a base of denser texture is present in which the crystals lie embedded distinctly. These greenstones, which, pure in one place, pass in another into serpentine, from the laminæ of diallage which they include (Fichtelgebirge), are occasionally found lying in beds upon the old stratification clefts of the green clay slate, and penetrating them; but they more frequently make their way through the rock in the manner of veins, or they present themselves as greenstone balls, analogous in all respects to balls of basalt and porphyry(225).

HYPERSTHENE ROCK—a granular mixture of Labrador felspar and hypersthene.

EUPHOTIDE and SERPENTINE, occasionally containing crystals of augite and uralite, instead of diallage, and thus nearly allied to a more common rock, and, I might add, one that indicates a still higher degree of eruptive activity, viz., augitic porphyry(226).

MELAPHYRE, AUGITIC, URALITIC, and OLIGOCLASSIC PORPHYRY. To the last belongs the true verd antique, so celebrated as a material employed in the arts.

BASALT, with olivine and constituents becoming gelatinous with acids, PHONOLITE (porphyritic slate), TRACHYTE and DOLERITE. The second of these rocks always divides into thin tables; the first only shows this structure partially, which, however, gives them both an appearance of stratification over extensive districts. According to Girard, mesotype and nepheline form important elements in the composition and intimate texture of basalt. The nepheline of basalt reminds the geologist of the miascite of the Ilmengebirge in the Ural chain(227), which frequently replaces granite, and occasionally contains zircon, as well as of the pyroxenic nepheline discovered by Gumprecht near Löbau and Chemnitz.

To the second class of fundamental forms, the SEDIMENTARY ROCKS, belongs the greater portion of the formations which used to be arranged under the old systematic, but by no means correct, designation of Transition and Floetz, or secondary and tertiary formations. Had the igneous rocks exerted nothing of an uplifting, and, with simultaneous quaking of the earth, of a concussive influence upon these sedimentary formations, the surface of our planet would have consisted of a series of uniform strata horizontally disposed one upon another. Without mountains, on whose acclivities the progressive diminution in the temperature of the air is picturesquely reflected, not only in the luxuriance of vegetation, but in the kinds of plants that are produced, the monotonous surface would only have been broken here and there by ravines eroded by water-courses or by small collections of drift, the effect of masses of fresh water thrown into gentle undulations; the several continents from pole to pole, and under every variety of climate, would have presented the dreary uniformity of the South American Llanos or of the Northern Asiatic steppes. As in the greater portion of these, we should then have seen the vault of heaven resting on the plain, and the stars rising and setting as if they emerged from the bosom of the ocean, and dipped into it again. But such a state of things even in the primitive world could never have been of any considerable duration as regards time, nor of any thing like general prevalence in respect of space; the subterraneous powers, at every epoch in the history of nature, have been at work striving to subvert and to change it.

Sedimentary strata are precipitated or deposited from liquids, according as the matter before the formation was either held chemically dissolved, as in the case of lime, or merely suspended and mixed, as in the case of clay-slate, mica-slate, &c. But even when earthy matters are thrown down from fluids impregnated with carbonic acid, the descent of the matter during its precipitation and accumulation into strata, must be regarded as a mechanical element in the process of formation. This view is of some importance in connection with the envelopment of organic bodies in petrifying calcareous tuffs. The oldest sediments of the transition and secondary formations have apparently taken place from waters more or less elevated in temperature, and at a period when the heat of the upper crust of the earth was still very considerable. In this way, therefore, a Plutonic influence was also at work to a certain extent in connection with the sedimentary strata, particularly the oldest of them; these strata, however, appear to have become hardened from the state of mud into the schistose structure, under great pressure; not like the rocks that have risen up from the interior (granite, porphyry, basalt), to have been consolidated by cooling. As the primitive waters of the globe cooled by degrees, they became capable of holding a larger and larger quantity of carbonic acid gas in solution, which they may have attracted from the atmosphere, surcharged with this gas in the earlier epochs of creation, and so of holding dissolved a larger quantity of calcareous earth.

The SEDIMENTARY STRATA, from which we here separate all the other exogenous purely mechanical precipitates of sand or fragmentary rocks, are these:

SCHISTS or SLATES of the inferior and superior transition rocks, consisting of the Silurian and Devonian formations; from the lower Silurian, or as they were once designated, Cambrian, strata, to the uppermost bed of the Old red sandstone or Devonian formation, where it comes in contact with the Mountain limestone;

CARBONIFEROUS deposits—Coal formation;

LIMESTONES, interstratified in the transition and coal formations; Zechstein, Muschelkalk, Jura formation and Chalk, also the portion of the tertiary formation which does not present itself to us as sandstone and conglomerate;

TRAVERTINE, fresh-water limestone, the silicious sinter of hot springs—formations that have originated not under the pressure of great pelagic coverings of water, but almost in contact with the air in shallow pools and rivulets;

INFUSORIAL STRATA, a geological phenomenon, the vast significance of which, as proclaiming the influence of organic activity upon the formation of the solid constituents of the earth, was discovered in very recent times, by my intellectually-gifted friend and fellow-traveller, Ehrenberg.

If in this short but comprehensive survey of the mineral constituents of the crust of the earth, we do not immediately refer to numbers of simple sedimentary rocks, the various conglomerate and sandstone formations, partly deposited from liquids, that are so variously intermingled with the schists and the limestones both of the floetz and transition series, this is only because these, besides fragments of erupted and sedimentary rocks, also contain pieces of gneiss, mica-schist, and other metamorphic masses. The obscure process of transformation (metamorphosis), and the influence it exerts, must, from this showing, constitute the third class of fundamental forms.

The endogenous or eruptive rocks (granite, porphyry, and melaphyre), exert an influence, as already oftener than once observed, not merely of a dynamical kind, shattering or upheaving, erecting or pushing strata aside; by their presence they farther produce changes in the chemical composition of their constituents, as well as in the nature of their intimate texture. New species of rocks are produced, gneiss and mica slate, and granular or saccharoidal limestone (Carrara and Parian marble). The old Silurian or Devonian transition schists, the belemnitic limestone of Tarantaise, the grey unlustrous macigno or cretaceous sandstone of the Northern Apennines, with its included sea-weed, are difficult of recognition after their transformation into new and frequently-sparkling textures. The belief in the metamorphosis, indeed, has only been confirmed since we have succeeded in following the several phases of the transformation, step by step, and have come to the assistance of inductive conclusions with the results of direct chemical experiments, the employment of different fusing heats, degrees of pressure, and rates of cooling. When the study of chemical combinations is extended under the guidance of leading ideas(228), we find that from the narrow confines of our laboratories, we can diffuse a clear light over the wide field of geology, over the great subterraneous rock-composing and rock-transforming workshop of Nature. The philosophical inquirer escapes being deceived by seeming analogies, by limited views of the natural processes, when he keeps steadily in his eye the complication of circumstances which, in the intensity, the immeasurableness of their force, were competent, in the primitive world, to modify the reciprocal influences of individual substances familiarly known to us at the present day. The simple or undecomposed bodies have unquestionably obeyed the same forces of affinity at all times; and where contradictions seem to meet us now, it is my most intimate persuasion, that chemistry will herself, for the most part, come upon the traces of conditions not fulfilled in like or due measure, as causes of these contradictions.

Accurate observations, embracing extensive districts of mountainous country, satisfy us that the eruptive rocks do not intervene as any disorderly or lawless power. In the most distant countries of the world, we frequently see granite, basalt, or diorite, exerting their transformative force, in every the most minute particular, alike upon strata of clay-slate, on thick beds of limestone, and on the grains of quartz of which sandstone consists. As the same kind of endogenous rock almost everywhere exerts the same kind of influence, different kinds of rocks belonging to the same class of endogenous or eruptive formations, exhibit, on the contrary, very different characters. Intense heat, above all, has exerted an influence in the whole of the phenomena; but the degree of molten fluidity attained—perfect mobility of particles, or a more viscid or glutinous adhesion among them—has been very different in granite and basalt; in different geological epochs, indeed (phases in the transformation of the crust of the earth), along with the eruptions of granite, basalt, porphyritic greenstone or serpentine, various other substances dissolved in vapours have arisen from the interior laid open. And this is the place to remind the reader anew, that in the rational views of modern geology, the metamorphosis of rocks is not limited to the mere phenomena of contact, to the apposition of two different kinds of rock; but that genetically it comprises all that has accompanied the protrusion of a particular ejected mass. In situations where no immediate contact has taken place, the mere vicinity of such a mass causes modifications in the induration, the silicification, the granulation, the crystallization of adjacent rocks.

All eruptive rocks penetrate the sedimentary strata, and other likewise endogenous masses, as veins; but the distinction that is apparent between the Plutonic rocks(229)—granite, porphyry, serpentine—and those which, in a more restricted sense, are called volcanic (trachyte, basalt, lava), is of especial importance. The rocks which our present volcanoes, as remnants of the activity of the body of the earth, produce, appear in narrow streams, which, however, may still form sufficiently wide beds when several of them meet in hollows or basins. Basaltic eruptions, where they have been traced deeply, have been repeatedly seen to terminate in slender taps. Flowing from narrow openings, as in the Pftasterkaute, near Marksuhl, two miles from Eisenach, in the blue knolls, near Eschwega (banks of the Werra), and at the Druid's-stone, on the Hollert ridge (Siegen), to cite three examples indigenous to Germany, the basalt breaks through the red sand-stone and greywacke schist, and spreads out above, like the cap of a mushroom, into knolls, which in one place appear split into columnar groups, in another are thinly stratified. Not so granite, syenite, quartzose, porphyry, serpentine, and the entire series of unstratified massy rocks, which, from an attachment to mythological nomenclature, have been called Plutonic. These, with the exception of a few veins, have been ejected, not in a molten liquefied state, but in one merely tenacious and softened, and not from narrow crevices, but from wide valley-like chasms, and extensive gorges. They have been forced, they have not flowed out: they present themselves not in streams, like lava, but spread out in immense masses[230]. Among the dolerites and trachytes, some groups give indications of a certain basalt-like fluidity; others, expanded into vast bell-shaped elevations and craterless domes, appear to have been merely softened when they were protruded. Other trachytes, again, those of the Andes among the number, which I frequently found very closely allied to the greenstones and syenitic-porphyries, so rich in silver, and then without quartz, lie in beds like granite and quartzose-porphyry.

Experiments upon the alterations which the structure and chemical constitution of rocks undergo through fire[231], have showed that the volcanic masses, diorite, augitic porphyry, basalt, and lava from Ætna, according to the degree of pressure under which they were melted, and the rate of their cooling, were either, when quickly cooled, brought to the state of a black glass of an even fracture, or when slowly cooled made to assume the appearance of a stony mass having a granular crystalline texture. The crystals in such cases were either produced on the sides and cavities, or embedded in the general basic mass. The same material—and this consideration is of great importance as regards the nature of the eruptive rock, or the transformations it has undergone—yields the most dissimilar-looking products. Carbonate of lime, melted under high pressure, does not lose its charge of carbonic acid; the cooled mass is granular limestone, saccharoidal marble. So much for crystallization in the dry way; in the moist way, calcareous spar as well as Aragonite is produced, the former under a moderate, the latter under a higher, degree of heat[232]. According to diversities of temperature, the consolidating particles of crystals in process of formation arrange themselves variously and in particular determinate directions; the very form of the crystals, indeed, varies with the temperature under the influence of which they are produced[233]. There is, moreover, under certain relations, and without the intervention of any fluid state, a transposition[234] of the minute particles of a body, which is proclaimed by optical effects. The phenomena presented by devitrifaction, by the production of cemented and cast steel, by the transition of the fibrous structure of iron into one that is granular, under the influence of elevated temperature[235], perhaps even of very insignificant but equable and long-continued concussions, all conduce to throw light upon the processes of geological metamorphosis. Heat can even induce opposite effects at the same time upon crystalline bodies; for Mitscherlich's beautiful experiments show that calcareous spar, without altering its state of aggregation, expands in the direction of one of its axis of crystallization, and contracts in another[236].

If from these general considerations we pass on to particular examples, we first observe schists turned into black-blue roofing slate by the vicinity of Plutonic ejected rocks. The clefts of stratification are then interrupted by another system of clefts which cut the former almost perpendicularly, and indicate the operation of a later influence[237]. By the penetration of silicic acid, clay slate, traversed by fragments of quartz, is partially changed into whetstone slate (Wetzschiefer, whitestone or Eurite?) and silicious slate (Kieselschiefer, quartzite?), the latter frequently carboniferous, and then galvanic in its effects on the nerves. The highest degree of silicification of the schists[238], however, is found in a precious material employed in the arts, ribboned jasper, produced in the Ural Mountains by the contact of eruptive augitic porphyry (Orsk), dioritic porphyry (Auschkul), or hypersthene rock (Bogoslowsk); in the Island of Elba (Monte Serrato), according to Fr. Hoffmann, and in Tuscany, according to Alexander Brongniart, by contact with euphotide and serpentine.

The contact and Plutonic influence of granite cause clay-slate to become granular, changing it into a granitic-looking mass—into a mixture of felspar and mica, in which again larger plates of mica lie embedded[239]—a fact which Gustavus Rose and I observed within the fortress of Buchtarminsk[240]. "That the whole of the gneiss lying between the Icy Sea and the Gulph of Finland has been formed and transformed by the agency of granite out of Silurian strata of the transition series, may now, as Leopold von Buch has said, be assumed as an hypothesis familiar to all geologists, and accepted by the greater number as demonstrated. In the Alps of St. Gothard cretaceous marl is met with transformed by granite, first into micaceous schist, and then into gneiss"[241]. Similar phenomena in respect of gneiss and mica slate formations, under the influence of granite, are presented: in the Oolitic group of Tarantaise[242], where belemnites have been found in rocks that already lay claim to the denomination of mica schist; in the schistose group of the western portion of the Island of Elba, not far from Cape Calamata, and in the Fichtelgebirge of Bayreuth, between Lomitz and Markleiten[243].

Precisely as jaspar, a substance employed in the arts, which was inaccessible to the ancients in large masses[244], is the product of volcanic agency upon augitic porphyry, so is the other artistic material, so variously and so successfully employed by them, granular marble, to be regarded as a sedimentary stratum altered by the heat of the earth and the vicinity of an eruptive rock. Careful observation of phenomena of contact, and the remarkable experiments of Sir James Hall on the fusion of rocks, now more than half a century old, in addition to the diligent study of granitic veins, which contributed so essentially to the early foundations of our present geology, warrant such a conclusion. The protruded rock has occasionally changed the dense calcareous deposit into granular limestone to a certain thickness only, or in a certain zone from the line of contact. We find a partial transformation, like a half-shadow, at Belfast in Ireland, where basaltic dykes penetrate the chalk; in the same way, in the compact floetz-limestone near the bridge of Boscampo, and by the waterfall of Canzocoli in the Tyrol, celebrated by Count Marzari Pencati, the strata have been partially bent where they come in contact with a syenitic granite[245]. Another kind of transformation is that in which the whole of the beds of compact calcareous rock are changed into granular limestone through the influence of granite, syenite, or dioritic porphyry(246).

Let me be allowed to refer particularly in this place to the Parian and Carrara marbles, which have become so necessary to the noblest efforts of the sculptor, and which have served but too long in our geological collections as principal types of *primitive* limestone. The effects of the granite here reveal themselves partly under immediate contact, as in the Pyrenees(247), partly, as in the continent of Greece and the islands of the Ægean Sea, through interposed strata of gneiss and mica slate. In both cases the process of transformation of the calcareous rock is contemporaneous, but dissimilar. It has been observed at Cubæa, in Attica, and in the Peloponnesus, "that the rule is, that the limestone which rests upon mica slate is by so much the more beautiful and crystalline as the schist is purer, that is, as it is freer from argillaceous admixture." Mica slate, as well as gneiss strata, present themselves at many deep points of Paros and Antiparos(248). If marine productions were discovered [in ancient times] in the quarries of Syracuse, and the "impression of a small fish" was seen in the deepest of the rocks of Paros, as we may infer from a notice in Origen, of the old Eleatic philosopher, Xenophanes of Colophon (249), who conceived the whole of the world to have been formerly covered by the sea, we might believe in the remains of a floetz stratum in this situation which had not undergone complete metamorphosis. The marble of Carrara (Luna), which was employed before the Augustan age, and was the principal source of the material for statues so long as the quarries of Paros remained closed, is a stratum of the same cretaceous sandstone (macigno) altered by Plutonic agency, which presents itself in the insulated Alpine height, Apuana, lying between gneiss-like micaceous and talcose schists(250). Whether or not granular limestone, formed in the interior of the earth, and filling fissures in the manner of veins (Auerbach on the Bergstrasse), have ever been forced to the surface by gneiss and syenite(251), I cannot, through want of personal observation, take it upon me to decide.

The most remarkable metamorphoses of compact calcareous strata, however, according to Leopold von Buch's able observations, are to be seen in the Southern Tyrol, and among the Italian slopes of the Alps, effected principally by the intrusion of dolomitic masses. The metamorphosis of the calcareous rock here proceeds from fissures, which traverse it in all directions. The clefts are everywhere covered with rhomboids of magnesian spar; the whole formation indeed, without stratification, and without a vestige of the fossils which it formerly included, then consists exclusively of a granular aggregation of dolomitic rhomboids. Talc leaves and transverse fragments of serpentine lie here and there dispersed through the new-fashioned rock. In Fassathal, the dolomite rises perpendicularly in the form of smooth walls of dazzling whiteness to the height of several thousand feet. It forms pointed conical hills, which stand side by side in great numbers without touching one another. Their physiognomical character brings to mind that sweetly fantastical mountain landscape with which Leonardo da Vinci has ornamented the back-ground of his portrait of Mona Lisa.

The geological features which we are here portraying excite the imagination as well as reflection; they are the work of an augitic porphyry, which has intruded and produced its effect, by upheaving, shattering, and transforming(252). The dolomitizing process is by no means regarded by the gifted inquirer who first pointed it out as an imparting of magnesian earth by the black porphyry, but as a change effected contemporaneously with the protrusion of the injected rock into extensive fissures filled with vapour. It remains for future inquiries to determine in what way the transformation is effected when the dolomite occurs in beds between limestone strata, without contact with the endogenous rock, where the conduits of Plutonic influences lie concealed. But it is not perhaps necessary, even here, to take refuge in the old Roman saying, according to which "much that is like in nature has been produced in totally different ways." If in an extensive district of country, two phenomena, viz., the protrusion of melaphyre, and the alteration in crystalline texture and chemical constitution of a compact calcareous rock, always go together, then may we, with some reason, conjecture, that in cases where the second phenomenon presents itself without the first, the seeming contradiction in the non-fulfilment is connected with certain conditions accompanying the occult principal cause. Should we question the volcanic nature of basalt and its state of liquefaction through fire, because a few rare instances have been met with in which dykes of this substance traverse carboniferous sandstone and cretaceous strata, without the coal being deprived of its bitumen, the sandstone reduced to the state of frit, or the chalk being turned into granular marble? Where we meet with even a twilight glimmer, with the faintest trace of a way in the obscure region of mineral formations, we must not thanklessly reject both, because there is still much unexplained in the relations of transition from one rock to another, and in the isolated interposition of altered between unaltered strata.

Besides the transformation of compact calcareous carbonate into granular limestone and into dolomite, there is a third metamorphosis of the same deposit, which must here be adverted to, and which is attributable to the volcanic eruption of sulphuric-acid-vapours, in primeval epochs. This transformation of limestone into gypsum is connected with the penetration of rock-salt and sulphur (the latter precipitated from watery vapour charged with the mineral). In the lofty chain of the Andes of Quindiu, far from all volcanoes, I have myself observed this precipitate within fissures in gneiss, whilst the sulphur, gypsum, and rock-salt of Sicily (Cattolica, near Girgenti) belong to the newest secondary strata, or to the chalk formations(253). I have farther seen fissures filled with rock-salt, in quantities that sometimes tempt the people to engage in an illicit traffic in the article, in the edge of the crater of Vesuvius. On both slopes of the Pyrenees it is impossible to doubt the connection of dioritic (and pyroxenic?) rock, with the appearance of dolomite, of gypsum, and of rock-salt([254]). Everything in the phenomena here referred to proclaims the influence of subterraneous forces upon the sedimentary strata of the ancient ocean.

The beds of pure quartz of enormous magnitude, which are so characteristic of the Andes of South America([255])—and I may here state that I have seen such beds between 7 and 8,000 feet in thickness, in the route from Caxamarca to Guangamarca, descending towards the southern ocean — are of enigmatical origin. They rest in one place upon quartzless porphyry, in another upon dioritic rocks. Have they been produced from sandstone, as M. Elie de Beaumont conjectures has been the case in regard to the quartz strata of the Col de la Poissonnière, to the east of Briançon?([256]) In the diamond districts of Minas Geraes and St. Paul, in Brazil, which have been lately so carefully examined by Clausen, Plutonic forces acting upon dioritic veins have developed in one place common mica, in another ferruginous mica, in the quartzose itacolumite. The diamonds of Grammagoa are contained in layers of solid silicic acid; occasionally they lie enveloped by plates of mica, exactly like the garnets formed in mica-slate. The most northern of all the diamonds which have been found since the year 1829, under the 58th parallel of north latitude, on the European declivity of the Ural mountains, also stand in geological relation to the black carboniferous dolomite of Adolfskoi([257]), as well as to augitic porphyry, which have not yet been made the subject of sufficiently accurate observations.

Among the most remarkable contact-phenomena, finally, are comprised the formation of garnets in clay-slate, under the influence of basalt and dolerite, instances of which occur in the county of Northumberland and in the island of Anglesea; and for the production of a great number of beautiful and very dissimilar crystals—garnets, Vesuviane, augite, and Ceylanite —which make their appearance upon the contingent surfaces of eruptive and sedimentary rocks, on the boundary of the syenite of Monzon with dolomite and compact limestone([258]). In the island of Elba, the masses of serpentine, which nowhere, perhaps, present themselves so conspicuously as eruptive rocks, have caused the sublimation of iron glance and red iron stone into the fissures of a cretaceous sandstone([259]). The same iron glance is still seen every day, sublimed from vapour, upon the edges of open fissures in the craters of Stromboli, Vesuvius, and Ætna, and in cracks of the recent lava streams of these volcanoes([260]). As we here perceive the materials of veins produced under the influence of volcanic forces before our eyes, and where the neighbouring rock has already attained to a state of solidity, we conceive how mineral and metallic veins may have been produced during the earlier revolutions of the crust of the earth; when the solid, but still thin shell of the planet, repeatedly shaken by earthquakes, shattered and rifted by alterations in its volume occasioned by cooling, presented numerous communications with the interior, numerous outlets for vapours laden with earthy and metallic substances. The stratified arrangement of the mineral matters parallel with the surfaces bounding veins, the regular repetition of similar layers on both sides, on the roofs and on the floors of veins, and the druses or elongated cavities of their middles, indeed, frequently bear immediate testimony to a Plutonic process of sublimation in metalliferous veins. As the matter traversing is of more recent origin than the matter traversed, we learn from the relations of position of the porphyry to the silver-ore formations of the Saxon Erzgebirge, that these, in the mountains which are richer in mineral treasures than any others in Germany, are at least younger than the trunks of trees of the coal formation and than the lower new red sandstone (Rothliegendes) ([261]).

All our geological speculations, in regard to the formation of the crust of the earth and the metamorphosis of rocks, have had unexpected light thrown on them, by the happy idea of assimilating the production of scoriæ in our smelting furnaces, to the formation of natural minerals, and of reproducing these artificially from their elements([262]). The same affinities, determining chemical combinations, come into play in all these operations, whether they be conducted in our laboratories or in the bosom of the earth. The most important simple minerals, characterizing the very generally distributed Plutonic and volcanic rocks, as well as those that have suffered metamorphosis through them, have been found in our artificial mineral formations in the crystalline state, and like the natural ones in all respects. We distinguish those that have arisen accidentally in scoriæ, from those that have been produced intentionally by chemists. To the former belong felspar, mica, augite, olivine, blende, crystalline oxide of iron (iron glance), octohedral magnetic iron, and metallic titanium([263]); to the latter garnet, idokras, ruby (equal in hardness to the Oriental stone), olivine, and augite([264]). The minerals now named form the principal constituents of granite, gneiss, and mica schist, of basalt, dolerite, and numerous porphyries. The artificial production of felspar and mica, in particular, is of signal geological importance for the theory of the formation of gneiss by the transformation of clay slate. This contains the elements of granite, potash not excepted([265]). It would not, therefore, be any thing very extraordinary, as an acute geologist, M. von Dechen, has observed, were we one day to find a piece of gneiss produced upon the walls of a smelting-furnace built of clay-slate and greywacke.

In these general considerations on the solid crust of the earth, and after having indicated three original forms of production in reference to its mineral masses, viz., Eruptive, Sedimentary, and Metamorphosed rocks, there still remains a fourth class, the Conglomerated, or fragmentary, to wit. This title of itself brings to mind the denudations or destructions which the surface of the earth has suffered; but it also farther reminds us of the process of cementation or agglutination that has been effected by oxide of iron, and by argillaceous and calcareous [and silicious] cements, by which in one case rounded, in another angular, fragments have been again united. Conglomerates and breccias, in the widest sense of these words, reveal the character of a two-fold mode of origin. The materials of which they are mechanically composed have not always been accumulated by the waves of the sea, or by streams of fresh water in motion ; there are fragmentary rocks in the production of which the shock or the action of water has had no part. "When basaltic islands, or trachytic mountains, arise upon fissures, the friction of the rock as it ascends against the sides of the fissure causes basalt and trachyte to become surrounded by agglomerates of their own masses. The grains of which the sandstones of many formations consist have been more detached by the attrition of outbreaking volcanic, or Plutonic rocks, than produced by the motion of a neighbouring ocean. The existence of such attrition-conglomerates, which are encountered in enormous masses in both divisions of the globe, bear witness to the intensity of the force with which the eruptive masses have been forced to the surface from the interior. The waters then obtained power over the smaller detached granules, and spread them out in layers which they themselves covered"(266). Sandstone formations are found intercalated among all the strata, from the lower Silurian transition series, to this side the chalk in the tertiary formation. On the edges of the vast plains of the New World, both within and without the tropics, they are seen as walls or bulwarks, indicating, as it seems, the coasts against which the mighty waves of a former ocean once dashed themselves into foam.

If we venture a glance at the geographical distribution of rocks, and their relations in point of place in that portion of the crust of the earth which is accessible to our observation, we perceive that the most generally distributed chemical material of all is silicic acid, either in the transparent and colourless state, or opaque and variously tinged. After solid silicic acid comes carbonate of lime ; then follow in order the compounds of silicic acid with alumina, with potash and soda, with lime, magnesia, and oxide of iron. If the masses which we call *rocks* be definite associations of a small number of minerals, to which a few others, but also of determinate kinds only, are added as parasites ; if, in the eruptive rock granite, the association of quartz (silicic acid), felspar and mica be the essentials, so do these minerals, either isolated or in pairs, present themselves in many other strata. With a view of illustrating by an example how quantitative relations distinguish a felspathic rock from another abounding in mica, I here remind the reader, as Mitscherlich has done, that if three times more alumina, and one third more silicic acid than belong to it naturally be added to felspar, we have the composition of mica. Potash is contained in both, a substance the existence of which in many mineral masses reaches far beyond the commencement of everything like vegetation on the surface of the earth.

The succession, and with this the age of the several formations, are ascertained or determined by the reciprocal position of the Sedimentary, Metamorphic, and Conglomerate strata, by the nature of the formations up to which

L

the Eruptive masses ascend, but most certainly and safely by the presence of organic remains and the diversities of their structure. The application of Botanical and Zoological knowledge to the determination of the age of rocks, the *chronometry* of the crust of the earth, which Hook's great spirit had already divined(267), marks one of the most brilliant epochs in the progress of geology, now finally abstracted, on the Continent at least, from semitic influences. Palæontological studies have, as with a vivifying breath, given grace and the charms of variety to the doctrine of the solid materials of the globe.

The fossiliferous strata present us with the entombed floras and faunas of bygone millenniums. We ascend in time, whilst, penetrating downwards from layer to layer, we determine the relations in space of the several formations. An animal and vegetable existence that has passed away is brought to light. Wide-spread revolutions of the globe, the upheaval of mighty mountain chains, whose relative ages we are in a condition to determine, denote the destruction of old organic forms, the appearance of new. A few of the older still show themselves for a time among the newer forms. In the narrowness of our knowledge of original production, in the figurative language with which this circumscription of view is concealed, we designate as *new creations* the historical phenomena of change in the organisms, as in the tenancy of the primeval waters, and of the uplifted dry land. These extinct organic forms are in one case preserved entire, even to the most minute details of covering and articulated parts ; in other instances nothing more remains of them than their footsteps imprinted on the wet sand or mud which they traversed when alive, or their coprolites (petrified dejections), containing the unassimilated portions of the food upon which they fed. In the lower Jura formation (the Lias of Lyme Regis), the preservation of the ink-bag of the sepia(267), is so wonderfully perfect, that the same material which the animal employed myriads of years before to preserve itself from its enemies, has been made to serve as the colour wherewith to paint its likeness. In other strata there is sometimes nothing more than the faint impression of a bivalve shell, and yet will this suffice, when brought by a traveller from a far distant country, if it be a *characteristic* shell (Leitmuschel, a guiding-shell) (268) to inform us of the material formations which there exist, and of the other organic remains with which it was associated ; it tells the history of the district whence it came.

The anatomical study of the ancient animal and vegetable life of the globe extends in a two-fold direction. The one is purely morphological in its bearings, and is especially devoted to the description and physiology of the organisms ; it fills up with extinct forms the gaps encountered in the series that still exist. The other direction is geological, and considers fossil organic remains in their relations to the superposition and relative age of the sedimentary formations. The first was long the usual direction taken, and in its imperfect and superficial comparisons of petrifactions with living species led off into erroneous ways, traces of which are still to be discovered in the extraordinary denominations of certain natural bodies. There was the constant disposition to recognize a living in every extinct species; just as, in the 16th century, false analogies led naturalists to confound the animals of the Old World with those of the New Continent. Camper, Soemmering, and Blumenbach, had the merit, by the scientific application of a better comparative anatomy, of first illustrating the osteological portion of Palæontology (the Archæology of organic life), in so far at least as the larger fossil vertebrate animals are concerned; but for the proper geological consideration of the Science of Petrifactions, for the happy combination of the zoological character with the age and order of deposition of strata, we are indebted to the great work of George Cuvier and Alexander Brongniart.

The oldest sedimentary formations, those to wit of the transition series, in the organic remains which they include, present a mixture of forms which assume very different places in the scale of development, gradually attaining to greater and greater perfection. Of vegetables, they contain indeed but a few Fuci, Lycopodiaceæ which perhaps were arborescent, Equisetaceæ, and tropical Ferns; but of animal remains, we find, strangely associated together, Crustacea (trilobites with reticular eyes, and calymene), Brachiopoda (spirifer, orthis), the elegant Spheronites, which are nearly allied to the crinoideæ(269) and orthoceratites from among the Cephalopoda, Stone-corals, and with these lower organisms, Fishes of singular forms in the upper Silurian strata. The heavily-armed family of Cephalaspidans, fragments from whose genus Pterychthys were long regarded as trilobytes, belong exclusively to the Devonian, or old red sandstone formation, and show, as Agassiz says, as peculiar a type in the series of fishes as Ichthyosauri and Plesiosauri do among the reptiles(270). The Goniatites, belonging to the group of Ammonites, likewise begin to show themselves in the transition limestone and greywacke of the Devonian, and even in the later members of the Silurian system(271).

The dependence of physiological gradation upon the age of the formation, which has hitherto been but little observed in the position of the invertebrate order of animals(272), is exhibited in the most regular manner in connection with the vertebrate series. The oldest of these, as we have just seen, are the fishes; and then, following the series of formations from the inferior to the superior, we come to reptiles and mammalia. The first reptile encountered, a saurian or lizard, and, according to Cuvier, a monitor, which had already attracted the attention of Leibnitz(273), makes its appearance in the copperslate floetz of the Zechstein [lower new red, or magnesian limestone formation], of Thuringia; and with this, and of the same age, according to Murchison, the palæosaurus and the codontosaurus of Bristol. The Saurians go on increasing in numbers in the Muschelkalk, in the Keuper, and in the Jura formation, in which they attain their maximum(274). Contemporaneously with this formation lived Plesiosauri, with long swan-like necks, containing thirty vertebræ; the Megalosaurus, a crocodilian monster, 45 feet long, and with bones of the feet like those of a heavy mammiferous land animal; eight species of large-eyed Ichthyosauri; the Geosaurus or Soemmering's Lacerta gigantea; finally, seven singularly hideous Pterodactyles(275), or Saurians furnished with membranous wings. In the chalk, the number of crocodilian Saurians falls off, yet the epoch which this deposit characterizes is distinguished by the Mæstricht crocodile, as it is called, the Mososaurus of Conybeare, and the colossal, perhaps herbivorous, Iguanodon. Other animals that belong to the present race of crocodiles Cuvier has met with ascending into the tertiary formations(276); and Scheuchzer's "Man attesting the deluge (homo diluvii testis)," a great salamander, allied to the axolotl, which I brought with me from the Mexican lakes, belongs to the newest freshwater formations of Oeningen.

The relative ages of organisms determined by the position of the rocky strata in which their remains are found, has led to important conclusions as to the relations that can be traced between extinct and still existing families and species (the latter, the species, in very small numbers). Older and newer observations agree in showing that the floras and faunas are by so much the more unlike the present forms of plants and animals, as the sedimentary formations in which their remains lie buried belong to the inferior; in other words, to the older strata. The numerical relations presented by these grand successive changes in the forms of organic life, first pointed out by Cuvier, have yielded decisive results through the meritorious labours of Deshayes and Lyell, in connection more especially with the various groups of the tertiary formations, which contain a considerable mass of carefully-studied forms. Agassiz, who has cognizance of 1700 species of fossil fishes, and who estimates the number of living species that have been described, or that are preserved in museums, at 8000, speaks out quite decisively, in his master-work. He says: "With the single exception of one small fossil fish, peculiar to the clay-geodes of Greenland, I have found no animal of this class in all the transition, floetz, and tertiary strata, which is specifically identical with any fish now living;" and he adds the following important observation: "In the inferior tertiary formations, the coarse limestone and London clay, for example, one-third of the fossil fishes even belong to genera that are wholly extinct; below the chalk there is not one of the genera of fishes of the present time to be found, and the extraordinary family of the Sauroids (or fishes with scales covered with enamel, which in structure almost approach reptiles, and ascend from the coal formation, in which the largest species lie embedded, to the chalk, where single individuals are still encountered), stand related to the two families Lepidosteus and Polypterus, which now inhabit the rivers of America and the Nile, in the same way as our present elephants and tapirs to the Mastodons and Anaplotheriums of the primeval world"(277).

The chalk-beds, however, which still contain two of these sauroid fishes, and gigantic reptiles, and which present themselves as an entire world of extinct corals and shells, are composed, according to Ehrenberg's beautiful discovery, of microscopic Polythalamia, many of which are still to be found in our seas, particularly in the middle latitudes of the North Sea and Baltic. The first group of the tertiary formation lying over the chalk, a group which it has become customary to designate by the name of the strata of the Eocene period, would appear by no means rightly to deserve this title — "inasmuch as the morning dawn of the nature that still exists with us extends far more deeply into the history of the earth than was until lately believed"(278).

As fishes, the oldest of all vertebrate animals, already show themselves in Silurian transition strata, and then occur without interruption in all subsequent formations up to the strata of the tertiary epoch; as we have seen the Saurians begin with the zechstein or magnesian limestone, so are the first mammiferous animals, the Thylactotherium, Prevostii, and T. Bucklandii, which Valenciennes regards as nearly allied to the Marsupialia(279), found in the Stonesfield slate, a lower member of the Jura or Oolitic formation, and the first bird occurs in the older cretaceous deposits(280). These, according to our present knowledge, are the inferior limits of fishes, saurians, mammalia, and birds.

But if, among the members of the invertebrate series of animals, stone corals and serpulites are found making their appearance in the oldest formations simultaneously with highly developed cephalopods and crustacea, the most different and dissimilar orders being, therefore, associated without distinction, we, on the other hand, discover very determinate laws in connection with the distribution of particular groups of the same orders. Fossil shells of the same kinds, goniatites, trilobites, and nummulites, compose entire mountains. Where different genera are mingled, it often happens that not only is there a determinate sequence of organisms recognizable, according to the relations of superposition in the several systems of strata, but the association of certain genera and species has also been observed in the subordinate strata of the same formations. By his happy discovery of the Law of Estimates (Lobenstellung), Leopold von Buch has been enabled to distribute the vast multitude of ammonites into well-characterized families, and shown how the ceratites belong to the muschelkalk, the arietes to the lias, the goniatites to the transition limestone and greywacke(281). Belemnites have their inferior limits in the Keuper which covers the Jura limestone, their superior limits in the chalk(282). The waters of countries far remote from one another were inhabited at the same epochs by testaceous animals, which partly at least, as is now known for certain, are identical with those that occur fossilized in Europe. Leopold von Buch has shown us exogyri and trigonia from the southern hemisphere (the volcano Maypo, in Chili), and d'Orbigny ammonites and grypheæ from the Himalaya mountains and the plains of Cutch in India, which are identical in kind with those left behind by the old Jurassic sea of France and Germany.

Strata characterized by determinate species of fossils, or by determinate rolled masses which they inclose, form a geognostical horizon, by means of which the geologist, when at a loss, can always ascertain his place, and pursuing which, he arrives at safe conclusions as to the identity and relative age of certain formations, the periodical recurrence of particular strata, their parallelism, and their total suppression or failure. If we will thus embrace the type of the sedimentary formation in its greatest simplicity and most general distribution, we find its members in the following order, proceeding from below upwards.

1st. The so-called Transition rocks, in the two divisions of inferior and superior greywacke, or Silurian and Devonian systems, the latter formerly designated the Old Red Sandstone formation;

2d. The inferior Trias(283)—Mountain limestone, the Coal measures together with the Red conglomerate (Todtliegendes), and Zechstein or Magnesian limestone;

3d. The superior Trias — Variegated-sandstone, Muschelkalk and Keuper(284);

4th. Jura limestone (Lias and Oolite);

5th. Massive sandstone, Inferior and Superior chalk, as the last of the floetz strata, which begin with the mountain limestone;

6th. Tertiary formations, in three divisions, which are indicated by the Coarse limestone, Brown coal or Lignite, and Sub-Apennine gravel.

In the alluvium or drift follow the gigantic bones of the extinct mammalia—the Mastodons, Dinotheriums, Missuriums, Megatheriums, Owen's Sloth-like Mylodon, 11 feet long, &c. With these primæval genera are associated the fossilized remains of many animals that still exist—the elephant, rhinoceros, ox, horse, deer, &c. The plain near Bogota, filled with the bones of Mastodons (the Campo de Gigantes, in which I had some careful digging performed) (285), lies 8,200 feet above the level of the sea, and the bones of extinct species of true elephants are found still higher in the lofty plateaus of Mexico. Like the chain of the Andes, which has certainly been upheaved at very different epochs, the advances of the Himalaya, the Sewalik hills (which Captain Cautley and Dr. Falconer have so carefully examined), besides the extinct Mastodon, Sivatherium, and gigantic land tortoise, the Colossochelys, 12 feet long and 6 feet high, contain remains of genera that still exist—elephants, rhinoceroses, giraffes; and this, which is very much to be regarded, within a zone which enjoys the same tropical climate at the present day which we may be permitted to conjecture prevailed during the epoch of the Mastodons(286).

After having thus compared the series of inorganic formations composing the crust of the earth, with the animal remains which lie buried in them, we have still to write another chapter in the history of the organic life of the globe—that, namely, which refers to vegetables; and to trace the epochs of vegetation, the floras varying with the increasing dimensions of the dry land, and the modifications which the atmosphere underwent.

The oldest transition strata, as already remarked, present us with nothing but cellular-leaved marine plants. It is in the Devonian strata that a few cryptogamic forms of vascular vegetables, calamites and lycopodiaceæ, are first encountered(287). Nothing seems to testify, as, on theoretical views *On the simplicity of the first forms of organic life*, it has been assumed, that vegetable life was awakened sooner than animal life, upon the face of the old earth, and that this was brought about or determined by that. The existence of races of men in the very northern polar zones, who subsist on the flesh of fish, and seals and whales, is enough of itself to assure us of the possibility of living without vegetable matter of any kind. After the Devonian strata and the mountain limestone, comes a formation, the botanical anatomy of which has made such brilliant progress in recent times(288). The Coal Formation comprises not only fern-like cryptogamic plants, and phanerogamous monocotyledons—grasses, yucca-like liliaceous vegetables and palms; it further contains gynospermic dicotyledons—coniferæ and cycadeæ. Nearly 400 species from the flora of the coal formation are already known. I here mention only arborescent calamites and lycopodiaceæ; scaly lepidodendrons; sigillariæ of 60 feet long, and occasionally found standing erect and rooted, and distinguished by a double vascular fasciculate system; cactus-like stigmariæ; a host of ferns now arborescent, and again mere fronds, and by their quantity proclaiming the still entirely insular character of the dry land(289); cycadeæ(290); and particularly palms(291) in small numbers, asterophyllites with verticillate leaves, allied to the Najades; araucaria-like coniferæ(292) with slight indications of annual rings. The diversity in character of a vegetation which flourished luxuriantly on the uplifted and dry-laid portions of the old red sandstone, from the vegetable world of the present time, still continues through the later phytological periods on to the last layers of the chalk(293); but with a great degree of strangeness in the forms, the flora of the coal formation still exhibits a very remarkable uniformity in the distribution of the same genera (if not always of the same species), over every part of the then surface of the earth; in New Holland, Canada, Greenland, and Melville Island, the genera are still the same.

The vegetation of the former world presents us with forms the affinities of which with various families of the present age remind us that with them many intermediate members in the series of organic developments have perished. To quote two instances only: the Lepidodendra, according to Lindley, stand between the Coniferæ and the Lycopoditeæ(294); the Araucarita and Pinita, on the other hand, in the combination of their vascular fascicles, exhibit something that is foreign and peculiar. But confining our views to the present order of things, the discovery of Cycadeæ and Coniferæ in the flora of the old coal measures in juxtaposition with Sagenaria and Lepidodendra, is still of great significance. The Coniferæ, to wit, have not only relationships with the Cupuliferæ and the Betulineæ, by the side of which we encounter them in the brown-coal formation, but they are further connected with the Lycopoditeæ. The family of the sago-like Cycadeæ approaches the Palms in external appearance, whilst agreeing essentially with the Coniferæ in the structure of the flowers and fruit(295). Where several series of coal strata lie over one another, the genera and species are not always mixed; they are rather and for the major part generically arranged, so that only Lycopodites and certain Ferns occur in one series of beds, and Stigmariæ and Sigillariæ in another.

In order to form an idea of the luxuriance of vegetation in the former world, and of the masses of vegetable matter accumulated by running water, and which have very certainly been converted into coal in the humid way(296), I remind the reader that in the Saarbrück coal field there are 120 seams of coal lying one over another, exclusive of a host of smaller seams less than a foot in thickness; that there are single seams of coal of 30 and even of more than 50 feet thick, as at Johnstone in Scotland, and Creuzot in Burgundy; whilst in the forest regions of our temperate zone, the carbon which the trees of a certain superficial extent of ground contain, would not cover this surface with a layer of much more than half an inch in thickness (7 lines) in the course of one hundred years(297). Near the mouth of the Mississippi, and in the *wood hillocks*, as they have been called, of the Siberian Icy Sea, described by Admiral Wrangel, however, there is at the present time such an accumulation of trunks of trees, such a quantity of *drift wood*, washed down by land streams, and brought together by ocean currents, that the phenomena remind us at once of the events which took place in the inland waters and insulated bays of the primeval world, and gave occasion to the production of the coal formations which we now discover hundreds of feet below the surface of the ground. It is also well to remember that these coal measures are indebted for no inconsiderable portion of their materials not to the trunks of mighty trees, but to small grasses, and to frondiferous and low cryptogamic vegetables.

The association of palms and cone-bearing trees which we have just signalized in the coal fields, continues through almost all the formations onwards to far into the tertiary period. In the present world they seem rather to fly each other's vicinity. We have, in fact, although improperly, habituated ourselves so much to regard the cone-bearers as northern forms, that I myself, ascending from the shores of the South Sea towards Chilpansingo and the elevated valleys of Mexico, was somewhat amazed when I found myself between Venta de la Moxonera and the Alto de los Caxones, 3,800 feet above the level of the sea, riding for a whole day through a dense forest of the Pinus occidentalis, in which this cone-bearing tree, so like our Lord Weymouth's or white pine, was associated with a fan-leaved palm—the Corypha dulcis, covered with flights of gay coloured parrots(298). Southern America produces oaks, but not a single species of pine; and the first time that I again encountered the familiar form of a fir-tree, it met me in the estranging presence of a palm with its fan-like leaves. In the north-east end of the Island of Cuba, too, and so within the tropics, but scarcely raised above the level of the sea, Christopher Columbus in the course of his first voyage of discovery observed coniferous trees and palms associated in their growth(299). This gifted and all-observing man speaks of the circumstance in his journal as a singularity; and his friend Anghiera, secretary to Ferdinand the Catholic, says, with evident astonishinand that "in the newly discovered oountry This ound palmeta and pineta growing togeth astonis is of the greatest interest in a geological point of view to contrast the present distroountr of plants upon the surface of the earth with that which the floras of the former world unfold to us. The temperate zone of the southern hemisphere, abounding in water and in islands, and in which tropical forms of vegetation mingle so strangely with the forms that belong to colder regions of the earth, presents us, according to Darwin's beautiful, animated description, with the most instructive examples for both the old and the new, the past and the present geography of plants(300). The primeval is in the strictest sense of the word a portion of the history of phytology.

The Cycadeæ, which, to judge by the number of species, played a much more important part in the world that has passed away than in that which now exists, accompany the allied Coniferæ from the epoch of the coal formation upwards. They are almost entirely wanting in the period of the variegated sandstone, in which Coniferæ of singular formation (Voltzia, Haidingera, Albertia) have grown luxuriantly; the Cycadeæ, however, attain their maximum in the Keuper strata and the lias, where about twenty different forms make their appearance. In the chalk the prevailing forms are those of marine and fresh-water plants (Fuci and Najades). The cycadean forests of the Jura formation have by this time been long exhausted, and even in the older tertiary formations they remain deep behind the cone-bearing tribes and palms(301).

The lignitic or brown-coal strata, which are present in every one of the divisions of the tertiary period, amongst the earliest forms of cryptogamic land plants, exhibit a few palms, many conifers with distinct annual rings, and frondiferous trees, of more or less decided tropical character. In the middle tertiary period we observe the complete recurrence of the palms and cycadeans, and in the last members of this epoch, at length, strong resemblances to our present flora. We come suddenly upon our pines and firs, our cupuliferous tribes, our planes, and our poplars. The dicotyledonous stems of the lignite are frequently distinguished by gigantic thickness and vast age. A trunk was found near Bonn, in which Nöggerath counted 792 annual rings. In the peat-moss of the Somme, at Yseux, not far from Abbeville, in the north of France, oaks have been found that are 14 feet in diameter, a size which, in the old hemisphere, is very remarkable beyond the tropics(302). Goeppert's excellent researches, which it is hoped will soon appear illustrated with plates, inform us, "that all the Baltic amber is derived from a coniferous tree, which, as proclaimed by the extant remains of the wood and bark, were obviously of different ages, came nearest to our white and red pine timber, but still constituted a particular species. The amber-tree of the former world (Pinites succifer) had a richness in resin with which none of the coniferous tribes of the present world will bear comparison, inasmnch as great masses of amber are contained not only within and upon the bark, but also between the rings of the wood and in the direction of the medullary rays, which, as well as the cells, are seen under the microscope to be filled with ambreous resin of a whiter or yellower colour in different places. Amongst the vegetable matters inclosed in amber we find both male and female flowers of indigenous, acicular-leaved, and cupuliferous trees; but distinct fragments of Thuja, C pressus, Ephedera, and Castania vesca, mingled with others of Junipers and Firs, indicate a vegetation which is different from that of the present coasts and plains of the Baltic Sea."

In the geological portion of our Representation of Nature, we have now gone over the whole series of formations, from the oldest eruptive rocks, and the oldest sedimentary strata, to the newest alluvium, upon which lie the great erratic blocks, the causes or means of whose distribution has long been matter of discussion, but which for my own part I am less disposed to ascribe to icebergs, than to the eruption and tumultuous descent of great masses of pent-up water suddenly let loose by the upheaval of mountain chains(303). The oldest members of the transition formation with which we are acquainted, are the schists and greywacke, which inclose some few remains of seaweed from the Silurian, formerly the Cambrian Sea. Upon what did these *oldest* rocks, as they are called, repose, if gneiss and mica-slate are to be regarded but as metamorphosed sedimentary strata? Shall we venture a conjecture in regard to that which cannot be the object of actual geological observation? According to an ancient Indian Myth, it is an elephant that supports the earth; and the elephant himself, that he may not sink, is borne by a gigantic tortoise. Whereon the tortoise stands, it is not allowed to the believing Brahmin to inquire. We make bold to attempt a problem of the sort, although prepared for variety of blame in its solution. On the first formation of the planets, as we have made it probable in the astronomical portion of our Picture, vaporous rings circulating about the sun became aggregated into spheres, and gradually consolidated from without inwards. What we call the older Silurian strata are only the upper portions of the solid crust of the earth. The eruptive rocks which we see breaking through, pushing aside, and heaving up these, arise from depths that are inaccessible to us; they exist, consequently, under the Silurian strata, composed of the same association of minerals which are familiar to us under the name of granite, augite, and quartz-porphyry, at the points where, by breaking through, they become visible. Resting on analogies, we may safely assume that that which at one and the same time fills extensive fissures in the manner of veins, and bursts through the sedimentary strata, can only be an offset from an inferior bed. The active volcanoes of the present day carry on their processes at the greatest depths; and from the strange fragments which I have found included in streams of lava in different quarters of the globe, I also hold it as more than probable that a primogenial granitic rock is the foundation of the great systems of stratification which are filled with such variety of organic remains(304). If basalts, containing olivine, first make their appearance in the cretaceous period, and trachytes show themselves still later, the eruptions of granite, on the contrary, belong (as metamorphic productions also assure us) to the epochs of the oldest sedimentary strata of the transition series. Where knowledge cannot take its rise from the immediate scrutiny of the senses, it is fairly allowable, even on grounds of pure induction, as also after a careful comparison of facts, to advance a conjecture which restores to the olden granite a portion of its threatened rights, and its distinction of primordiality.

The late advances of geology, the extended knowledge of the geological epochs, which are characterized by the mineralogical diversity of their rocks or mineral masses, by the peculiarities and succession of the organic remains which they contain, by the position, the erection or the undisturbed horizontal lie of the strata, all these considerations lead us, following the intimate causal connection of phenomena, to the division, in respect of space, of the solid and the fluid, of the continents and the seas which constitute the surface of our planet. And here we indicate a point of union between that which is historical in geology with reference to the earth—cosmographical geology, and geographical geology, or the general consideration of the form and partition of continents. The limitation of the Solid by the Fluid, and the relations in respect of area between the one and the other, have been very different at different times in the long succession of geological epochs, according as the sedimentary carboniferous strata were deposited horizontally on the upright strata of mountain lime- and old red sand-stone; as lias and oolite were laid on banks of kuper and muschelkalk; and as chalk was accumulated on the acclivities of the green sand and Jura limestone. If, with M. Elie de Beaumont, we designate the waters under which the Jura limestone and the chalk were precipitated in the shape of mud or slime, as the Jurassic and cretaceous seas, then will the contour of the two formations just mentioned give us the boundary for two epochs, between the ocean still engaged in forming rocks, and the land already laid dry. The happy idea has even been conceived of forming maps of these physical elements of primeval geography; and these maps are perhaps more accurate than those which have been composed in illustration of the wanderings of Io and the Homeric narratives. The latter give graphic representations of opinions and mythical images; the former exhibit facts in the positive science of formation.

The result of investigations into the extent of exposed area, or dry land, is this: that in the earliest times, in the Silurian or Devonian transition epochs, as also in the first floetz period, throughout its tripartite division, the dry land, the surface occupied by land plants, was limited to separate islands; that these islands united at later epochs, and inclosed numerous inland lakes by the sides of deeply-indented bays of the sea; that finally, when the mountain chains of the Pyrenees and Apennines and Carpathians arose—towards the time of the older tertiary strata, therefore—extensive continents, having almost the dimensions of those of the present day, had appeared. In the times of the ofstrrian world, as well as in the epoch of tertiary est luxuriance of the Cycadeæ and gighavi str.urians, the quantity of dry land from pole to pole might very possibly have been even less than it is in the Pacific and Indian Oceans at the present time. How this preponderating mass of water, in common with other causes, conduced to elevation of temperature, and to greater equality of climate, will be the subject of consideration by and by. Here it must only be farther remarked, in considering the gradual augmentation (agglutination) of the uplifted dry land, that shortly before the revolutions, which, after shorter or longer pauses in the diluvial period, occasioned the sudden extinction of so many gigantic vertebrate animals, portions of the present continental masses were still completely separate from one another. In South America and the Australasian lands, there is a great prevailing resemblance between the existing animals and those that have become extinct. In New Holland, the fossil remains of kangaroos have been discovered, and in New Zealand the semifossil bones of a gigantic struthious bird, Owen's Dinornis, closely allied to the existing Apterix, but having little affinity to the so lately extinguished Dronte or Dodo of the island of Rodriguez.

The outline of former continents was perhaps indebted in principal measure for its elevation above the surrounding sea-level to the eruption of quartzose porphyry, an event which so powerfully shook the first great vegetable covering of the dry land, from which were derived the materials of the coal measures. What we call plains or flats (in continents), are no more than the broad backs of hills and mountains whose feet are at the bottom of the sea. Each plain, in its submarine relations, is, in fact, a lofty plateau or table-land, whose inequalities have been concealed by new sedimentary depositions in horizontal beds, as well as by alluviums spread over its surface by floods.

Among the general considerations which belong to a Picture of Nature, the foremost place must be given to the quantity of TERRA FIRMA projecting, uplifting itself above the level of the sea; such a determination of continental areas includes the consideration of their individual forms in point of horizontal extension (segmentary relations), and of perpendicular elevation (the hypsometrical relations of mountain chains). Our planet has two coverings or envelopes: one general, the Atmosphere, as elastic fluid, and one particular, only locally distributed, bounding the Solid, and thereby giving it its figure, the Sea. These two coverings, the air and the ocean, form a natural whole which gives the surface of the earth its climate, diverse according to the relative extent of the sea and of the land, of the division and geographical position of the land, and of the direction and elevation of its mountain chains.

From this knowledge of the reciprocal influences of the air, ocean, and land it appears that great meteorological phenomena, severed from geological considerations, cannot be understood. Meteorology, like the geography of plants and animals, first began to make some progress since observers have become persuaded of the mutual interdependence of the phenomena to be investigated. The word Climate implies in the first instance a specific constitution of the atmosphere; but this constitution depends on the ceaseless reciprocal influences exerted between an ever and deeply-agitated ocean, crossed in different directions by currents of totally dissimilar temperatures, and the heat-radiating dry land, variously partitioned, elevated, coloured, naked or covered with lofty trees or lowly herbs.

In the present condition of the surface of our planet, the area of the dry to that of the fluid is as 1 : 2¾; according to Rigaud(305), as 100 : 270. The islands form at present scarcely $\frac{1}{23}$ of the continental masses. The latter are so unequally divided, that in the northern hemisphere they offer a three times greater extent of surface than they do in the southern hemisphere. The southern hemisphere is consequently most especially oceanic in its prevailing character. From 40° S. latitude on towards the antartic pole, the crust of the earth is almost entirely covered with water. Even as predominating, and only broken here and there by insignificant clusters of islands, is the fluid element between the east coasts of the Old and the west coasts of the New World. The learned hydrographer, Fleurieu, by way of distinguishing this extensive sea basin from other seas, has very well entitled it the Great Ocean. Within the tropics it includes a breadth of as many as 145 degrees of longitude. The southern and western hemispheres, beginning the reckoning from the meridian of Teneriffe, are thus the regions of the earth's surface that most abound in water.

These are the principal points in the consideration of the relative quantities of the land and sea, a relation which exerts so vast an influence upon the distribution of temperature; the variation of atmospheric pressure; the direction of winds, and the hygrometric state of the air which particularly and so essentially determines the force of vegetation. When we think that nearly three-fourths of the surface of the earth are covered with water(306), we are less astonished at the imperfect state of meteorology up to the commencement of the present century—an epoch when a considerable mass of accurate observations on the temperature of the sea, under different parallels of latitude and at different seasons of the year, was first obtained and numerically contrasted.

The horizontal figure of the land, in its most general relations of extension, was already an object of ingenious consideration at an early period in the history of the Greek civilization. It was sought to ascertain the greatest extension from east to west, and Dicearchus, according to the testimony of Agathemerus, found this to lie in the latitude of Rhodes in a direction from the Pillars of Hercules to Thinæ. This is the line which was called the Parallel of the Diaphragm of Dicearchus, the astronomical accuracy of whose position (which I have myself examined in another place) must ever be the subject of admiration(307). Strabo, led apparently by Eratosthenes, appears to have been so thoroughly persuaded that as this parallel of 36°, the maximum extension of the world, as known to him, was intimately connected with the figure of the earth, that he fixes the place of the continent which he prophesied must exist in the northern hemisphere, between Iberia and the coast of Thinæ, as also falling under the same degree of latitude(308).

If, as already remarked, considerably more land has been raised above the level of the sea in the one hemisphere than in the other—and this is the case whether the globe be halved in the line of the equator, or in that of the meridian of Teneriffe—the two great masses of land, true islands surrounded by the sea on every side, which we designate the Eastern and Western continents, the Old and New Worlds, beside the most striking contrasts in configuration at large, or rather in the position of their greater axes, still present many points of resemblance in the details of their configuration, particularly in the extent and outline of their opposite coasts. In the Eastern division, the prevailing direction or position of the longer axis is from east to west (more correctly, from south-west to north-east); in the Western continent, however, it is meridional, or from north to south (more accurately, from south-south-east to north-north-west). Both masses are cut off towards the north in the line of the same parallel of latitude—generally in that of 70°; and to the south they both run out into pyramidal points, which have mostly a submarine extension in the shape of islands and shoals. This is proclaimed by the archipelago of Terra del Fuego; the Lagullas bank, to the south of the Cape of Good Hope, and Van Diemen's Land, separated from New Holland by the Bass Straits. The Northern Asiatic coast exceeds, or runs up beyond the parallel of 70° mentioned above, about Cape Taimura (78° 16′ N. lat. according to Kreusenstern), whilst from about the embouchure of the great Tschoukotschja river eastward, in the direction of Behring's Straits, the north-eastern promontories of Asia (Cook's East-Cape) do not reach higher than 66° 3′ according to Beechey(309). The northern shore of the New Continent follows the 70th parallel pretty closely; as both south and north of Barrow's Straits, from Boothia-felix and Victoria-land, all the land consists only of detached islands.

The pyramidal figure of all the southern terminations of continents belongs to the "Similitudines physicæ in configuratione mundi," to which Bacon had already directed attention in the Novum Organum, and with which Cook's companion in his second voyage round the world, Reinhold Forster, has connected some very acute and interesting considerations. Proceeding from the meridian of the Island of Teneriffe eastward, we observe the southern extremities of three great continents, namely, of Africa (the extreme of the old world), Australia, and South America, approaching the south-pole successively nearer and nearer. New Zealand, which is fully twelve degrees of latitude in length, forms a very regular intermediate member lying between Australia and South America, and also ending with an island—New Leinster—to the south. Another remarkable feature in the configuration of our present continents is this: that almost under the same meridians under which the most southern stretches of the land are made, the northern coasts also shoot out and reach the highest latitudes towards the arctic pole. This appears on comparing the Cape of Good Hope and the Lagullas bank with the North Cape, and the peninsula of Malacca with Cape Taimura in Siberia(310). Whether the poles are girded with terra firma, or surrounded by an ocean covered with horizontal strata of ice (consolidated water), we know not. The North-pole has been approached as high as 82° 55′ N. latitude; the South-pole not higher than 78° 10′ S. latitude.

In the same way as the great continental masses terminate pyramidally towards the south, the like configuration is variously and almost everywhere repeated on a smaller scale, not only in the great Indian Ocean (the peninsulas of Arabia, Hindostan, and Malacca), but also, as observed by Eratosthenes and Polybius, in the Mediterranean, where the Iberian, Italian, and Hellenic peninsulas present corresponding sensible configurations(311). Europe, with an area of but one-fifth that of Asia, is in like manner but a western, many-membered peninsula of the Asiatic and almost undivided portion of the globe; and the climatic peculiarities of Europe also show that it stands to Asia very much in the same relationship as the peninsula of Brittany does to the rest of France(312). The influence which the subdivisions of a continent, the higher development of its form, exerts at once upon the manners and whole civilization of a people, is obviously particularly alluded to by Strabo(313), when he commends the "greatly diversified form" of our small division of the globe, as an especial advantage. Africa(314) and South America, which in other respects exhibit such similarities in their configuration, are those among the great masses of land which have the simplest outlines of coast. It is only the eastern seaboard of Asia, broken in upon by the currents of the east sea (fractas ex æquore terras), that shows variety and irregularity of outline(315). Peninsulas and a succession of islands there alternate from the equator to 60° of N. latitude.

Our Atlantic Ocean bears every feature of a great valley. It is as if floods had directed their shocks successively to the north-east, then to the north-west, and then to the north-east again. The parallelism of the opposite coasts northward from 10° of S. latitude, their advancing and retreating angles, the convexity of the shores of Brazil opposite those of the Gulf of Guinea, the convexity of Africa under the same parallels of latitude as the deep indentation formed by the Gulf of Mexico, all vouch for this apparently bold view(316). In this Atlantic valley, as almost everywhere else in the configuration of great masses of land, indented and isle-studded shores stand opposite to unindented coasts. It is long since I directed attention to the circumstance how remarkable in a geological point of view was the comparison of the west coasts of Africa and South America within the tropics. The deep bay-like inward sweep of the African coast by Fernando Po (4½° N. lat.), is repeated on the American continent under 18¼° S. lat. at the tropical point near Arica, where (between the Valle de Arica and the Morro de Juan Diaz) the Peruvian coast suddenly changes its course from south to north into a north-western direction. This change of direction extends in like measure to the lofty chain of the Andes, which here proceeds in two parallel connected lines; and not only to the lofty plateaus near the coast(317), but also to the eastern plains, the earliest seat of human civilization in the South American continent, where the little alpine lake of Titicaca is bounded by the colossal mountains, Sorata and Illimani. Farther towards the south, from Valdivia and Chiloe (40° to 42° S. lat.), through the Archipelago de los Chonos on to the Terra del Fuego, the curious Fiord-formation, the complication of narrow, deeply-penetrating bays or arms of the sea, is repeated, which, in the northern hemisphere, we find characterizing the west coasts of Norway and of Scotland.

Such are the most general considerations that suggest themselves on the configuration of continents (the extension of the dry land in a horizontal direction), as a survey of the surface of our planet offers them at the present time. We have here placed facts in juxtaposition, analogies in form occurring in remote districts of the earth, which, however, we do not venture to speak of as Laws of Form. When on the flanks of a still active volcano, of Vesuvius for example, we observe the not uncommon phenomenon of partial upheavings of the soil, in which small portions of the solid earth, either before or in the course of an eruption, permanently change their level by several feet, and rise in penthouse-like ridges or flat elevations, we perceive how it must depend on trifling accidents of intensity in the force of subterraneous vapours, and in the amount of resistance to be overcome, that the upheaved parts assume this or that form and direction. Even so may slight disturbances of the equilibrium in the interior of our planet have determined the upheaving elastic forces to operate towards the Northern in a greater degree than towards the Southern hemisphere; to throw up the Eastern hemisphere as a broad continuous mass with its principal axis running nearly parallel to the equator, the Western and more oceanic hemisphere, again, as a narrower band, with its axis nearly in the plane of the meridian.

On the ætiological connection of such grand incidents in the production of the dry land, of similarity and contrast in the configuration of continents, there is little to be made out empirically. We only know one thing: that the efficient cause is subterraneous; that the present fashion of continents and islands has not been obtained at once; but, as has been already observed, that from the epoch of the Silurian formation (Neptunian separation), on to that of the tertiary deposits, there have been many alternate elevations and depressions of the surface, which, on the whole, has gradually increased in extent, and, from numerous smaller divisions, has coalesced into the larger masses which we now behold. The present configuration is the product of two causes, which exerted their influence in succession, one after another: firstly, a subterraneous manifestation of force, whose measure and direction we call accidental, because we have no means of determining them; because, to our understanding, they are abstracted from the circle of necessity; secondly, powers that are efficient on the surface, among which, volcanic eruptions, earthquakes, the upheaval of mountain chains, and ocean currents, have played the principal part. How totally different would have been the state of the earth, in reference to temperature, and, along with this, how dissimilar the state of vegetation, of agriculture, and of human society, had the principal axis of the new continent lain in the same direction as that of the old—had the Andes, instead of being uplifted in the plane of the meridian, been raised from east to west—had there been no extensive tropical land radiating heat to the south of Europe (Africa)—had the Mediterranean, which once communicated and made one with the Caspian and Red seas, and has proved so essential a means in promoting the civilization of mankind, had no existence—had its bottom been raised to the same level as the plains of Lombardy and Cyrene!

The alterations in the respective levels of the solid and fluid portions of the earth's surface—alterations which, at one and the same time, determine the outlines of continents, and leave dry or overflow districts of low-lying land, are to be ascribed to a variety of causes operating at different times. The most powerful have unquestionably been: the force of elastic vapours, which the interior of the earth encloses: the sudden change of temperature of great mountain chains([318]); the unequal secular loss of heat by the crust and core of the earth, which has occasioned the wrinklings or zigzag foldings conspicuous on many occasions in the solid surface; local modifications of the force of gravitation([319]), and, as a consequence of these, altered curvature of a portion of the fluid element.

That the elevation of continents has been an actual, not a seeming one only, attributable to the form of the surface of the sea, appears to follow from views now adopted by geologists generally, and from the long observation of connected facts, as well as from the analogy of the more important volcanic phenomena. The merit of this view also belongs to Leopold von Buch, who announced it in the account of his remarkable travels through Norway and Sweden, in the years 1806 and 1807, when it was first introduced to science([320]). Whilst the whole of the coasts of Sweden and Finland, from the limits of north Scania (Sölvitsborg), through Gefle, to Torneo, and from Torneo to Abo, is rising (the rise, in the course of a century, amounts to four feet), south Sweden, on the contrary, according to Nilson, is sinking([321]). The maximum of the upheaving power appears to lie in north Lapland. The upheaval falls off gradually towards the south as far as Calmar and Sölvitsborg. Lines of what were old sea-levels within historical times, are indicated along the coasts of the whole of Norway, from Cape Lindesnaes to the extreme north Cape, by beds of shells of the present ocean([322]), and have lately been most accurately measured by Bravais, during the long winter residence at Bosekop. These shores lie as many as 600 feet above the present mean sea-level, and, according to Keilhau and Eugenius Robert, the same thing extends nor-nor-west to the coasts of Spitzbergen, opposite the North-cape. Leopold von Buch, who was the first to direct attention to the raised bed of shells near Tromsoe (69° 40′ N. lat.), has, however, shown that the old upheavals along the line of the North Sea belong to another class of phenomena than the smooth and gradual rising of the Swedish coasts of the Gulf of Bothnia. The last phenomenon, vouched for by sure historical testimony, must not, therefore, be confounded with that alteration in the level of the surface which accompanies earthquakes, as in the case of the coasts of Chili and of Cutch. It has very recently given occasion to precisely similar observations in other countries. To the rising there occasionally corresponds, as a consequence of the folding of strata, an obvious sinking, as in West-Greenland (according to Pingel and Graah), in Dalmatia and in Scania.

If we regard it as extremely probable, that in the earlier ages of our planet the oscillating movements of the soil, the alternate elevations and depressions of the surface, were greater than they are at present, we shall be less surprised at finding single spots on the face of the globe, in the interiors of continents, that lie deeper than the present uniform level of the ocean. Examples of this kind are presented by the Natron lakes, described by General Andreossy, the small bitter lakes of the Isthmus of Suez, the Caspian Sea, the Sea of Tiberias, and, above all, the Dead Sea([323]). The level of the Sea of Tiberias is 625 feet, and that of the Dead Sea no fewer than 1230 feet lower than that of the Mediterranean mirror. Could the drift and alluvium that cover the rocky strata in so many parts of the earth be all at once removed, it would then be obvious how much of the rocky foundation lies actually lower than the present sea-level. The periodical, although irregular, alternate rise and fall in the waters of the Caspian Sea, of which I have myself seen unquestionable traces in the northern parts of this basin([324]), appear, like the observations of Darwin in the Coral Ocean([325]), to proclaim, that without any proper shock or concussion, the surface of the earth is still susceptible of the same smooth and progressive undulations which in primeval times, and when the thickness of the consolidated crust was much less than it is at present, were much more general [and extensive] than they are now.

The phenomena to which we here direct attention remind us of the instability of the present order of things, in the changes which, at far distant intervals of time, the outline and configuration of continents have in all probability undergone. Incidents that are scarcely recognizable to successive generations of men, accumulate in periods of the length of which the movements of the heavenly bodies supply the measure. In the course of 8000 years the east coast of the Scandinavian peninsula has risen to the extent perhaps of about 320 feet; after the lapse of 12,000, if the motion prove continuous and equable, parts of the bottom of the ocean that lie near the peninsula, and at the present day are covered with 100 feet of water, and more, will have come to the surface, and begun to be laid dry. But what is the brevity of these intervals compared with the length of the geological periods, which the succession of strata in the several formations, and the host of extinct and totally different organisms which they inclose, reveal to us! We have here considered the phenomenon of upheavement only; but we can readily, resting on the analogies of facts observed, in like measure figure to ourselves the possibility of the sinking or submersion of whole districts of country. The mean height of the level, or non-mountainous portions of France is not quite 480 feet. Contrasted with former geological periods, in which more extensive changes went on in the interior of the earth, we perceive that no very long period of time were requisite to have considerable portions of the north-west of Europe permanently overflowed, and presenting in its sea-board a very different outline from that which now distinguishes it.

Risings and fallings of the solid, or of the fluid—in their several effects so evenly balanced that the rise of the one occasions the seeming fall of the other—are the cause of every change in the configuration of continents. In a general Picture of Nature, in a liberal, not one-sided, presentment of the phenomena of nature, the possibility at least of a diminution in the mass of waters, of a true sinking in the mean sea-level, must therefore be indicated. That with the former high temperature of the surface of the earth, with the greater water-engulfing fissuration of its crust, with a totally different constitution of its surrounding atmosphere, great variations in the level of the sea may have taken place in connection with the increase or decrease of the liquid element, there is no room left for doubt. In the actual condition of our planet, however, we are totally without any direct evidence of an actual progressive decrease or increase of the sea; we are also without any proof of change in the mean height of the barometer at the sea level of the same points of observation. From Daussy's and Antonio Nobile's researches, it appears that an increase in the height of the barometer would of itself be accompanied with a depression of the sea-level. But as the mean pressure of the atmosphere at the level of the sea, in consequence of meteorological causes—direction of the wind, moistness of the air—is not the same under every parallel of latitude, the barometer of itself can supply no certain evidence of change in the liquid level of our globe. The remarkable phenomenon which was observed in the beginning of the present century, when several harbours of the Mediterranean were repeatedly left completely dry for many hours, appears to indicate that alterations in the direction and strength of currents, without any actual diminution in the quantity of water, without any general depression of the level of the ocean, may give rise to local recessions of its waters, and to permanent exposures of small portions of its shores. From the knowledge lately obtained of these complicated phenomena, it seems that we must be particularly cautious in interpreting them, inasmuch as effects may very readily be ascribed to one of the "old elements," the water, which belong of right, and in fact, to two others, the earth and the air.

As continents, which we have hitherto delineated in their horizontal extension, by their configuration, by their external distribution and their variously indented coasts, exert a beneficial influence upon climate, commerce, and the progress of civilization, so is there another kind of internal subdivision effected by perpendicular elevations of the surface—by mountain chains and lofty table lands—which have consequences that are not less important. All that occasions change, variety of form and feature, in the surface of the planet—the dwelling-place of the human family—besides mountain chains, great lakes, grassy steppes, and even deserts surrounded by wooded regions as by coasts, impresses a peculiar character on communities. Lofty ridges covered with snow interrupt communication, interfere with traffic; but a mixture of less elevated mountain members lying apart(326), and of low lands, such as the West and South of Europe, present in such happy interchange, occasion variety in the meteorological processes, as well as in the products of the vegetable kingdom; and further beget wants, as every district even under the same degree of latitude then falls under the dominion of a different kind of husbandry, the satisfaction of which arouses the activity of the inhabitants. Thus have the dreadful convulsions that have ensued upon the reactions of the interior against the exterior, upon sudden upheavals of portions of the oxidized crust of the earth, upon the elevation of vast mountain chains, still proved conducive, with tranquillity restored, with the revival of the slumbering might of the organizing forces, to cover the dry land of either half of the globe with a beautiful abundance of individual forms, and to free at least the greater portion of it from the blank of uniformity which appears to cramp and impoverish both the physical and the intellectual powers of man.

To each system(327) of these mountain chains there is, according to the grand views of Elie de Beaumont, a relative age to be assigned: the upheaval of the range must necessarily fall between the times when the erupted strata were deposited, and those in which the horizontal beds, that stretch up to the very foot of the mountains, were laid down. The furrowings of the crust of the earth, in other words, the erections of strata which are of like geological age, appear, moreover, to attach themselves to one and the same direction. The line of strike, or heaving of the strata, is not always parallel to the axis of the chain, but sometimes cuts it through; so that, according to my views(328), the phenomenon of erection of strata which is even found repeated in the neighbouring level, must be older than the elevation of the chain. The principal direction of the whole of the dry-land in Europe (south-west to north-east) is opposed to the great fissure or valley which runs from north-west to south-east, from the mouths of the Rhine and the Elbe, through the Adriatic and Red Sea, across the mountain system of Puschti-Koh in

Luristan, towards the Persian Gulf and the Indian Ocean. Such a nearly rectangular intersection of geodetical lines has exerted a vast influence on the commercial relations of Europe with Asia and the north-west of Africa, as well as on the march of civilization along the once more fortunate shores of the Mediterranean Sea(329).

If vast and lofty mountain chains appear to our imagination as evidences of great revolutions undergone by the surface of the earth, as boundaries of climates, as dividers and determiners of the courses of rivers, as bearers of another vegetable world, it is the more necessary, by accurate numerical estimates of their volumes, to show how insignificant, on the whole, is the quantity of the upheaved masses in contrast with the areas of entire continents. The mass of the Pyrenees, for example, a chain the mean height of whose ridges, and the extent of surface of the base which it covers, have been ascertained by accurate measurements, if distributed evenly over the area of France, would raise the surface of that country by no more than about 108 feet. The mass of the eastern and western Alps, spread in the same way over the area of Europe, would only raise the land by about 20 feet. By a laborious calculation(330), which from its nature can only give an extreme superior limit, in other words, a number which may be less, but cannot be greater than the truth, I have found that the centre of gravity of the volume of the countries which in Europe and North America rise above the level of the sea, lies at a height of 630 and 702 feet, and in Asia and South America, at an elevation of 1062 and 1080 feet. These estimates show the slight elevation of the northern regions: the vast steppes of the Siberian levels are compensated by the enormous rise of the Asiatic soil from 28½° to 40° N. lat. between the Himalaya, the north Thibetic Kuen-luen, and the Thianschan or Celestial Mountains. We read, to a certain extent, in the numbers found, where the Plutonic forces of the interior of the earth have put forth their greatest strength in uplifting continental masses.

There is nothing to assure us that these Plutonic powers may not in the course of future centuries add new members to the mountain systems of different ages and having different directions, which have been enumerated by Elie de Beaumont. Wherefore should the crust of the earth have lost the property of folding on itself? Almost the last of the mountain systems that appeared, the Alps and the Andes, have reared colossuses in Mont-Blanc and Monte Rosa, in Sorata, Illimani and Chimborazo, that do not allow us to infer any falling off in the intensity of the subterranean forces. Geological phenomena of all kinds indicate alternating periods of activity and repose(331). The repose we now enjoy is only apparent. The shocks which the surface experiences under every variety of climate, and along with every description of rock, Sweden rising in its level, and the appearance of new eruptive islands, bear no testimony to quiescence in the internal life of the globe.

The two coverings of the solid crust of our planet—the liquid and the gaseous, the ocean and the atmosphere, besides the contrasts which arise from the great diversities in their states of aggregation and elasticity—also present numerous analogies by reason of the mobility of their particles, of their currents, and their relations to temperature. The depth of the sea and of the aërial ocean are both of them unknown to us. In some places under the tropics no bottom has been found to the sea with 25,300 feet of line (more than a [German] geographical mile); and the atmosphere, supposing it, as Wollaston will have it, to be limited and so subject to undulations, may be inferred, from the phenomena of twilight, to have a nine-times greater profundity. The aërial ocean rests partly on the solid earth, whose mountain chains and lofty table-lands, as already said, rise up like green and wood-crowned shoals; partly on the ocean, whose surface forms the fluctuating bottom upon which the inferior denser and moister strata repose.

From the limits of both the atmosphere and the ocean upwards and downwards, the aërial and liquid strata are alike subjected to certain laws of decrease of temperature. In the atmosphere this decrease is much slower than in the ocean. Under every zone the tendency of the sea is to preserve the temperature of its surface in equilibrium with that of the stratum of air which rests immediately upon it, inasmuch as the chilled particles [supposing the temperature of the air to be the lower] sink, [and the warmer particles, *vice versâ*, keep their place on the surface]. A vast series of careful observations on temperature, teach us that in the usual and mean state of its surface, the ocean, from the equator to 58° of north and south latitude, is somewhat warmer than the stratum of air that rests immediately upon it(332). On account of the decrement of temperature with the increasing depth, fishes and the other inhabitants of the sea, which, by reason perhaps of the nature of their branchial and cutaneous respiratory systems, love deep water, are able to find the lower temperatures, that agree particularly with them in higher latitudes, under the temperate and colder zones. This circumstance, analogous to the temperate, even to the cold alpine atmospheres of the lofty plateaus of the torrid zone, exerts an essential influence on the migrations and geographical distribution of many marine animals. The depths in which fishes live, by the increase of pressure they occasion, modify in like measure the cutaneous respiration and the contents in oxygen and azote of the air in the swimming bladder.

As fresh and salt water do not attain their maximum density at the same temperature, and the saline contents of the sea cause the thermometrical indication of greatest density to descend, water was obtained from the abyss of the ocean in the voyages of Kotzebue and Dupetit-Thouars, which indicated the low degrees of 2·8° and 2·5° C. This icy temperature of the water also prevails in the depths of the tropical sea, and its discovery gave the first information of the existence of inferior polar currents, proceeding from either pole towards the equator. Without such *under-sea* currents, the abyss of the tropical ocean could only have a temperature equal to the maximum of cold which the particles of water descending locally from the surface radiating heat, and cooled by the contact of the atmosphere, could acquire in a tropical region. In the Mediterranean Sea, as Arago acutely observes, a corresponding great depression of temperature in the inferior strata is only not observed, because the influx of the deep polar stream by the Straits of Gibraltar, through which the Atlantic is flowing from west to east, is encountered by a westward under-current of the Mediterranean towards the Atlantic.

The fluid-covering of our planet, equalizing and tempering climates in general, where it is not intersected by pelagic currents of colder or warmer water, and far from the coasts of tropical countries, particularly between 10° north and 10° south latitude, may be said to exhibit a truly wonderful equality and steadiness of temperature over areas that are thousands of square miles in extent(333). It has, therefore, been said with reason(334), that a long-continued and careful investigation of the thermal relations of the tropical seas would give us information in the simplest manner on the grand and much discussed problem of the constancy of climates, and of the temperature of the earth. Great revolutions in the luminous disc of the sun, were they of long continuance, would be simultaneously reflected in the altered mean temperature of the sea still more certainly than in the mean temperature of the land.

The zones in which the maxima of density (saline contents) and temperature lie, do not coincide with the equator. The two maxima are distinct from one another, and the warmest water appears to form two not completely parallel bands to the north and south of the geographical equator. The maximum of saline contents was found by Lenz, in his voyage round the world, in the Pacific, in the two parallels of 22° north and 17° south latitude. The zone of least density, again, was found to lie a few degrees to the south of the line. In the region of the Calms, the heat of the sun cannot occasion any great amount of evaporation, because a stratum of air saturated with saline vapour there sleeps unmoved and unrenewed upon the surface of the ocean.

The surface of all the seas that communicate one with another, must be regarded as generally perfectly equal in respect of mean elevation. Local causes, mostly prevailing winds and currents, have, however, in particular extensively land-locked seas—the Red Sea, for example, produced permanent, though still inconsiderable differences of level. At the isthmus of Suez the level of the Red Sea is from 24 to 36 feet above that of the Mediterranean at different hours of the day. The form of the canal, (the Straits of Babelmandel), by which the Indian Ocean communicates with the Red Sea, being such, that the waters find a readier access than outlet, appears to assist in producing this remarkable permanent superior elevation of the surface of the Red Sea, which was already known to the Ancients(335). The admirable geodetical operations of Coraboeuf and Delcros along the chain of the Pyrenees, have shown that there is no appreciable difference in the surface of equilibrium, in the sea-level, on the north coast of Holland and at Marseilles, of the ocean and the Mediterranean(336).

Disturbances of the Equilibrium and motions of the mass of waters consequent on these, sometimes irregular and transient, depending on winds and producing Waves which in the open ocean and far from land mount during a storm to a height of 35 feet and more; in other instances, regular and periodical, occasioned by the position and attraction of the sun and moon—the Tides; in still other instances, permanent, but of unequal force, as Oceanic currents. The phenomena of ebb and flow, which extend over every sea with the exception of those that are very small and much land-locked, in which the tidal wave is either little or not at all observable, are completely explained by the Newtonian natural philosophy, *i. e.* referred to the circle of necessary effects. Each of these periodically recurring oscillations of the ocean, is somewhat longer than half a day. In the open ocean they scarcely rise to the extent of a few feet; but in consequence of the position and configuration of coasts and estuaries which meet the coming tidal wave they rise in some places to extraordinary heights—in St. Malo to 50 feet, and in Acadia, Nova-Scotia, to from 65 to 70 feet. "Under the supposition that the depth of the ocean is inconsiderable when contrasted with the semi-diameter of the earth, the analysis of the great geometrician Laplace, has shown how the stability in the equilibrium of the ocean requires that the density of its fluid should be less than the mean density of the earth." And indeed, as we have seen above, the density of the oarth is five times greater than that of water. The high lands of the earth, therefore, can never be *overflowed*, and the remains of marine animals found on mountains can by no means have been brought into such situations by former floods or deluges produced by the position of the sun and moon(337). It is no trifling tribute to analysis, which in the unscientific circles of society is presumptuously held so cheap, that Laplace's perfected Theory of the Tides has made it possible to predict in our astronomical ephemerides or nautical almanacks, the height of the spring-tide to be expected at each new and full moon, and so to forewarn the inhabitants of the coasts of the increased danger with which they are threatened at these seasons, particularly when the moon is in her perigee.

Oceanic currents, which exercise so considerable an influence on the intercourse of nations and on the climatic relations of coasts, are almost simultaneously dependent on a multitude of very dissimilar, now greater, now apparently more insignificant causes. To the number of these belong: the progressive time of appearance of the ebb and flow of the tidal wave in its course round the world; the duration and force of prevailing winds; the density and specific gravity of the watery particles modified under different parallels of latitude by their temperature and saline impregnations(338); the horary variations of the atmospheric pressure, which proceed successively from east to west with such regularity within the tropics. The currents of the ocean present this remarkable spectacle: that they cross it of definite breadths in different directions, in the manner of rivers, neighbouring unmoved watery strata, forming the banks, as it were, of these streams.

This distinction between the portion which is moved and that which is at rest, is most remarkable where large quantitics of sea-weed carried along with the current permit us to estimate its velocity. We occasionally observe similar phenomena of limited currents in the inferior strata of the atmosphere: after tempests that have swept over dense forests, it sometimes happens that the trees are only found shattered and blown down in the course of narrow strips.

The general motion of the sea between the tropics from east to west, entitled the equatorial current, is regarded as a consequence of the advancing times of the tides and of the trade winds. It alters its direction in consequence of the resistance of the east coasts of the continents which it encounters in its progress. The new results which Daussy has obtained from the motion of bottles thrown out on purpose by navigators (10 French sea miles, of 925 toises each, every 24 hours), argees to within $\frac{1}{18}$th of the velocity which I had ascertained from a comparison of earlier data([339]). In the log-book of his third voyage (the first in which he sought to make the tropics in the meridian of the Canaries), Christopher Columbus says: "I hold it as certain that the waters of the sea move with the heavens (*las aguas van con los cielos*)," that is to say, from east to west, like the apparent motion of the sun, moon, and stars([340]).

The narrow currents, true oceanic rivers, which take their way through the sea, run warmer water in higher, colder water in lower latitudes. To the first class belongs the celebrated Gulf-stream([341]), which was known to Anghiera([342]), and particularly to Sir Humfrey Gilbert in the 16th century. The commencement and first impulse of this mighty current is to be sought for southward from the Cape of Good Hope, and it debouches from the Caribbean Sea and the Gulf of Mexico, through the Straits of Bahama; running from south-south-west to north-north-east, getting farther and farther from the shores of the United States of America, it turns off eastward by the banks of Newfoundland, crosses the Atlantic, and frequently throws the seeds of tropical plants (Mimosa scandens, Guilandina bonduc, Dolichos urens), upon the coasts of Ireland, the Hebrides and Norway. The north-eastern prolongation of the Gulf-stream contributes to moderate the cold of the sea-water and also of the climate about the north Cape of Scandinavia. The warm Gulf-stream, after it has turned eastward from the banks of Newfoundland, at no great distance from the Azores, sends off a branch to the south, and it is here that the Sargasso-sea, as it has been called, the great bank of sea-weed, is met with, which made so lively an impression on the imagination of Columbus, and which Oviedo called the sea-weed meadow (Praderias de Yerva). A host of small marine animals inhabit this ever-verdant mass of Fucus natans, one of the most widely diffused of the social plants of the ocean, which is constantly drifted hither and thither by the tepid winds that blow across its surface.

In contrast to the Gulf-stream, which belongs almost exclusively to the northern hemisphere of the Atlantic valley, and runs between America, and Europe and Africa, is the great current of the Pacific Ocean, the inferior temperature of whose waters has an appreciable influence on the climate of the sea-boards along which it sweeps, as I first observed in the autumn of 1802([343]). This current, in fact, brings the colder water of high southern latitudes to the coast of Chili, runs along the shores of this country and those of Peru, first from south to north, and then (from the bay of Arica) from south-south-east to north-north-west. In the middle of the tropics at certain seasons of the year the water of this cold ocean stream is not higher than 15° 6 C. (60° 0 F.), whilst the motionless water beyond its limits is as high as from 27° 5 to 28° 7 C. (81° 5 to 84° 6 F.). Where the sea-board of South America, southward from Payta, advances farthest to the west, the stream turns suddenly in the same direction from off the land, and takes a course from east to west; so that he who sails northward [by crossing the stream] comes suddenly from a colder to a warmer sea.

It is not known to what depth the oceanic currents, whether hot or cold, extend, how near they run to the bottom. The deviation of the South African current produced by the Lagullas bank, where the water is full 70 or 80 fathoms deep, appears to indicate a considerable extension in depth. Sand-banks and shoals outside the streams are mostly recognizable, as the excellent Benjamin Franklin discovered, by the coldness of the water over them. This depression of temperature appears to me to be connected with the circumstance, that with the communication of motion to the neighbouring ocean, deep cold water is made to rise over the edges of the banks and to mix with the upper warmer water. My immortal friend, Sir Humphrey Davy, on the other hand, ascribed the phenomenon, from which the seaman can frequently draw practical inferences conducive to his safety, to the descent of the superficial strata of water cooled in the course of the night: these remain nearer the surface, because the shoal prevents them from sinking to a greater depth. The thermometer was turned by Franklin into a plumb-line; fogs are frequent upon banks and shoals: their colder water causes precipitation of the vapour that is dissolved in the sea air. I have observed such fogs to the south of Jamaica, and also in the Pacific, indicating the outline of shoals sharply and quite distinctly from a distance. They present themselves to the eye like air-pictures, in which the fashion of the sub-marine bottom is reflected. A still more remarkable influence of these cold shallows is this, that they produce an obvious effect upon the superior strata of the atmosphere, almost in the same way as low coral or sandy islands. Far from all land, in the high seas, when the air is elsewhere quite clear, clouds are frequently seen hovering over the spots where shoals occur. In such cases their bearings can be taken by the compass, precisely as if they were lofty mountains or isolated peaks.

Without the variety of external forms that characterize the surface of continents, the ocean, when its interior is narrowly scanned, presents a greater mass of organic life than is perhaps to be found collected together in any other portion of the earth's surface. Charles Darwin observes with justice, in the interesting Journal of his extensive sea-voyage, that our woods on shore do not harbour so many animals as the woody regions of the ocean, where the sea-weed groves, rooted to the bottom of the shallows, or the fuci detached by waves and currents, supported by air-cells and swimming free, unfold their delicate arms and branches. The use of the microscope increases still farther, and in the most remarkable manner, the impression of the universal life of the ocean, the astounding assurance that here sensibility is everywhere diffused and active. In depths that surpass the height of our most lofty mountains, every one of the several superposed strata of waters, is animated with its own Polygastric worms, Cyclidia, and Ophrydia. Here swarm, turning each wave into luminous foam, and attracted to the surface by particular weather-influences, the innumerable host of small light-flashing Mammaria from the Orders of the Acalephæ, Crustacea, Peridinia, and Nereides moving in circles.

The abundance of these small animals, and of the animal matter which their rapid destruction supplies, is so immeasurable, that the sea-water at large becomes a nutritious fluid for much larger creatures. If this exuberance of living forms, these myriads of dissimilar microscopical, and yet in part extremely perfect organisms, engage and pleasantly excite the fancy, this is appealed to in a more earnest, I might say a more solemn manner, by the sense of the Limitless and the Immeasurable, which every sea-voyage presents to our contemplation. He who is awakened to a spiritual self-activity, and who delights to build up a world within himself, fills the amphitheatre of the boundless ocean with the lofty image of the INFINITE and the ENDLESS. His eye is fixed especially by the far horizon, where indefinitely and as in mist, the ocean and the air meet bounding one another, in which the stars set and rise anew before the eyes of the beholder. But still, with the eternal play of this interchanging scene, as everywhere else with human happiness, there comes the breath of sadness, of ungratified longing, to mix itself with the joy.

A peculiar predilection for the sea, grateful remembrances of the impressions which the mobile element between the tropics, in the peace and silence of the night, or roused and at war with the natural forces, has left upon my mind, could alone have induced me to speak of the individual enjoyment of the prospect, before referring to the beneficial influence which contact with the ocean has had on the development of the intelligence and character of various nations; on the multiplication by its means of the bonds that ought to embrace the whole of the human family; on the possibility it has afforded of attaining to a knowledge of the configuration of the earth and its parts; lastly, on the improvement it has led to in astronomy, and in the mathematical and natural sciences at large. A portion of this influence was originally confined to the waters and the shores of the south-western parts of Asia; but from the 16th century onwards it has extended far and wide, and even attained to nations that live in the interior of continents remote from the sea. Since Christopher Columbus was "sent forth to unchain the ocean"[344] (for so was he addressed in a dream by an unknown voice whilst he lay on a sick-bed by the river Belem), man, too, mentally more free, has ventured with greater boldness into unknown regions.

The second and most external and universally diffused of the coverings of our globe, the ATMOSPHERE, on whose depths, or shoals, which are lofty table-lands and mountains, we live, present six classes of natural phenomena, connected in the most intimate manner with one another; these are: chemical composition; alterations in the transparency, polarization, and colour; in the density or pressure; in the temperature, humidity, and electricity. If in its oxygen the air contains the first element of physical animal life, another excellence, it might almost be said of a higher order, must be indicated in its constitution. The air is the "carrier of sound," and so also the bearer of speech, the means of communicating ideas, of maintaining social intercourse among men. The earth, robbed of its atmosphere, like the moon, presents itself to the imagination as a desert brooded over by silence.

The relations of the substances which belong to the strata of the atmosphere that are accessible to us, have, since the beginning of the 19th century, been made the object of researches, in which Gay Lussac and I took an active part; it is but very recently, however, through the admirable labours of Dumas and Boussingault, that the chemical analysis of the atmosphere, pursued in new and trustworthy ways, has been advanced to a high degree of perfection. From this analysis dry air appears to contain per volume 28·8 oxygen, and 79·2 azote; besides from 2 to 5 ten thousands of carbonic acid, a still smaller quantity of carburetted hydrogen[345], and from the important experiments of Saussure and Liebig, traces of ammoniacal vapours[346], which may supply plants with their azotized constituents. That the quantity of oxygen may vary in a trifling but still appreciable degree according to season, situation of a place—upon the sea or in the interior of a continent—has been rendered probable by some observations of Lewy. It is conceivable that changes in the quantity of oxygen held in solution by water, induced by microscopical animal organisms, may be followed by changes in the strata of air that lie in immediate contact with its surface[347]. The air collected by Martins on the Faulhorn at a height of 8226 feet, did not contain more oxygen than the air of Paris[348].

The admixture of carbonate of ammonia in the atmosphere may probably be held as older than the existence of organic beings on the surface of the earth. The sources of the carbonic acid of the atmosphere are extremely numerous[349]. We may here mention the respiration of animals, which receive the carbon they exhale from the vegetable food they consume, as vegetables themselves derive it from the atmosphere; the interior of the earth in the country of extinct volcanoes and thermal springs; the decomposition of the slight admixture of carburetted hydrogen contained in the atmosphere, by the electrical discharges of the clouds, so frequent in intertropical countries.

Besides the substances which have just been mentioned, and which may be held proper to the atmosphere under all circumstances and in all situations, there are other accidental matters associated with it, which occur especially near the ground, and of which several, designated miasms and contagions, affect the animal system prejudicially. The chemical nature of these substances has not yet been made known by any immediate analysis; but, considering the putrefactive processes which proceed incessantly on the surface of our planet, covered as it is with animal and vegetable matters, and led as well by combinations and analogies derived from the domain of pathology, we may fairly conclude on the existence of such injurious local admixtures. Ammoniacal and other azotized vapours, sulphuretted hydrogen, combinations, indeed, resembling the multibasic, (ternary and quarternary), compounds of the vegetable kingdom(350), may form miasmata, which, in a variety of shapes, and by no means only on naked swampy bottoms, or on sea-coasts strewed with putrifying molluscs, or covered with under-growths of mangrove (Rhizophora), and Avicenniæ, may produce fevers of aguish or typhoid types. Fogs which diffuse a peculiar smell, remind us at certain seasons of the year of such accidental contaminations of the lower strata of the atmosphere. Winds and ascending currents of air occasioned by the heating of the surface, raise even solid, though of course finely pulverized substances, to considerable heights. The dust, which makes the air misty over a great area, and falls about the Cape de Verd Islands, to which Darwin has so properly directed attention, is found from Ehrenberg's observations to contain an infinity of silicious shelled infusory animalcules.

As principal features in a general physical picture of the atmosphere, we may distinguish, 1st. In the variations of the air's pressure: the regular, and between the tropics, so readily appreciable hourly oscillations, a kind of ebb and flow of the atmosphere, which cannot be ascribed to the attraction of the mass of the moon(351), and which is so different according to the latitude, the season of the year, and the height of the place of observation above the level of the sea. 2d. In the climatic distribution of heat: the influence of the relative position of the transparent and opaque masses—the fluid and solid superficial areas, as well as of the hypsometrical or perpendicular configuration of continents, relations which determine the geographical position and curvature of the isothermal lines* in the horizontal or vertical direction, in the ground-plane, or in the aerial strata lying one above another. 3d. In the distribution of the moisture of the atmosphere: the consideration of the quantitative relations according to diversity in the solid and oceanic surfaces, distance from the equator, and height above the level of the sea; the forms in which precipitation of the watery vapour takes place, and the connection of this precipitation with the changes of temperature, and the direction as well as the succession of the winds. 4th. In the relations of the aerial electricity, whose primary source, when the air is serene, is still much disputed: the relation of ascending vapours to the electrical charge and the fashion of clouds according to the time of the day and the season of the year, the colder or hotter zones of the earth, the lower or higher-lying plains; the frequency and rarity of storms; their periodicity and occurrence in summer and winter; the casual connection of electricity with the extremely rare occurrence of hail-showers by night, as also with water-spouts and sand-spouts, which have been so ably investigated by Peltier.

The horary variations of the barometer, in which within the tropics the instrument is twice in the course of the day at its highest, viz., at 9 or 9¼ A. M. and 10 or 10¾ P. M., and twice at its lowest, viz., at 4 or 4¼ P. M., and 4 A. M., nearly the hottest and coldest hours in the round of the twenty-four, consequently, long formed the subject of my most careful daily and nightly observations(352). The regularity of these is so great, that the time, especially in the day, may be ascertained by the height of the column of mercury, without an error on the average of more than from fifteen to seventeen minutes. In the torrid zone of the New Continent, on the coasts as well as on heights of more than 12,000 feet above the level of the sea, where the mean temperature falls to 7° C (43° 8 F.), I have not found the regularity of this ebb and flow of the atmosphere to be disturbed either by tempests of thunder or of wind, by rain or by earthquakes. The amount of the daily fluctuation diminishes from the equator on to 70° N. latitude (a parallel under which we possess very accurate observations made by Bravais at Bosekop) (353), from 1·32 line, to 0·18 line. That, much nearer the pole, the mean height of the barometer is actually less at 10 A. M. than at 4 P. M., so that the times of the maxima and minima are severally interchanged, is by no means to be concluded from Parry's observations at Bowen Harbour (73° 14′ N. latitude).

The mean height of the barometer, by reason of the ascending current of air, is somewhat less under the equator, and especially under the tropics, than in the temperate zone(354); it appears to attain its maximum, in the West of Europe, in the parallels of 40° and 45°. If, with Kæmtz, we connect those places which present the same mean differences in their monthly barometrical extremes by isobarometrical lines, curves are engendered, the geographical position and direction of which yield us important conclusions in regard to the influence of the configuration of continents, and the expanse of seas upon the oscillations of the atmosphere. Hindostan, with its lofty mountain ranges and triangular-shaped peninsula, the East coasts of the New Continent, at the point where the warm gulf-stream turns eastward by Newfoundland, show greater isobarometrical fluctuations than the West India Islands, and the Western parts of Europe. Prevailing winds exert the most especial influence on the diminution of the atmospheric pressure, and with this, according to Daussy, as we have already

---

* Lines of equal mean temperature.

observed, the mean height of the sea is increased(355).

As the whole of the most important variations in the weight or pressure of the atmosphere—whether they occur regularly at certain hours and seasons, or are accidental and excessive, when they are often accompanied with danger(356)—like all the rest of what are called weather phenomena, have their principal cause in the heating power of the sun's rays ; so the directions of the wind (partly on Lambert's proposition) were at an early period compared with the state of the barometer, with variations in temperature, and with differences in the hygrometric state of the atmosphere. Tables of the pressure of the atmosphere along with particular winds, designated by the title of *barometrical wind-cards*, have given a deep insight into the connection of meteorological phenomena(357). With wonderful acumen, Dove perceived, in the laws of the rotation of the winds of both hemispheres, which he discovered, the cause of many grand variations (processes) in the atmospheric ocean(358). The thermal difference between countries lying near the equator and those situated near the pole, engenders two opposite currents in the upper regions of the atmosphere and on the surface of the earth. In consequence of the diversity of the rotatory velocity in the parts lying nearer the pole, or nearer the equator, the air which is streaming from the pole acquires an eastern, that which is pouring along from the equator a western direction. From the struggle between these two currents, the place of descent of the higher, the alternating displacements of the one by the other, depend the most important phenomena of atmospheric pressure, of the heating and cooling of the aerial strata, of the precipitation of moisture, and, indeed, as Dove has correctly shown, of the formation of clouds and their configuration. The forms of clouds, those all-enlivening ornaments of the landscape, are faithful indications of what is going on in the upper regions of the air ; and in calms, and floating in the warm summer's sky, they are also the "projected image" of the heat-radiating surface of the ground.

Where the influence of the radiation of heat is conditional on the relative position of great continental and oceanic surfaces, as betwixt the East coast of Africa and the West coast of the peninsula of Hindostan, regular periodical changes in the direction of the winds accompany the changes in the declination of the sun, and constitute the Indian monsoons(359), the Hippalos of the Greek navigators. These winds must have been amongst the earliest regular winds recognized and taken advantage of by mankind. In this knowledge of the monsoons, which has certainly been spread over China and Hindostan, the Eastern, Arabian, and Western Malayan Seas, for thousands of years, as well as in the still older and more generally diffused observation of the sea and land breeze, lies the hidden germ of the fast-advancing meteorological science of the present day. The long series of magnetic stations which have now been established from Moscow to Pekin, through the whole of Northern Asia, as they have it also in charge to observe meteorological phenomena in general, will soon become of great importance in establishing the LAW OF THE WINDS. The comparison of observations made simultaneously at places many hundreds of miles apart, will determine whether or not the same east wind blows from the barren table-lands of Gobi to the interior of Russia, or whether, and at what point in the line of stations, the direction of the current becomes changed through a descent of air from the higher regions. We shall then, in the true sense of the phrase, learn " whence the wind cometh." If we would base the required result on observations continued for not fewer than twenty years, Mahlman's careful notifications assure us that in the middle latitudes of the temperate zone in both continents the west-south-west is the prevailing wind.

Our knowledge of the DISTRIBUTION OF HEAT in the atmosphere has gained, in some respects, in clearness, since attempts have been made to connect the points that indicate the mean temperature of the year, of the summer and of the winter, by different orders of lines. The system of Isothermal, Isotheral, and Isochimenal lines, which I first proposed in 1817, may, perhaps, when it has been gradually perfected by the united efforts of natural philosophers, be found to supply a general and grand basis for a comparative Climatology. Terrestrial magnetism first acquired a scientific shape when scattered partial results were connected graphically with one another by lines of equal variation, of equal dip, and of equal intensity.

The expression CLIMATE, in its most general acceptation, indicates every change in the atmosphere which sensibly affects our organs—temperature, humidity, alteration of barometrical pressure ; calms or storms of wind from various quarters ; amount of electrical tension ; purity of atmosphere, or its contamination with gaseous exhalations more or less pernicious ; finally, degree of habitual transparency and serenity of the sky, which is not merely important in connection with the amount of radiation from the ground, the organic evolution of plants, and the ripening of fruits, but also with the feelings and whole mental estate of mankind.

Were the surface of the earth composed of one and the same homogeneous fluid mass, or of rocky strata of like colour, like density, like smoothness, like capacity of absorption for the sun's rays, and like power of radiation into planetary space, then would the Isothermal, Isotheral, and Isochimenal lines run parallel to one another, and to the Equator. In such an hypothetical condition of the earth's surface, the power of absorbing and of emitting light and heat would be the same in the same parallel of latitude all round tho globe. And it is, in fact, from such a mean, and, as it were, primary condition, which neither excludes the transmission of heat to the interior of the earth, nor towards the atmosphere involving it, nor the communication of heat by currents of air, that the mathematical consideration of climates sets out. All that alters the absorbing and radiating powers of the surface in particular parts lying in the same parallels of latitude, produces inflections in the Isothermal lines. The nature of these inflections, the angle under which the isothermal, isotheral, and isochimenal lines cut the parallel circles, the portion of the convexities or concavities of these lines in respect of the pole of the corresponding hemisphere, are the effects of calorific or frigorific causes which show themselves possessed of more or less power under different geographical longitudes.

The progress of Climatology has been favoured in a remarkable manner by the spread of European civilization from two opposite seaboards, by its extension from our Western European coast to an Eastern coast on the other side of the great Atlantic valley. When the British, after the temporary establishments which had proceeded from Iceland and Greenland, had founded the first permanent colonies on the shores of the United States of America, where religious persecution, fanaticism, and love of freedom, soon swelled the ranks of the settlers, the bold adventurers must have been amazed at the severity of the winters which they encountered, from North Carolina and Virginia to the River St. Lawrence, in comparison with those which prevail under corresponding parallels of latitude in Italy, France, and Great Britain. Such climatic observations, however exciting they must have been, still only bore fruits when they could be based on numerical results of mean annual temperatures. If, between the parallels of 58° and 30° N. latitude we compare Nain, on the coast of Labrador, with Gottenburg, Halifax with Bordeaux, New York with Naples, St. Augustin in Florida with Cairo, we find the differences in mean annual temperature between the East of America and the West of Europe, under similar parallels of latitude, progressing from north to south, from 11°·5, 7°·7 and 3°·8 to almost 0 Cent. The gradual decrease of difference in the above series, through 28 degrees of latitude, is very remarkable. Still farther to the south, and within the tropics, the isothermal lines in almost every part of both divisions of the globe run parallel with the equator. From the examples here given, it is obvious that the questions we hear so constantly repeated in our social circles, as to how many degrees America—and without any distinction of East or West coast—is colder than Europe? and how many degrees the mean annual temperature in Canada and the United States of America is lower than under corresponding parallels of latitude in Europe? when taken as general expressions, are totally without meaning. The difference under each particular parallel is different from what it is under every other parallel; and without special comparisons of the winter and summer temperatures of the opposite coasts, no right conception can be formed of the several particular climatic relations in so far as they influence agriculture, trade, and the feelings of comfort and convenience, or the contrary.

In enumerating the causes that may produce disturbances in the form of the isothermal lines, I distinguish the causes tending to exalt, and the causes tending to depress temperature. To the first class belong: the vicinity of a west coast in the temperate zone; the configuration of a continent cut up into numerous peninsulas; deep bays, and far-penetrating arms of the sea; the right position of a portion of dry land—*i. e.* its relations either to an ocean free from ice which extends beyond the polar circle, or to another continent of considerable extent which lies between the same meridional lines under the equator, or, at all events, in part within the tropics; farther, the prevalence of southerly and westerly winds on the western confines of a continent in the northern temperate zone; mountain chains, which serve as screens against winds from colder countries; the rarity of swamps, which continue covered with ice through the spring, and even some way into summer; the absence of forests on a dry sandy soil; finally, the constant serenity of the heavens in the summer months, and the neighbourhood of a pelagic stream of running water of a higher temperature than that of the surrounding sea.

To the second class of causes, or those that tend to depress the mean annual temperature by exciting cold, I enumerate: the elevation of a place above the sea level, without any thing like remarkable elevated plains surrounding it; the vicinity of an eastern coast in high and middle latitudes; the massive or unbroken outline of a continent without indentation of its coasts and deep sea bays; the wide extension of the land towards the poles up to the region of eternal ice (without the intervention of a sea open in winter); a geographical position in longitude of such a kind that the equatorial and tropical regions belong to the ocean—in other words, the absence of a heating, radiating tropical country between the same meridian lines as the country whose climate is to be determined; mountain chains whose form and direction are such that they prevent the access of warmer winds; or the neighbourhood of isolated summits down whose slopes cold currents of air descend; extensive forests, which hinder the sun's rays from reaching the ground, whose appendicular organs (the leaves), by their vital activity, throw off large quantities of watery vapour, and vastly increase the amount of radiating or cooling superficial surface, and so act in a threefold manner—by shading, by evaporating, and by radiating; great swamps, which, up to the middle of summer, in the north, form a kind of subterraneous glacier in the flats; a misty or overcast summer sky, which diminishes the effect of the sun's rays by intercepting them in their passage to the earth; finally, a very clear winter's sky, by which radiation is favoured(360).

The simultaneous activity of disturbing, whether heating or cooling causes, determines as a total effect the inflexions of the isothermal lines projected upon the surface of the earth, their course being especially influenced by the relations of extent and configuration between the opaque continental and the fluid oceanic masses. The perturbating causes engender convex or concave summits of the isothermal curves. But there are disturbing causes of different orders, each of which must first be separately considered; subsequently, in order to ascertain the whole effect upon the motion (direction or local curving) of the isothermal lines, it must be discovered which of the several influences in their combinations modify, annul, or strengthen each other, as happens in the case of other small oscillations that meet and intersect each other. Such is the spirit of the method, by which I flatter myself it will one day become possible to connect immeasurable series of apparently isolated facts with one another, by empirical numerically expressed laws, and to demonstrate the necessity of their mutual dependence.

As we find westerly or west-south-westerly winds in both temperate zones as the prevailing counter-currents to the trades or east winds of the tropics, and as these, to a country with an eastern sea-board, are land winds, and to a country with a western sea-board again are sea winds (*i. e.* as they blow over a level, which by reason of its mass and the descent of the cooled particles of water is susceptible of no great degree of chilling); so comes it that, where oceanic currents running near the shore do not influence the temperature, the east coasts of continents are colder than the west coasts. Cook's junior companion in his second voyage, the gifted George Forster, whom I have to thank for urging me on to various extensive undertakings, was the first who directed particular attention to the difference of temperature of the east and west coasts in both hemispheres, as well as to the correspondence between the temperature of the west coasts of North America in the middle latitudes, with that of the west of Europe within the same parallels([361]).

Accurate observations show a striking difference even in pretty high northern latitudes between the mean annual temperature of the east and west sea-boards of America. At Nain in Labrador (57° 10′ N. lat.) this temperature is 3°·8 C [5°·16 F.] under the freezing point of water [*i. e.* 26°·8 F.], whilst at New Archangel on the north-west shore of Russian America (57° 3′ N. lat.) it is still 6°·9 C. [12°·4 F.] above the freezing point [*i. e.* 44°·4 F.]. At the first named place the mean summer temperature scarcely reaches 6°·2 C. [43°·1 F.], whilst at the second it is as high as 13°·8 C. [56°·5 F.]. The mean winter temperature of Pekin (39° 54′ N. lat.) is at least 3° C. below the freezing point; whilst in the west of Europe, even at Paris (48° 50′ N. lat.), it is fully 3°·3 C. above this point. The mean winter cold of Pekin is thus lower by 2°·5 C. than that of Copenhagen, which lies 17 degrees of latitude farther to the north.

We have already spoken of the extreme slowness with which the great masses of the ocean follow alterations in the temperature of the air, and how in virtue of this property the ocean acts as an equalizer of temperature. It tempers at once the rudeness of the winter's cold and the fervour of the summer's heat. From hence a second important contrast: the difference between the insular or sea-board climates which all deeply indented continents abounding in bays and peninsulas enjoy, and the climates of the interior of great masses of terra firma. This remarkable contrast, in the variety of its phenomena, in its influence on the power of vegetation, and the improvement of agriculture, on the transparency of the atmosphere, the radiation of the earth's surface and the height of the line of perpetual snow, was first fully developed in the writings of Leopold von Buch. In the interior of the Asiatic continent, Tobolsk, Barnaul on the Obi and Irkutsk, have summers like those of Berlin, Munster and Cherbourg in Normandy; but these summers are followed by winters in which the coldest month reaches the fearful mean temperature of from —18° to —20° C. [0°·4 to —4° F.]. In the summer months, again, the thermometer for weeks together is seen standing at 30° and 31° C. [86° and 87°·8 F.]. Such *continental climates* are therefore well and properly characterized as excessive by Buffon, who was so well versed both in mathematics and in physics; and the inhabitants of the countries where they prevail, seem doomed, like the unfortunates in Dante's Purgatory([362]),

—— "a soffrir tormenti caldi e geli."*

In no quarter of the globe, not even in the Canary Islands or in Spain, or the South of France, have I met with more delicious fruit, particularly more beautiful grapes, than in Astrachan, near the shores of the Caspian Sea (46° 21′ N. lat.). With a mean annual temperature of about 9° C. [about 48½° F.], the mean summer temperature rises to 21°·2 C. [70°·1 F.], equal to that of Bordeaux; whilst not only there, but still farther to the south, at Kislar on the mouth of the Texel, in the latitudes of Avignon and Rimini, the thermometer in the winter season sinks to —25° and —30° C. [—13° and —22° F.]

Ireland, Guernsey and Jersey, the Peninsula of Brittany, the coasts of Normandy, and the South of England, in the mildness of their winters and the low temperature and overcast sky of their summers, present the most remarkable contrasts with the continental climate of the interior of the east of Europe. In the north-east of Ireland (54° 56′ N. lat.), under the same parallel as Königsberg in Prussia, the myrtle grows as vigorously as it does in Portugal. The month of August, the temperature of which in Hungary is 21° C., is scarcely 16° C. in Dublin, which stands on the same isothermal line of 9½°; and the mean winter temperature, which sinks in Buda to —2°·4 C., in Dublin (with its mean annual temperature, lower by 9° C.) is still 4°·3 above the freezing point of water; *i. e.*, it is 2° C. higher than in Milan, Pavia, Padua, and the whole of Lombardy, where the mean annual temperature is fully 12°·7 C. At Stromness in the Orkneys, not half a degree further to the south than Stockholm, the mean winter temperature is 4° C., higher consequently than that of Paris, and nearly equal to that of London. Even in the Faro Islands in 62° N. latitude, the influence of the westerly winds and of the ocean is such, that the water of the inland lakes never freezes. On the pleasant coasts of Devonshire, where Salcombe, by reason of its mild climate, has been called the Montpellier of the North, the Agave Mexicana has been seen flowering in the open air, and Oranges, trained as espaliers, and scarcely protected for a few weeks with mats, have borne fruit. There, as well as at Penzance and Gosport, and Cherbourg on the Norman coast, the mean winter temperature is as high as 5°·5 C., that is to say, but 1°·3 below the temperature of the corre-

[* "From beds of raging fire to starve in ice."
Milton, after Dante, though the English poet lays the scene in his Hell.—Tr.]

sponding season in Montpellier and Florence(363). The relations now indicated show how important for vegetation, agriculture, the growth of fruit, and the feeling of climatic comfort is the distribution of the same annual mean temperature over the different seasons of the year.*

The lines which I have entitled isochimenal and isotheral (lines of like mean winter and summer heat) are by no means parallel with the isothermal lines (lines of like mean annual heat). If in places where the Myrtle grows untended, and the ground in winter is never permanently covered with snow, the temperature of the summer and autumn is still just sufficient—nay, it might be said, is barely sufficient to bring the apple to perfect ripeness; if the vine, when it yields drinkable wine, flies islands, and almost all sea-boards, even those with a western exposure; the cause of this does not alone reside in the lower summer temperature of the coasts, which our thermometer in the shade proclaims; it lies in the hitherto so little considered, and yet in other phenomena (such as an explosion of a mixture of chlorine and hydrogen gas) so important distinction between direct and diffused light with a clear or clouded state of the heavens. It is long since I directed the attention of the observers of natural phenomena and of botanical physiologists to these distinctions, as well as to the unestimated heat locally developed in the vegetable cell under the influence of direct light(364).

If we descend in the thermal scale of husbandry of different kinds(365), beginning with the hottest climates, where Vanilla, Cacao, the Banana, Plantain, and Cocoanut Palm are successfully cultivated, to the regions in succession of the Pine-apple, Sugar-cane, Coffee, Date, Cotton-tree, Citron, Olive, true Chestnut, and Vine yielding drinkable wine, the careful geographical consideration of the limits of each of these species of culture, respect being had at once to the plain and to the mountain slope, assures us that other climatic relations than those connected with the mean annual temperature here come into play. To take the single instance of the vine, I remind my reader, that in order to have palatable wine(366), not only must the mean annual temperature exceed 9½° C. [49°·55 F.], but that the mean winter cold must not fall quite to the freezing point (0°·5 C., 33°·4 F.), and this must be followed by a mean summer heat of at least 18° C. [64°·4 F.]. At Bordeaux, in the valley of the Garonne (North latitude 44° 50′), the temperature of the year, of the winter, of the summer, and of the autumn, are respectively 13°·8; 6°·2; 21°·7; and 14°·4. In the plains of the Baltic, where wine is grown that is not palatable, though it is nevertheless consumed, the corresponding numbers are 8°·6; —°7; 17°·6; and 8°·6. If it seem strange that the great differences which the cultivation of the vine, favoured or opposed by climate, exhibits, are not more conspicuously shown by our thermometrical numbers, this strangeness will be lessened by the consideration, that a thermometer set for observation in the shade, and as effectually as possible protected from the effects of direct insolation and nocturnal radiation, does not by any means give the true superficial temperature for every division of the year, under periodical variations of the heat of the ground, exposed to the whole amount of insolation [and of radiation].

In the same way as the milder, more equable climate of the peninsula of Brittany stands related to the climate of the rest of the compact continent of France, colder in winter, hotter in summer, so to a certain extent does the climate of Europe stand related to that of the general continent of Asia, to which Europe forms, in fact, a kind of western peninsula. Europe owes its milder climate: to the geographical position of Africa, which in its vast extent, favouring the ascending current of air, presents a solid radiating surface within the tropics, whilst southward from Asia the equatorial region is mostly oceanic; to its partitions and vicinity to the sea—its forming the western boundary of the northern part of the Old World; to the existence of a sea free from ice, where it extends towards the north. Europe from this would become colder were Africa to be overflowed by the sea and to disappear(367); were the Mythical Atlantis to arise and connect Europe with North America; were the gulf-stream to cease from flowing and pouring its tepid current into the northern sea, or were another continent, raised by volcanic forces, to intervene between the Scandinavian peninsula and Spitzbergen. If we see the mean annual temperature of Europe sinking as we proceed along the same parallel of latitude from the shores of the Atlantic, from France, through Germany, Poland, and Russia, towards the Ural Mountains, from west to east, therefore, the principal cause of the phenomenon is to be sought for in the progressively less and less subdivided or more compact form of the land as the longitude increases, in the increasing remoteness of the tempering ocean, as in the feebler influence of the west wind. Beyond the Ural chain the west becomes the chilling land-wind, for then it is blowing over extensive tracts of country covered with ice and snow. The intense cold of Western Siberia is greatly connected with such relations of configuration in the land and of currents of air(368), nowise, as Hippocrates and Trogus Pompeius presumed, and as distinguished travellers in the 18th century have gone on fancying, with great elevation of the country above the level of the sea.

If we pass on from the consideration of diversities of temperature in the plains, to inequalities in the polyhedral configuration of the surface of our planet, we contemplate the mountains either according to their influence on the climate of the neighbouring low lands, or according to the influences which they exert, in consequence of hypsometrical relations, upon their own summits, frequently spread out into lofty plateaus or table-lands. The grouping of mountains into chains divides the surface of the earth into different basins, sometimes into narrow circular valleys surrounded by lofty walls—circus-like cauldrons, which (as in Greece and a portion of Asia Minor) give in-

[* For a great deal of interesting information on temperature the reader is referred to an excellent "Thermometrical Table," by Alfred S. Taylor, published by Willatt, 98 Cheapside. It is a complete Encyclopedia of Thermotics. —TR.]

dividual local characters to the climate in respect of warmth, dampness, frequency of winds and storms, and transparency of atmosphere. These circumstances have from time immemorial exerted a powerful influence upon the nature of the productions of the soil, and on the manners, forms of government, and likings and dislikings of neighbouring races for one another. The character of the *geographical individuality* reaches its maximum, as it were, where the diversities in the configuration of the surface, both in the vertical and the horizontal direction, in the relief and the partitioning of continents, are the greatest possible. With such relations of the soil are contrasted the steppes of Northern Asia, the grassy plains (Prairies, Savannas, Llanos, and Pampas) of the New Continent, the heaths or moors of Europe, and the sandy and rocky deserts of Africa.

The law of the decrement of temperature according to the height above the sea under different parallels of latitude, is one of the most important particulars in connection with the knowledge of meteorological processes, with the geographical distribution of plants, the theory of terrestrial refraction, and the various hypotheses which bear upon the determination of the height of the atmosphere. In the course of the numerous mountain expeditions I have undertaken, both within and without the tropics, the determination of this law has always been one of the principal objects of my observations and experiments(369).

Since the true relations of thermal distribution over the surface of the earth, *i. e.*, the inflections of the isothermal and isotheral lines, and the unequal distances of these from each other in the several systems of eastern and western temperature of Asia, mid-Europe, and North America, have been studied and made more generally known, we must not any longer inquire, even in a general way, what fractional part of the mean annual or summer temperature corresponds to a change of one degree of geographical latitude? In each system of isothermal lines of like curvature there prevails an intimate and necessary connection between three elements: the decrease of temperature in the perpendicular direction from below upwards; the difference of temperature in changing the place of observation by 1° of latitude; the equality of the mean temperature of a mountain station, and the polar distance of a point laid down on the level of the sea.

In the East American system, the mean annual temperature changes from the coasts of Labrador to Boston for every degree of latitude by 0°·88 C.; from Boston to Charleston by 0°·95 C.; from Charleston to the tropic of Cancer in Cuba onwards, the change, however, becomes less—there it is only 0°·66 C. Within the tropics the change is still smaller, the variation from Havanna to Cumana, corresponding to a degree of latitude, being no more than 0°·20 C.

It is quite different in the system of isotherms of mid-Europe. Between the parallels of 38° and 71° I find the decrease of temperature to coincide very accurately with half a degree (0°·5 C.) for each degree of latitude. But, as in this country, the fall in temperature is 1° C. for every 480, or 522 feet of perpendicular rise, it follows that here a rise of from 240 to 262 feet above the level of the sea corresponds, in respect of temperature, to one degree of latitude. The mean annual temperature of the Convent on Mount St. Bernard, 7,668 feet above the sea-level, in latitude 46° 50′, would thus be met with again in the plain, in latitude 75° 50′.

In that part of the chain of the Andes which lies within the tropics, my observations, which have been carried out to an elevation of 18,000 feet, indicate a fall of 1° C. for 96 toises, or 576 feet; my friend Boussingault, thirty years later, found 90 toises, or 540 feet, as the mean corresponding to the same fall. On comparing the places which stand among the Cordilleras at equal heights above the sea, whether on the slopes themselves, or on the extensive plateaus which they form, I found an increase of from 1°·6 to 2°·3 C. in mean annual temperature of the latter over the former. Without the cooling effects of nocturnal radiation, the difference would be still greater. As the climates are there stratified, as it were, superposed in layers from the Cacao groves of the lowlands up to the line of perpetual snow, and as the temperature in the tropical zone varies but very slightly in the course of the whole year, a tolerably fair idea is formed of the relations in respect of temperature to which the inhabitants of the great cities of the Andes are exposed, when these relations are compared with the temperature of particular months in the plains of France and Italy. Whilst the temperature of the day on the wooded banks of the Orinoco is such that it exceeds, by 4° C., that of the month of August at Palermo, we find when we have ascended the mountains to Popayan (911 toises), that we are in the temperature of the three summer months at Marseilles; in Quito, again (1492 toises), the temperature is that of the end of the month of May at Paris, and when we have attained the Paramos or mountain wilds, overgrown with dwarf Alpine plants, still bearing large flowers (1800 toises), we meet with the temperature of the beginning of the month of April at Paris.

The acute Peter Martyr de Anghiera, one of the friends of Christopher Columbus, was the first who perceived (in the expedition of Rodrigo Enrique Colmenares, Oct. 1510), that the snow-line always rises higher the nearer the equator is approached. I find these words in the beautiful work, De Rebus Oceanicis(370): "The River Gaira comes from a mountain (in the Sierra Nevada de Santa Marta), which, from the reports of the companions of Colmenares, is higher than any mountain yet discovered. It must undoubtedly be so, if, in a zone which is at most 10° from the equinoctial line, it retains its covering of snow continually." The inferior limit of the eternal snow in a given latitude is the summer limit of the snow-line; that is, the maximum height to which the snow-line recedes in the course of the entire year. From this summer limit of the snow-line, three other phenomena must be distinguished: Annual fluctuations of the snow-line; occasional or sporadic falls of snow; and glaciers, which appear to be peculiar to the temperate and frigid zones, on which Saussure's immortal work on the Alps, and in later years, the labours of Venetz, of Charpentier, and of Agassiz, endowed with perseverance that set danger at naught, have thrown much interesting and new light.

We know only the inferior, not the superior, boundary of the eternal snow; for the mountains of the earth do not rise into the ethereal or olympic empyrean, into the thin dry strata of the atmosphere, which we may presume with Bouguer no longer contain any vesicular vapour turned into crystals of ice, and thus made visible. The lower snow-limit, however, is not merely a function of the geographical latitude, or the mean annual temperature; the tropics, even the equator itself, is not the situation, as was long believed and taught, where the snow-limit attains its highest elevation above the level of the sea. The phenomenon which we here advert to is, in fact, an extremely complicated one, and depends generally on various relations of temperature, moisture, and mountain configuration. If these relations themselves be subjected to a more special analysis, as a great number of new measurements permit us to do([371]), we discover as coefficient causes determining the snow-line: Differences in temperature of the different seasons of the year; direction of the prevailing winds, and their contact with the sea and land; the degree of dryness or moistness of the upper strata of the atmosphere; the absolute magnitude or thickness of the deposited and accumulated snow; the relation of the snowy summit to the total height of the mountain; the relative position of the particular mountain considered in the chain; the steepness of the declivities; the vicinity of other mountains likewise capped with perpetual snow; the extent, lay, and height of the plain or level from which the snowy mountain rises isolated, or as one in a group or chain, and which may be a sea-coast, or the interior of a continent, covered with wood, or with a thick short turf, which may be sandy, barren, and strewn with naked rocks, or a wet mossy bottom.

While the snow-line in South America reaches a height under the equator which equals that of the summit of Mont Blanc, and in the high lands of Mexico, near the northern tropic, in 19° North latitude, according to recent measurements, descends from that by a quantity equal to about 960 feet, it rises, according to Pentland, in the southern tropical zone (lat. 14½° to 18° south), and in the western or Chilian Andes, not in the eastern chain, to more than 2500 feet higher than it is under the equator, on Chimborazo, Cotopaxi, and Antisana, not far from Quito. Dr. Gillies states, indeed, that much farther to the south, namely, on the declivity of the volcanic mountain Penguenes (33° S. lat.), he found the snow-line at an elevation between 2270 and 2350 toises above the level of the sea. The evaporation of the snow, in consequence of the radiation into an atmosphere which is excessively dry in summer, into skies which are scarcely obscured by a cloud, is so rapid, that the volcano of Aconcagua, to the north-east of Valparaiso (lat. 32½° south), which was found by the Expedition of the Beagle to be more than 1400 feet higher than Chimborazo, was once seen without snow([372]).

In almost the same parallel of North latitude (30¾° to 31°), the snow-limit of the southern slopes of the Himalaya is found nearly at the elevation which various combinations and comparisons might lead us to expect, viz., 12,180 feet; on the northern slopes, however, under the influence of the lofty table-land of Thibet, the mean height of which appears to be 10,800 feet, the snow-limit is only met with at an elevation of 15,600 feet. This phenomenon, which has often been the subject of discussion both in Europe and in India, on the cause of which I have myself made known my views in several papers([373]), possesses more than a merely physical interest; it has had an important influence upon the state of numerous tribes of mankind. Meteorological processes fit or unfit extensive districts of a continent for agriculture or pasturage.

As with the temperature the quantity of vapour contained in the atmosphere increases, this, which is so important an element for the whole of the organic creation, varies with the hour of the day, the season of the year, the degree of latitude, and the height above the level of the sea. The recent experience so generally obtained through the use of August's Psychrometer, according to the ideas of Dalton and Daniell, for the determination of the relative moistness of the air by means of the difference between the dew-point and the temperature of the air,* has considerably increased the extent of our knowledge of the hygrometrical relations of the surface of the earth. Temperature, atmospheric pressure, and quarter of the wind, all stand in most intimate connection with the vivifying moisture of the air. This vivification, however, is not so much a consequence of the quantity of vapour held dissolved under different latitudes, as of the manner and frequency of its precipitation in the shape of dew, fog, rain, or snow, which moistens the ground. From the deduction of the gyratory law of winds by Dove, and the views of this distinguished philosopher([374]), it appears that in our northern zone "the elasticity of vapour is greatest with south-west, least with north-east winds. On the west side of the wind-card it diminishes, and on the contrary it rises on the east side. On the west side, viz., the colder, heavier, drier current, forces back the warmer, lighter, much moister air; whilst on the east side the former is overcome by the latter. The south-west current is the penetrating equatorial stream; the north-east the sole prevailing polar current."

The beauty and fresh verdure of many trees which grow in countries within the tropics, where for five, six, or seven months together there is never a cloud to be seen on the face of the heavens, where no visible dew or rain ever falls, inform us that the appendages of the stem or the leaves have the power, in virtue of a peculiar vital process, which perhaps is not one merely producing cold by radiation, of

[* Now very conveniently obtained by the different readings of two thermometers, as like each other as possible, one of which has its bulb dry, the other its bulb wet. The instrument is commonly sold under the name of Mason's Hygrometer in England.—TR.]

withdrawing water from the atmosphere. With the parched levels of Cumana, Coro, and Ceara, in North Brazil, the deluges of rain which fall in other districts of tropical countries contrasts strongly: for example, in Havana, where observations carried on for six years by Ramon de la Sarga show the mean annual fall of rain to amount to 102 Parisian inches—four or five times as much as it is in Paris or Geneva[375]. On the slopes of the Andes, the quantity of rain that falls, like the temperature, diminishes with the height[376]. It was found by my companion in my South American journey, M. Caldas, of Santa Fe de Bogota, not to exceed 37 inches at a height of 8200 feet, which is but little more than the quantity that falls on some of the west coasts of Europe. At Quito, when the temperature was from 12° to 13° C., Boussingault sometimes saw Saussure's hygrometer recede to 26°; and in his great aerostatic ascent Gay Lussac saw the same instrument at 25°·3, his elevation at the time being 6600 feet, and the temperature of the air 4°·6 C. The greatest degree of dryness yet observed in a low country was seen by Gustavus Rose, Ehrenberg, and myself, between the valleys of the Irtisch and Obi, in Northern Asia. In the Platowskaja Steppe, after the south-west wind had been long blowing from the interior of the continent, the temperature of the air being 23°·7 C., we found the dew-point 4°·3 below the freezing-point. The air only contained $\frac{16}{100}$ of watery vapour[377]. Several able observers, Kaemtz, Bravais, and Martins, have of late years called in question the great degree of dryness of the mountain air, which seemed to follow from Saussure's observations among the Alps, and my own among the heights of the Cordilleras. The relative moistness of the air in Zurich was contrasted with that of the air of the Faulhorn, a mountain which indeed could only be called high in Europe[378]. The moisture with which the peculiar species of large-flowered, myrtle-leaved Alpine shrubs are almost perpetually bedewed in the region of the Paramos of the tropical Andes, between 11,000 and 12,000 feet above the sea level, and not far from the line where snow begins to fall, does not, however, necessarily imply a great absolute moistness of the air in this region; like the frequent fogs in the beautiful plateau of Bogota, it only proclaims the frequency of precipitations. Banks of fog at these heights form and disappear several times in the course of an hour when the air is calm; such rapid changes characterize the lofty plateaus and paramos of the Andes.

The ELECTRICITY OF THE ATMOSPHERE, whether considered in the lower regions or in the cloudy canopy aloft, viewed problematically in its silent periodical diurnal progression, or in the brilliant and noisy explosions of the thunder-storm, stands in manifold relationship with all the phenomena of thermal distribution, of atmospheric pressure and its disturbances, of hydrometeors, and apparently also of the magnetism of the outer crust of the earth. It exerts a most powerful influence upon the whole of the animal and vegetable world, and this not merely through the meteorological processes, precipitations of watery vapour, of acids, or of ammoniacal compounds, which it occasions, but also immediately as the electrical force, that force which excites the nerves and occasions or assists the circulation of the juices. This is not the place to renew the contest in regard to the source of the electricity of the serene sky, which has at one time been ascribed to the evaporation of impure fluids, *i. e.* fluids loaded with earths and salts[379], at another to the growth of vegetables[380], or other chemical decompositions proceeding on the surface of the earth, to the unequal distribution of heat in the different strata of the atmosphere[381], finally, according to Peltier's able inquiries[382], to the influence of a constantly negative charge of the globe. Limited to the results which electrometrical observations, particularly those which the clever arrangement of an electro-magnetical apparatus, first proposed by Colladon, have given, Physical Cosmography ought to indicate the unquestionable increase of the general positive aerial electricity with the height of the station and freedom from surrounding trees[383], its daily ebb and flow (according to Clarke's Dublin experiments, in more intricate periods than Saussure and I had detected), and its differences according to season, distance from the equator, and the continental or oceanic nature of the surface.

If the electrical equilibrium, on the whole, be less disturbed where the atmosphere is resting on the sea than on the land, it is the more remarkable to observe how small clusters of islands surrounded by an extensive ocean act upon the state of the atmosphere and give occasion to thunder-storms. In fogs, and at the beginning of falls of snow, I have in the course of a long series of observations seen the previous permanent vitreous, change suddenly into the resinous electricity, and these alternate repeatedly, as well in the plains of the frigid zone as under the tropics in the Paramos or Alpine wildernesses of the Cordilleras between 10,000 and 12,000 feet high. The alternate transition was in all respects similar to that which the electrometer had shown shortly before during the continuance of a thunder-storm[384]. When the vesicles of vapour have become aggregated into clouds with determinate outlines, the electrical tension of the outer layer or surface[385] upon which the electricity of the insulated vesicular vapour overflows, increases with the measure of the condensation. Slate-gray coloured clouds, according to Peltier's Paris experiments, have resinous, white, rose, and orange-coloured clouds, have vitreous electricity. Thunder clouds not only involve the highest summits of the Andes, (I have myself observed the vitrifying effects of lightning on one of the rocky crags which rise from the crater of the Volcano of Toluca, 14,300 feet high), but storm clouds have been measured, which were floating over low lands in the temperate zone, at a vertical height of 25,000 feet[386]. Occasionally, however, the thundering and lightning stratum of cloud descends to an altitude of five, and even of three thousand feet from the ground.

According to Arago's experiments, the most comprehensive we yet possess upon this difficult portion of meteorology, there are discharges of lightning of three kinds: zig-zag or forked lightnings, sharply defined on their edges;

lightnings that illuminate whole clouds, which seem to open up at once; lightning in the form of fire-balls(387). If the two first of these scarcely last for the $\frac{1}{1000}$ of a second, the globular kind of lightning, on the contrary, moves much more slowly, and continues visible for several seconds. Occasionally—and late observations confirm the description of the phenomenon already given by Nicholson and Beccaria—single clouds show themselves high above the horizon, which, without audible thunder, without any appearance of a storm, continue steadily luminous for a long time both in the interior and around the edges; hail-stones, drops of rain, and flakes of snow, have also been observed, which were luminous as they fell, without any precursory thunder-storm.

In the geographical distribution of storms, the coasts of Peru, in which it never thunders or lightens, present the most remarkable contrast with all the rest of the tropical zones besides, in which at certain seasons of the year, four or five hours after the culmination of the sun, thunder-storms occur almost every day. From the concurring testimony of northern navigators — Scoresby, Parry, Ross, Franklin — which has been collected by Arago, it is impossible to doubt that in high northern latitudes, such as the parallels from 70° to 75°, electrical explosions are extremely rare(388).

The meteorological portion of our Delineation of Nature, which we here conclude, shows that all the processes—absorption of light, evolution of heat, alteration of elasticity, hygrometrical state and electrical tension, which the immeasurable atmospheric ocean presents, are so intimately connected, that each individual meteorological process is simultaneously modified by any one, or by all the others. These varied disturbances, which involuntarily remind us of those that the nearer and particularly the smaller of the heavenly bodies, the satellites, comets, and shooting stars, experience in their course through space, render the interpretation of the complex meteorological processes difficult; they circumscribe, and, for the most part, make impossible, the prediction of atmospherical changes, which for horticulture and agriculture, for navigation and the enjoyment and pleasure of existence, would be so important. Those who place the value of meteorology not in a knowledge of the subject itself, but in such problematical prognostications, are penetrated with the belief that this portion of natural science, on account of which so many journeys have been made into remote mountainous countries of the globe, cannot boast of any advance for centuries. The confidence which they refuse to natural philosophers, they yield to the changes of the moon, and to certain famous days in the calendar.

"Great departures from the usual distribution of mean temperature seldom occur locally; they are, for the most part, relatively shared in by extensive districts of country. The amount of departure is a maximum at one particular spot, from which it diminishes to the confines around. If these confines be exceeded, great variations in the opposite sense are forthwith discovered. Like constitutions of the weather are more frequently observed from South to North than from West to East. The maximum cold of the end of 1829, (when I concluded my Siberian travels), occurred at Berlin, whilst North America enjoyed an unusual mildness of season. It is an entirely arbitrary assumption to suppose that a hot summer follows a cold winter, or that a mild winter succeeds a cool summer"(389). The states of the weather in neighbouring lands—two corn and wine-growing countries, for example, often so varied and so opposite, produce the most beneficial effects in equalizing the prices of agricultural produce in each. It has been well observed, that the barometer alone informs us of what is going on in respect of alterations in the pressure of the whole of the strata of air above the place of observation, even to the extreme limits of our atmosphere, whilst the thermometer and hygrometer only report upon the local temperature and moistness of the lower stratum in contact with the surface. We only conclude as to the thermometrical and hygroscopical modifications of the upper strata, where immediate observations made on mountains or in aerostatic journeys are wanting, from hypothetical combinations, so that the barometer may likewise come to serve both as a thermometer and hygrometer. Important changes in the state of the weather are not owing to any merely local cause at the place of observation itself; they are usually consequences of a condition which has begun, through perturbation in the equilibrium of the currents of air, at a vast distance, and, for the major part, not at the surface of the earth, but in the highest regions: bringing hither cold or warm, dry or moist air, impairing or increasing the transparency of the atmosphere, changing the piled masses of cumulus-clouds into the light and feathery cirrus. Inaccessibility of phenomena thus allying itself to multiplicity and complexity of perturbations, it has always appeared to me, that meteorology must seek her welfare and her roots in the torrid zone; in that favoured region where the same winds always blow, where the ebb and flow of the atmospheric pressure, the course of hydrometeors, and the occurrence of electrical explosions, are periodically and regularly recurrent.

Having now passed through the entire circuit of the inorganic life of the earth, and delineated our planet with a few leading touches, in its configuration, its internal heat, its electromagnetical charge, its luminous processes at either pole, and its internal or volcanic reaction upon the solid and variously compounded crusts; having, finally, considered the phenomena of its double outer covering, the ocean and the atmosphere, our Picture of Nature, according to the older ideas of Physical Geography, might be held as finished. But when the philosophic view essays to reach a higher point, the delineation would seem to want its attractive features, did it not at the same time present the sphere of ORGANIC LIFE in the numerous grades of its typical developments. The idea of animation is so closely connected with the idea of the existence of the impelling, ceaselessly active, decompounding, compounding, and fashioning natural forces, which inhere in the terrestrial ball, that in the popular Mythus of the nations of antiquity, the production of plants and animals was always ascribed to these forces ; the state of the surface of our planet, when it was unoccupied by life, was even referred to the chaotic primeval ages of the conflicting elements. To the empirical domain of objective sensuous consideration, to the delineation of That which has become, or of the actual state and condition of our planet, the mysterious and unresolved problem of Things becoming does not rightfully belong.

Our description of the world, abiding soberly by reality, remains a stranger, not from timidity, but from the nature of the subject and its limits, to the obscure history of the beginning of organized things(390) ; the word history being here taken in its most usual acceptation. But in our account of the actual Universe, we may direct attention to the fact, that in the inorganic crust of the earth, the same elementary substances are present which enter into and compose the organic frames of plants and animals. We may say, that the same powers prevail in these as in those, which combine and separate bodies, which give consistency and fluidity in the organic tissues: but, subjected to conditions which, not yet fathomed, have been systematically grouped according to analogies more or less happily imagined, under the very indefinite titles of effects of the vital force. It is, therefore, felt as a want, in the frame of mind which leads us to look on nature with a contemplative eye, that we pursue the physical phenomena which present themselves to us on earth to their very farthest limits, to the evolution of vegetable forms, and the discovery of that which in the organisms of animals is endowed with self-motive force. In this way does the geographical distribution of things organically-animated, *i. e.*, plants and animals, connect itself with the delineation of the inorganic natural phenomena of the body of the globe.

Without pretending in this place to discuss the difficult question of "the self-motive" in animals, *i. e.*, the difference between animal and vegetable life, we must first direct attention to the circumstance that, were we endowed by nature with a microscopic eye, and were the integuments of plants completely transparent, the world of vegetation would not meet us with that aspect of immobility and repose in which it now presents itself to our senses. The interiors of the cellular structures of vegetables are ceaselessly animated by the most diversified currents, rotatory, rising and falling, dividing and ramifying, or altering their direction—as is made manifest by the movement of the granular sap-corpuscles in the leaves of several water-plants (Najades Characeæ, Hydrocharideæ), and in the hairs of phanerogamous land-plants; there is at the same time seen a confused, molecular movement, first observed by the distinguished botanist, Robert Brown, but which also occurs among finely-divided particles of matter of all kinds, the phenomenon not taking place only within organic cells ; the circular movement of the globules of the cambium, in a system of special vessels (cyclosis) ; lastly, the singular articulated filiform vessels of the anthers of the Chara and the reproductive organs of the liverworts and sea-weeds which have the faculty of uncoiling themselves, and in which Meyen, snatched too soon away from science, believed that he recognized the analogues of the spermatozoa of the animal creation. If to the multifarious excitements and movements we add those that belong to endosmose and the processes of nutrition and growth, and farther to the penetration [and exhalation] of air, we have a picture of the forces which, almost unknown to us, are active in the silent life of the vegetable world.

Since I first portrayed the universal life of the surface of the earth, and the distribution of organic forms, both in the line of the height and of the depth, in my "Views of Nature," our knowledge in this direction also has been surprisingly increased by Ehrenberg's brilliant discoveries, "on the demeanor of minute life in the ocean as well as in the ice of the polar lands," discoveries made not by the way of induction, but by that of simple accurate observation. The sphere of vitality, we might almost say the horizon of life, has extended itself before our eyes. "There is not only an invisibly small, or microscopical, incessantly active life in the neighbourhood of both poles, where larger organisms are no longer produced ; but the microscopical forms of life of the South Polar Sea, collected in the Antarctic Voyage of Sir James Ross, comprise a wonderful variety of entirely new and often extremely beautiful forms. Even in the remains of the liquefied rounded masses of ice that were picked up swimming about under the latitude of 78° 10′, more than fifty species of silicious-shelled polygastrica, coscinodisca, with their green ovaries, and consequently living and successfully struggling with the extreme of severe cold, were discovered. In the bay of the Erebus, in from 1242 to 1620 feet of water, sixty-eight silicious-shelled polygastrica and phytolitharia, and with them only a single calcareous-shelled polythalamium, were drawn up by means of the lead.

The oceanic microscopic forms have hitherto been in vastly preponderating proportion of the silicious-shelled kinds, although silica does not appear among the constituents of sea-water discovered by analysis, and the earth can only be well conceived as mixed with or suspended in the waters. The ocean, however, is not only in particular spots, and in arms and bays, or near the shore, thickly peopled with invisible, *i. e.*, by the unassisted eye, unseen living atoms ; it may be assumed, from the samples of water drawn to the south of the Cape of Good Hope, under 57° S. latitude, as well as from the middle of the Atlantic, under the tropics, by Schayer, in his return from Van Diemen's Land, that in its ordinary state, without showing any particular colour, without being filled with floating fragments of the silicious-shelled filaments of the genus Chætoceros, which so much resemble the Oscillatoria of our fresh waters, but when perfectly transparent to the naked eye, the ocean still contains numerous independent microscopical organisms. Several polygastrica from Cockburn Island, mixed with the excrements of Penguins and sand, appear to be spread over the whole earth ; others, again, are common to either pole(391).

From this (and all the more recent observations confirm the view) it appears that in the eternal night of the depths of ocean, animal life especially prevails, whilst upon continents, vegetable life, which requires the periodic stimulus of the sun's rays, is the more extensively diffused. Considered with reference to mass, the vegetable far exceeds the animal world on the face of the globe. What is the number of great cetaceans and pachydermatous tribes, in comparison with the bulk of the thick-set trunks of lofty trees, from eight to twelve feet in diameter, that grow in the forests of the tropical zone of South America, between the Orinoco, the Amazon's River, and the Rio de Madeira ! Even allowing the character of the several countries of the earth to depend on the aspect of external phenomena at large ; if the outline of mountains, the physiognomy of plants and animals, the blue of the sky, the contour of the clouds, and the transparency of the atmosphere produce the general impression, still it is not to be denied that the principal element in this impression is the vegetable covering of the surface. The animal kingdom wants mass, and the motions of individuals withdraw them frequently from our sight. The vegetable world works upon our imagination by the mere force of quantity ; its mass indicates its age ; and in vegetables alone are age and the expression of inherent power of renovation associated(392). In the animal kingdom—and this consideration is also the result of Ehrenberg's discoveries—it is precisely the life that we are wont to designate as the smallest in point of room, which by its subdivision and rapid increase(393) presents the most remarkable relations in respect of mass. The smallest of the Infusoria, the Monadæ, only obtain a diameter of $\frac{1}{3000}$th of a line, yet do these silicious-shelled organisms, in moist countries, compose, by their accumulation, subterraneous strata several fathoms in thickness.

The impression of an all-animated nature, so exciting and so salutary to feeling man, belongs to every zone ; but it is most powerfully produced towards the equator, in the peculiar zone of the palms, the bamboos, and the arborescent ferns—in regions where, from sea shores covered with molluscs and corals, the ground rises in stages to the line of eternal snow, and the relations of plants and animals, in respect of local position, embrace almost every height and every depth. Organic forms even descend into the interior of the earth, and occur not merely in places where, through the operations of the miner, great excavations have been made ; in natural cavities, also, which have been opened for the first time by blasting, and to which meteoric water alone could have penetrated through fissures, I have found the snowy stalactitic walls covered with the delicate reticulations of an Usnea. Podurellæ penetrate into the icy circles of the glaciers of Monte Rosa, of the Grindelwald, and the Upper Aar. Chionœa arenoides, described by Dalman, and the microscopic Discerea nivalis, or Protococcus, as it used to be called, lives among the snows of the polar regions as well as of our loftier mountains. The red colour of old snow was known to Aristotle, and was probably observed by him among the mountains of Macedonia(394). Whilst upon the lofty summits of the Swiss Alps, Lecideas, Parmelias, and Umbilicarias alone, and sparingly, tint the rocks left bare of snow, in the elevated regions of the tropical Andes, at the height of 14,000 and 14,400 feet above the level of the sea, single specimens of beautiful phanerogamous plants are still encountered—the tomentose Calcitium rufescens, Sida Pichinchensis, and Saxifraga Boussingaulti. Hot springs contain small insects — Hydroporus thermalis, Galionellæ, Oscillatoria, and Confervæ ; they even irrigate the roots of phanerogamous plants. As water, earth, and air are peopled by animated beings at the most dissimilar temperatures, so also is the interior of the most dissimilar parts in the bodies of animals inhabited. Animated organisms have been found in the blood of the frog, salmon, &c. According to Nordmann, the whole of the fluids of the fish's eye are often filled with a suctorial worm (Diplostomum) ; and in the gills of the brasse lives that extraordinary double animal, denominated by the naturalist just mentioned the Diplozoon paradoxum ; a creature consisting, as it seems, of two perfect animals, grown crosswise together, having two heads and two caudal extremities.

Granting the existence of meteoric infusoria, as they have been called, to be more than doubtful, still the possibility must not be denied, that as the pollen of the pine-tree has fallen year after year from the air, so may minute infusory animalcules be passively raised with the watery vapour, and floated for a season in the atmosphere(395). This circumstance deserves to be taken into serious consideration, in connection with the old dispute in regard to spontaneous generation(396), (generatio spontanea) ; all the more, since Ehrenberg, as already observed above, has discovered in the kind of dust-rain which navigators frequently encounter in the neighbourhood of the Cape de Verd Islands, at a distance of 380 sea miles from the coast of Africa, the remains of eighteen species of silicious shelled polygastric animalcules.

The exuberance of organisms whose distribution in space is studied in the geography of plants and animals, is considered either according to the diversity and relative number of the types of formation, according to the configuration of the existing genera and species, or according to the number of the individuals which each particular species presents upon a given superficial area. Among plants, as among animals, it is an important distinction in their mode of life, whether they are met with singly or living in company. The species which I have designated social plants(397) cover large tracts of country with one unvarying growth. To this class belong many species of sea-weed in the ocean, Cladoniæ and Musci in the waste levels of Northern Asia, Grasses and tubular looking Cactuses, Avicennia and Mangrove in the tropical world, forests of coniferous trees and birches in the Baltic and Siberian plains(398). This kind of geographical distribution of plants, along with the individual aspect of the species, their size, the form of their leaves and flowers, determines in an especial manner the physiognomical character of a country. The shifting image of animal life, so varied and attractive, appealing so immediately to our feelings of liking or disgust, remains almost wholly foreign, or at least is much less powerfully felt, in connexion with the members of the vegetable kingdom. Agricultural nations increase artificially the domain of various social plants, and so increase the aspect of uniformity presented by nature in several districts of the temperate and northern zones; they also root out and destroy various wild-growing plants, and unintentionally propagate others that follow man in his wanderings. The luxuriant zone of the tropical world resists more powerfully this forcible metamorphosis of creation.

Observers who have perambulated extensive districts of country in short intervals of time, who have ascended mountain ranges in which the climates lie stratified one over another, must soon have been awakened to the regular distribution of vegetable forms. They collected the raw material of a science whose name was not yet pronounced. The same zones or regions of plants which Cardinal Bembo(399), in the 16th century, when yet a youth, described as occurring on the slopes of Etna, were found repeated on Mount Ararat, by Tournefort, who acutely compared the Alpine floras with the floras of plains under different latitudes; and who first remarked that the elevation of the ground above the level of the sea in mountainous districts influences the distribution of plants in the same way as distance from the pole in plains. Menzel, in an unedited Flora of Japan, incidentally used the expression GEOGRAPHY OF PLANTS. This phrase again recurs in the fantastic but pleasant "Studies of Nature" of Bernardin de St. Pierre. But the scientific treatment of the subject commenced when the distribution of plants was viewed in close connexion with the doctrine of the distribution of heat over the surface of the earth; when plants were arranged into natural orders, and it was thus made possible to distinguish numerically the particular forms which increase or diminish from the equator towards the poles, to perceive, in the different regions of the earth, in what numerical relationship each family stands to the whole of the mass of phanerogamous vegetables which are there indigenous. It is one of the fortunate events in my life, that at the time when I was giving my attention almost exclusively to botany, my studies should have been directed to the subject of inquiry just mentioned, by the spectacle of nature on the grandest scale, and offering the strongest contrasts in respect of climate.

The geographical distribution of animal forms, upon which Buffon first advanced general, and, for the major part, very accurate views, has in recent times had great assistance from the progress of vegetable geography. The curvatures of the isothermal, and particularly of the isochimenal lines, are displayed in the limits which certain species of plants, and of animals that do not roam far towards the north or towards the tops of snow-covered mountains, seldom exceed. The Elk, *e. g.*, lives in the Peninsula of Scandinavia, almost ten degrees farther to the north than in the interior of Siberia, where the lines of like winter temperature are so remarkably concave. Plants wander or migrate in the egg, in the seed. The seeds of many species are provided with peculiar organs for far journeys through the air. Once rooted, they are more dependent on the soil and the temperature of the atmosphere which surrounds them. Animals widen at will the circle of their presence from the equator towards the pole, and particularly in regions where the isotheral lines arch out towards the north, where hot summers succeed the severest winters: royal tigers, which do not differ from those of India, roam every summer in Northern Asia to the latitudes of Berlin and Hamburg, as Ehrenberg and I have shown in another place(400).

The groups or associations of vegetable species which we are accustomed to designate FLORÆ (spheres or domains of vegetation), appear to me, from what I have seen of the earth, by no means to reveal the prevalence of individual families to such an extent as authorizes us to establish geographical regions of the Umbellatæ, Solidagineæ, Labiatæ, or Scitamineæ. My particular views differ in this respect from those of several of my friends among the most distinguished botanists of Germany. The character of the Floras in the high lands of Mexico, New Granada, Quito, European Russia, and Northern Asia, consists, as I believe, not in the relatively larger number of species which one or two natural families exhibit, but rather in the much more complex relations of the aggregate life of many families, and the relative numerical value of their species. In meadow and steppe districts Gramineæ and Cyperaceæ are the prevailing families; in our northern woods we meet especially with Coniferæ, Cupuliferæ, and Betulineæ; but this prevalence of form is only apparent, and deceptive by reason of the mass of the social plants arresting the eye. The north of Europe, and Siberia in the zone northward from the Altai, no more deserve the title of a realm of grasses or cone-bearing trees, than the endles Llanos between the Orinoco and the mountain chain of Caraccas or the pine forests of Mexico. In the associated life of the vegetable forms which partly replace one another, in their relative numbers and grouping, lies the aggregate impression of richness and variety, or of poverty and monotony of vegetable nature.

In this brief consideration of the phenomena of organized beings, I have ascended from the simplest cell(401), and so, from the first breath of life, to higher and higher forms. "The aggregation of mucus-granules into a definitely formed cell-germ, around which a membrane in form of a vesicle being developed, it is connected into a closed cell," is either effected by a pre-existing cell, so that cell arises from cell(402), or the evolution of cells is involved in the obscurity of a chemical process, as in the case of the torula cerevisiæ, or yeast fungus. The most mysterious subject of Incipiency can only be lightly touched upon here. The geography of organized beings—plants and animals—treats of the germs already developed, of their habitats from migrations effected on purpose or accidentally, of their respective relations, and their aggregate distribution over the surface of the earth.

The general delineation of nature, which I here endeavour to present, would remain incomplete, were I not to yield to the disposition I feel, with a few touches, to portray the HU-

MAN KIND in its physical gradations, in the geographical distribution of its simultaneously existing types, in the influence which it derives from the forces of nature, and on the contrary, though in a less degree, the influence which it has exercised on these. Dependent, although not to the same extent as plants and animals, on the ground and the meteorological processes of the atmosphere, more readily escaping from under the dominion of some of the natural forces through activity of mind, and intelligence exalted by degrees, as well as through a wonderful pliability of constitution, which adapts itself to every climate, the human kind takes an essential part in the whole vitality of the earth. Through these relations we are brought into contact with the obscure and much agitated problem of the possibility of common descent in the circle of ideas which the physical cosmography embraces. The investigation of this problem, if I may so express myself, shall, through ennobled and purely human interests, be made the last aim of my work. The immeasurable realm of language, in the diverse organizations of which, the capacities of nations are foreshadowed, as it were, is most intimately connected with the subject of alliance of race; and what even slight diversity of race is competent to produce, is taught us by the Hellenic world in the bloom of its mental culture. The most important questions in the history of the progress of society connect themselves with ideas of descent, community of language, and immutability in an original direction of the affective and intellectual nature of man.

So long as extremes in diversity of colour and configuration were alone considered, and the first liveliness of sensible impression was yielded to, there might have been the disposition to consider races, not as mere varieties, but as originally different kinds of men. The permanency of certain types(403) even amidst the most inimical operation of external, particularly climatic influences, appeared to favour such an assumption, short though the time be through which historical information has come down to us. But vouching far more strongly, according to my views, for the unity of the human race, are the many middle tints(404) in colour of skin, and grades in form of skull, which the rapid spread of geographical knowledge in recent times has made known to us; the analogy of variety in other wild and domesticated classes of animals, and the sure experience which has been collected in regard to the limits of fruitful hybrids of different kinds(405). The greater number of the contrasts which in former times were believed to have been discovered, have been disposed of by the industrious work of Tiedemann, "On the Brain of the Negro and the European," and by the anatomical inquiries of Vrolik and of Weber, "On the Form of the Pelvis." If we embrace the dark-skinned African nations, on which Prichard's admirable work* has thrown so much light, in their universality, and compare them with the races of the South Indian and West Australian Archipelagos, with the Papuas and Alfourous (Haraforans, Endamenans), we see clearly that black colour of the skin, woolly hair, and negro-like features are by no means always conjoined(406). So long as but a small portion of the world was open to the western nations, they necessarily came to narrow or one-sided conclusions. Heat of sun in the tropical world, and dark colour of skin, seemed inseparable. "The Æthiopians," sings the old tragedian, Theodectes of Phaselis(407), "are dyed by the near sun-god in his course, with a dark and sooty lustre; the sun's heat crisps and dries up their hair." The expeditions of Alexander, which were so influential in exciting ideas of the physical cosmography, first fanned the dispute on the uncertain influence of climate upon races of men.

"The races of animals and plants," says one of the greatest anatomists of the age, Joannes Müller, in his very comprehensive "Physiology of Man,"* "undergo changes during their spread over the surface of the earth, within the limits prescribed to species and genera. But they are propagated organically as types of varieties of species. From the co-operation of different, as well internal as external conditions, not to be specified in individual instances, have the present races of animals proceeded, the most remarkable varieties of which are met with amongst those that are capable of the widest distribution over the face of the earth. All the races of men are forms of a single species, which are capable of fruitful union and of propagation; they are not different species of one genus; were they so, their mixed progeny would prove unfruitful. Whether the various races of men are descended from several or from a single primitive man, cannot be decided from experience"(408).

Geographical researches into the ancient seat, the cradle, as it has been called, of the human race, possess in fact a purely mythical character. "We know," says William von Humboldt, in a work yet unpublished, on the Diversity of Languages and of Nations, "we know, neither historically, nor by tradition that can be trusted, of any epoch in which the human race have not been collected together into tribes or communities. Whether this condition was the original one, or first arose at a later period, cannot be decided historically. Isolated traditions met with in many different places on the earth's surface negative the first assumption, and derive the whole of the human race from a single human pair. The wide diffusion of this belief has sometimes led to its being assumed as a primitive recollection among mankind. But this very circumstance much rather informs us, that nothing traditional, and nothing historical lies at the root of the persuasion, but merely similarity of the human faculty of conception which leads to the same explanation of the same phenomenon; many similar myths have very certainly arisen, without historical connection, out of the similarity of man's poetical and speculative constitution. These traditions also bear the entire stamp of human invention in this, that they explain the phenomenon of the first appearance of the hu-

* [Researches into the Physical History of Mankind, 3d and 4th edit., 4 vols. 8vo, 1841–44. The Natural History of Man, 1 vol. 8vo, 1843, 2d edit., 1845.—TR.]

* [Ably rendered into English, and copiously commented and illustrated, by Dr. Wm. Baly, 2 vols. 8vo, Lond., 1842. —TR.]

man race (a point which lies beyond the reach of all experience), in a way that accords with the experience of the day, and proceed to show how, in times when the human kind must already have existed for thousands of years, a desert island, or a sequestered valley, may have been peopled. It is in vain, however, that reflection attempts to dive into this problem of original production, seeing that man is so bound up with his kind and with time, that an individual without contemporaries, and without a past, can by no means be conceived in human existence. Whether, therefore, in this question, which can neither be decided by the way of reasoning nor of experience, this pretended tradition be the historical truth, or the human kind from its commencement has possessed the earth in the shape of tribes or communities, can neither be determined by Philology out of the elements of its science, nor, assuming the decision on other grounds, can it use the conclusion come to in illustration of its own propositions."

The distribution of the human kind is no more than a distribution into varieties, which have been designated by the somewhat indefinite word *races*. As in the vegetable kingdom, and in the natural history of birds and fishes, the system of grouping into many small families is more certain than that into a few divisions, embracing larger masses, so does it appear to me preferable, in the determination of races of men, to arrange them into smaller families. The old classification of my master, Blumenbach, into five races—the Caucasian, the Mongolian, the American, the Æthiopian, and the Malayan, may be followed ; or with Prichard, seven races may be assumed—the Tranian(409), the Turanian, the American, the Hottentot and Buschman, the Negro, the Papuan, and the Alfourousian ; still is there no typical rigour, no natural principle of classification, recognizable in such arrangements. The extremes in reference to configuration and colour are separated, without reference to the stocks that cannot be connected with one or other of these, and which have at one time been entitled Scythian, at another Allophylian races. Tranian, in reference to the European nations, is certainly a less objectionable name than Caucasian; but it may be maintained in a general way, that geographical designations as derivative points of races are very indeterminate, when the country which is chosen to confer the title on the race, for example, Turan (Maweran-nahr), has at different epochs been possessed by most dissimilar races — of Indo-Germanic and Finnish, but not Mongolian origin(410).

Languages, as mental creations of man, as closely intertwined with his spiritual development, inasmuch as they exhibit national forms, possess high importance in connection with the recognizable similarities and dissimilarities of races. They have this importance, because community of descent leads into the mysterious labyrinth in which the enchainment of physical (bodily) aptitude with mental power is exhibited in an endless variety of forms. The brilliant advances which philosophical philology has made in Germany, especially within the last half century, facilitate inquiries into the national character of languages, into that which descent appears to have added to them(411). As in all other regions of abstract speculation, however, the danger of being deceived is here set beside the hope of collecting a rich and assured booty.

Positive ethnographical studies, based upon solid historical knowledge, warn us that the greatest caution is necessary in all comparisons of nations with the languages which they have made use of at determinate epochs. Subjugation, living long together, the influence of a foreign religion, and intermixture of races, though often effected by a relatively small number of more powerful and more civilized intruders, have produced a phenomenon which has recurred in like measure, in both continents, viz. : that totally different families of languages are met with in use by one and the same race ; that among nations of very different descent, idioms of the same original tongue are encountered. Asiatic conquerors have had the greatest influence upon such phenomena.

Speech, however, is a portion of the natural science of mind ; and if the freedom wherewith the spirit in a state of happy independence steadily pursues the self-elected course under totally different physical influences, strives vigorously to withdraw it from the power of the earth, still the unfettering is never quite complete. There ever remains something of that which belongs to natural aptitude, to descent, to climate, to the bright blue sky, or to the cloudy atmosphere of the insular world. And since copiousness and grace in the structure of a language are unfolded from thought as from the most delicate blossom of the soul, so would we not, that in the intimacy of the bond which unites the two spheres—that of the physical nature, and that of the intellect and feelings—our delineation should be without the favourable light and colouring which it must derive from a consideration, here only indicated, it is true, of the relations of hereditary descent to language.

In maintaining the unity of the human kind, we at the same time repudiate all the unsatisfactory assumptions of higher and lower races of men(412). There are races of men more flexible, more highly polished, through mental culture more ennobled, but none naturally more noble. All are in equal measure ordained for liberty ; for liberty which in ruder conditions of society appertains to the individual, which in more polished states, in civil life and among men in the enjoyment of political institutions, is the right of the community. "If we would signalize an idea which is conspicuous throughout the entire current of history, and ever with a wider import, when any one assures us of the much-discussed, but still more extensively misapprehended perfectibility of mankind, it is the idea of HUMANITY : the effort to cast down the barriers which prejudice and one-sided views of every kind have hostilely raised betwixt man and man, and to treat mankind at large, without reference to religion, to nation, or to colour, as one great and nearly-related family—as a whole, that exists for the accomplishment of this single end, THE FREE DEVELOPMENT OF INTERNAL POWER. This is the extreme, the ultimate purpose of the social state, and at the same time it is the tendency infixed in the nature of man striving after indefinite extension of his being. He looks upon the ground, as it spreads out beneath his feet—on the heavens, as they arch over his head—on the stars that shed their light upon him, as intimately his own, as given to him for contemplation and for reality. The very child longs to get beyond the hills, the lakes that bound his narrow home ; and then, plant-like, he longs to return ; for it is a touching and a beautiful element in the nature of man, that all his desires for things agreeable and for things lost, still approve him exclusively attached to the moment : firmly rooted in the inmost nature of man, and at the same time commanded by his loftiest aspirations, this benevolent, this humane association of the entire race becomes one of the grand leading ideas in the history of mankind"(413).

With these words, which draw their charm from the depth of the feelings that gave them birth, be it allowed the Brother to conclude this general representation of the natural phenomena of the universe. From the farthest nebulæ of heaven, and from revolving double stars, we have come down to the smallest organisms of the animal creation that live by sea and land, and to the delicate vegetable germs that cover the rocks on the flanks of snow-clad mountain summits. Here we have found that the phenomena could be arranged according to laws which are partially known. Laws of another and a mysterious kind come into play in the higher circles of life in the organic world ; in those especially that are occupied by the races of mankind variously conformed, endowed with creative mental energies, and gifted with the faculty of inventing language. A *physical* Delineation of Nature indicates the boundary where the sphere of intelligence begins, and the far-piercing glance is lost in another world. It indicates, but does not pass the boundary.

NOTES TO PRECEDING SECTION.

1 (p. 29.)—The optical considerations on the differences which a single luminous point or a disc of measurable angle presents, in which the power of light remains the same at every distance, may be found discussed by Arago, Analyse des travaux de Sir William Herschel (Annuaire du Bureau des Long. 1842, p. 410—412, and 441.)

2 (p. 29.)—"The two Magellanic clouds, Nubecula major and minor, are highly remarkable objects. The larger cloud is an aggregation of stars, and consists of clusters of stars of irregular configuration, of globular clusters and nebulous stars of different sizes and densities. Between these lie large nebulæ which are not resolvable into stars, which apparently are star dust, and even with the 20-feet telescope present themselves only as a general luminousness of the field, as a brilliant back-ground, upon which other objects of very remarkable and incomprehensible configuration are scattered. In no other part of the heavens are so many clusters of nebulæ and of stars collected together within so small a compass as in this cloud. The Nubecula minor is much less beautiful; it shows a larger quantity of unresolvable nebular light, and the groups of stars it includes are fewer in number and smaller."—Letter from Sir John Herschel, dated Feldhuysen, Cape of Good Hope, June 13. 1836.

3 (p. 29.)—The fine expression, χόρτος οὐρανοῦ, which Hesychius borrows from an unknown poet, I have rendered as above by the phrase "garden of heaven;" χόρτος may, perhaps, rather signify an enclosed place, and would then be better translated, the celestial space. The connection of the word with the German garten, English garden, (Gothic gards, according to Jacob Grimm, from gairdan, cingere, to gird,) is, however, not to be overlooked, any more than the affinity with the Sclavonian grad, gorod, and, as remarked by Pott (Etymol. Forsch. Th. i. S. 144), with the Latin chors, (whence the modern words corte, court, cours), and the Ossetic khart. The northern gard, gard, a fence, an enclosure, and a country-seat; and the Persian gerd, gird, a circle, and also a princely country-seat, a castle, a town; as in the old names in Firdusi's Schahnameh: Siyawakschgird, Darabgird, &c.

4 (p. 30.)—For α Centauri, *vide* Maclear, in Trans. Astronom. Soc., vol. xii. p. 370. More probable mean error 0″·0640: for 61 Cygni, *vide* Bessel, in Schum. Jahrbuch, 1839, S. 47—49, and in Schum. Astronom. Nachr. Bd. 17, S. 401, 402. Mean error 0″·0141. On the relative distances of stars of different orders, as those of the third magnitude are probably three times more distant, and as to how we are to imagine the strata of stars in their bodily configuration, I find in Kepler's Epitome Astronomiae Copernicanae, 1618, tom. i. lib. i. p. 34—39, a remarkable passage: "Sol hic noster nil aliud est quam una ex fixis, nobis major et clarior visa, quia proprior quam fixa. Pone terram stare ad latus, una semidiametro viae lacteae, tunc haec via lactea apparebit circulus parvus, vel ellipsis parva, tota declinans ad latus alterum; eritque simul uno intuitu conspicua, quæ nunc non potest nisi dimidia conspici quovis momento. Itaque fixarum sphæra non tantum orbe stellarum, sed etiam circulo lactis versus nos deorsum est terminata."

5 (p. 31.)—"Si dans les zones abandonnées par l'atmosphère du soleil il s'est trouvé des molécules trop volatiles pour s'unir entre elles ou aux planètes; elles doivent en continuant de circuler autour de cet astre offrir toutes les apparences de la lumière zodiacale, sans opposer de résistance sensible aux divers corps du systéme planétaire, soit à cause de leur extrême rareté, soit parce que leur mouvement est à fort peu prés le même que celui des planètes qu'elles rencontrent."—Laplace, Exp. du Syst. du Monde, (5e éd.) p. 415.

6 (p. 31.)—Laplace, loc. cit. p. 396 and 414.

7 (p. 31.)—Littrow, Astronomie, 1825, Bd. ii. S. 107. Mädler, Astr. 1841, S. 212. (Laplace, loc. cit. p. 210.)

8 (p. 31.)—Kepler on the decreasing density and increasing volume of the planets, with their distance from the sun, which is described as the most dense of all bodies; *vide* his Epitome Astron. Copern. in vii. libros digesta, 1618—1622, p. 420. Leibnitz was also of Kepler's and Otto von Guericke's opinion, that the planets increase in volume in proportion to their distance from the sun. *Vide* his Brief an den Magdeburger Bürgermeister (Mainz, 1671), in Leibnitz, deutschen Schriften, herausg. von Guhrauer, Th. i. S. 264.

9 (p. 31.)—For a co ordination of the masses, see Encke, in Schum. Astr. Nachr. 1843, Nr. 488, S. 114.

10 (p. 32.)—If the semidiameter of the moon, according to Burckhardt's determination, be 0·2725, and its volume $\frac{1}{49 \cdot 09}$, its density would then come out 0·5596; nearly $\frac{5}{9}$. *Vide* Wilh. Beer und H. Mädler, der Mond, S. 1 und 10, wie Mädler, Astr. S. 157. The actual contents of the moon, acording to Hausen, equal $\frac{1}{54}$, according to Mädler, $\frac{1}{49 \cdot 6}$ of the material contents of the earth; its mass $\frac{1}{87 \cdot 73}$ that of the earth. In the case of the largest of all Jupiter's satellites, the third, the relations to the primary in volume are $\frac{1}{15370}$; in the mass, $\frac{1}{11300}$. On the oblateness of Uranus, *vide* Schum. Astron. Nachr. 1844, Nr. 493.

11 (p. 33.)—Beer und Mädler, l. c. § 185, S. 208, u. § 347, S. 332; also their Phys. Kenntniss der himml. Körper, S 4 und 69, Tab. i.

12 (p. 34.)—The four oldest comets whose orbits have been calculated—and this from Chinese observations—are those of the years 24 (under Gordian III.), 539 (under Justinian), 565 and 837. Whilst the last of these comets was less than 500,000 miles from the earth, and Louis the Pious, greatly alarmed, was seeking to avert the presumed danger by founding various monasteries, the Chinese astronomers were following quite scientifically the course of the star, whose tail, 60° in length, appeared first simple and then divided. The first comet which could be calculated from European observations alone, is that of 1456, (those of Halley, which were long but incorrectly regarded as the first accurate elements). *Vide* Arago, in Annuaire, 1836, p. 204. See, also, under Note 26.

13 (p. 34.)—Arago, in Annuaire, 1832, p. 209—211. In the same way as the tail of the comet of 1402 was seen in bright sunshine, so was the last great comet of 1843 seen both in its nucleus and tail between 1 and 3 o'clock on the 28th of February by J. G. Clarke, of Portland, Maine, U. S. Distances of the extremely dense nucleus from the sun's edge could be measured with great accuracy. The nucleus and tail presented themselves as a very pure white cloud; only between the tail and the nucleus there was a darker space.—American Journal of Science, vol. xlv. No. 1, p. 229. (Schum. Astr. Nachr. 1843, Nr. 491, S. 175.)

14 (p. 34.)—Phil. Trans. for 1808, pt. ii. p. 155, and for 1812, pt. i. p. 118. The diameters of the nuclei found by Herschel were 538 and 428 English miles. For the dimensions of the comets of 1798 and 1085, *vide* Arago, in Annuaire pour 1832, p. 203.

15 (p. 35.)—Arago des changemens physiques de la Comète de Halley du 15—23 Oct. 1835, in Ann. 1836, p. 218—221. The more usual direction of the tail or emanation was observed in Nero's time: "Comæ radios solis effugiunt." Seneca, Nat. Quaest. vii. 20.

16 (p. 35.)—Bessel, in Schum. Astr. Nachr. 1836, Nr. 300—302, S. 188, 192, 197, 200, 202, und 230. Also in Schum. Jahrb. 1837, S. 149—168. W. Herschel also believed that he had observed the rotation of the nucleus and tail in his observations on the beautiful comet of 1811. (Philos. Trans. 1812, pt. i. p. 140.) So, also, Dunlop in the third comet of 1825 at Paramatta.

17 (p. 35.)—Bessel, in Astr. Nachr. 1836, Nr. 302, S. 231, (Schum. Jahrb. 1837, S. 175). *Vide*, also, Lehmann über Cometenschweife in Bode's Astron. Jahrb. für 1826, S. 168.

18 (p. 35.)—Aristot. Meteor. i. 8, 11—14 and 19—21 (ed. Ideler t. i. p. 32—34). Biese, Phil. des Aristoteles, Bd. ii. S. 86. In the influence which Aristotle exerted on the whole of the middle ages, it is infinitely to be lamented that he showed himself so inimically disposed to the grand views of the structure of the universe espoused by the old Pythagoreans, and which approached the truth so closely. He declares the comets to be transient meteors belonging to our atmosphere, in the same book in which he quotes the opinion of the Pythagoreans to the effect that comets were planets with long periods of revolution. This doctrine of the Pythagoreans, which, however, from the testimony of Apollonius Myndius, appears to be much older, and to have been that of the Chaldeans, passed over to the imitative Romans. The Myndian describes the orbits of comets as passing far into the upper celestial spaces. Whence Seneca (Nat. Quaest. vii. 17): "Cometes non est species falsa, sed proprium sidus sicut solis et lunæ: altiora mundi secat et tunc demum apparet quum in imum cursum sui venit;" and (vii. 27): "Cometas æternos esse et sortis ejusdem, cujus cætera (sidera), etiamsi faciem illis non habent similem."

Pliny, too, evidently plays upon Apollonius' words, when he says, "Sunt qui et hæc sidera perpetua esse credant suoque ambitu ire, sed non nisi relicta a sole cerni."

19 (p. 35.)—Olbers, in the Astr. Nachr. 1828, S. 157, 184. Arago, de la constitution physique des comètes im Annuaire de 1832, p. 203—208. The ancients had already seen it as remarkable that we can see through comets as through a flame. The oldest testimony to stars having been seen through comets, is that of Democritus (Aristot. Meteorol. i. 6, 11). This statement leads Aristotle to the not unimportant observation, that he himself had seen the occultation of one of the stars of Gemini by Jupiter. Seneca very certainly refers to the translucency of the tail only (Nat. Quæst. vii. 18): "Non in ea parte qua sidus ipsum est spissi et solidi ignis, sed qua rarus splendor occurrit et in crines dispergitur. Per intervalla ignium, non per ipsos" (vii. 26). The last portion of the remark is superfluous, as we do positively see through a flame if it be not too thick, as remarked by Galileo (Lettera a Mons. Cesarini, 1619).

20 (p. 35.)—Bessel in den Astr. Nachr. 1836, Nr. 301, S. 204—206. Struve im Recueil des Mém. de l'Acad. de St. Pét. 1836, p. 140—143, and Astr. Nachr. 1836, Nr. 303. "For Dorpat, the star during the conjunction, was only 2″·2 from the brightest point of the comet. The star remained steadily visible, and was not sensibly weakened; whilst the nucleus of the comet appeared to be extinguished beside the brilliancy of the minute star (9th to the 10th magnitude)."

21 (p. 35.)—The first attempts of Arago to apply polarization to Comets were made on the 3d July, 1819, the evening of the sudden appearance of the great comet. I was present in the observatory, and along with M. Mathieu and the late M. Bouvard, was satisfied of the inequality of the strength of light in the polariscope when it received the light of the comet. With Capella, which was near the comet, and about the same altitude, the images were of like intensity. When Halley's comet appeared in 1835, the apparatus was so altered that it gave two images of complementary colours—green and red, according to the discovery by Arago of chromatic polarization. Annales de Chimie, t. xiii. p. 108. Annuaire, 1832, p. 216. "On doit conclure," says Arago, "de l'ensemble de ces observations que la lumière de la comète n'etait pas en totalité composée de rayons doués des propriétés de la lumiére directe, propre ou assimilée: il s'y trouvait de la lumière réfléchie spéculairement ou polarisée, c'est-à-dire venant du soleil. On ne peut assurer d'une manière absolue que les comètes brillent seulement d'un éclat d'emprunt. En effet en devenant lumineux, par eux-mêmes, les corps ne perdent pas pour cela la faculté de réfléchir des lumières étrangères."

22 (p. 36.)—Arago, in Ann. 1832, p. 217—220. Sir John Herschel, Astron. § 488.

23 (p. 36.)—Encke, in the Ast. Nachr. 1843, Nr. 489, S. 130—132.

24 (p. 36.)—Laplace, Exp. du Syst. du Monde, p. 216 and 237.

25 (p. 36.)—Littrow, Beschreibende Astr. 1835, S. 274. On the inner comet lately discovered by M. Faya, of the Parisian Observatory, whose excentricity is 0·551, perihelias distance 1·690, and apheliac distance 5·832, Schum. Astr. Nachr. 1844, Nr. 495.

26 (p. 37.)—Laugier, dans les Comptes, rendus des Séances de l'Acad. 1843, t. xvi. p. 1006.

27 (p. 37.)—Fries, Vorlesungen über die Sternkunde 1833, S. 262—267. A not very fortunate argument for the beneficent nature of comets occurs in Seneca, who speaks (Nat. Quæst. vii. 17 and 21) of the comet, "Quem nos Neronis principatu lætissimo vidimus et qui cometis detraxit infamiam."

28 (p. 38.)—One of my friends, accustomed to trigonometrical surveys, saw his chamber illuminated by a fire-ball at mid-day, and while the sun was shining, in the town of Popayan (N. lat. 2° 26′; 5520 feet above the sea level). He was standing with his back to the window, and when he turned round a great portion of the course traversed by the ball was still most brilliantly lighted. The titles for falling-stars are often extremely vulgar: the Germans speak of them as *star-snuffs:* according to the vulgar idea the lights of heaven want snuffing. In the woody country of the Orinoco, and on the solitary banks of the Cassiquiare, the shooting stars were designated by the natives *star-urine*, and the dew which lay on the beautiful leaves of the Heliconias like pearls, was *star-spittle*. The Lithuanian Mythus gives a more noble and imaginative interpretation of the nature and significance of falling stars: "The spinstress Werpeja begins to spin the thread of the destiny of the new-born child, and each of these threads ends in a star. And then when death approaches the man, the thread breaks and the star falls, quenching its light, to the earth."—Jacob Grimm, deutsche Mythologie, 1843, S. 685.

29 (p. 38.)—From the account of Denison Olmsted, Professor of Yale College, New Haven, Connecticut, U. S., vide Poggendorff's Annalen der Physik, Bd. xxx. S. 194. Kepler, who banishes falling stars from astronomy, they being, according to him, mere meteors, engendered by emanations from the earth, still expresses himself very cautiously in regard to them. "Stellæ cadentes," says he, "sunt materia viscida inflammata. Earum aliquæ inter cadendum absumuntur, aliquæ verè in terram cadunt, pondere suo tractæ. Nec est dissimile vero, quasdam conglobatas esse ex materia fœculentâ, in ipsam auram æthercam immixta: exque aetheris regione, tractu rectilineo, per aerem trajicere, ceu minutos cometas, occultâ causa motus utrorumque."—Kepler, Epit. Astron. Copernicanae, t. i. p. 80.

30 (p. 38.)—Relation historique, t. i. p. 80, 213, and 527. If we distinguish a head or nucleus and a tail in falling stars as in comets, we are made aware of the greater transparency of the atmosphere in tropical regions by the greater length and brilliancy of their trains. The phenomenon need not therefore be more frequent because it is more readily seen, and remains longer visible. The influence of the state of the atmosphere also shows itself occasionally in connection with falling stars even in our temperate zone, and at very short distances. Wartmann informs us, that on occasion of one of the November phenomena, the difference between the number of meteors seen at Geneva, and at Planchettes, two places very near to one another, was 1 : 7 (Mém. sur les Etoiles filantes, p. 17). The train of the falling star, upon which Brandes has made so many accurate and delicate observations, is by no means to be ascribed to the continuance of the impression of light upon the retina. It sometimes continues visible for a whole minute; in rare cases even longer than the light of the head of the falling star. The luminous track then usually remains motionless (Gilb. Ann. Bd. xiv. S. 251). This circumstance also proclaims the analogy between large shooting stars and fire-balls. Admiral Krusenstern, in his voyage round the world, saw the tail of a fire-ball that had long vanished, remain visible for an hour, with very little apparent motion (Reise, Th. i. S. 58). Sir Alexander Burnes gives a charming account of the transparency of the dry atmosphere of Bokhara (N. lat. 39° 43′, 1200 feet above the sea-level), so favourable formerly to the study of astronomy: "There is a constant serenity in its atmosphere, and an admirable clearness in the sky. At night, the stars have uncommon lustre, and the milky-way shines gloriously in the firmament. There is also a never-ceasing display of the most brilliant meteors, which dart like rockets in the sky; ten or twelve of them are sometimes seen in an hour, assuming every colour; fiery, red, blue, pale and faint. It is a noble country for astronomical science, and great must have been the advantage enjoyed by the famed observatory of Samarkand." Burnes' Travels into Bokhara, vol. ii. (1834) p. 158. We must not charge it upon the solitary traveller that he speaks of ten or twelve falling stars in an hour as *many;* it has but lately been discovered, from careful observation, that eight meteors per hour are the mean number that fall within the circle of vision of an individual (vide Quetelet, Correspond. Mathem. Nov. 1837, p. 447.) Olbers, the acute observer, reduces this number to from five to six. (Schum. Jahrb. 1838, S. 325.)

31 (p. 38.)—On meteoric dust, vide Arago, in Annuaire pour 1832, p. 254. I have very lately in another place (Asie centrale, t. i. p. 408) endeavoured to show how the Scythian myth of the sacred gold that fell glowing from heaven, and remained the property of the golden hordes of Paralatæ (Herod. iv. 5—7), may have been the obscure recollection of the fall of an Aërolite. The ancients also had their fables (Dio Cassius, lxxv. 1259) strangely enough of silver that fell from heaven, and with which attempts were made to plate the copper money under the Emperor Severus. Metallic iron was nevertheless recognized in meteoric stones by Pliny (ii. 56). The frequently recurring phrase lapidibus pluit, must not be always viewed as referring to aërolites. In Livy (xxv. 7) it is used in connection with the rejected masses—pumice, rapilli, of the not quite extinct volcanic Mons Albanus, Monte Cavo: vide Heyne, Opusc. Acad. t. iii. p. 261, and my own Relation historique, t. i. p. 394. To another circle of ideas belongs the conflict of Hercules against the Ligyes, on the way from Caucasus to the Hesperides. It is an attempt mythically to explain the origin of the rounded quartz blocks in the Ligyan stone field at the mouth of the Rhone, which Aristotle ascribes to an earthquake, Posidonius to the action of the waves of an inland sea. But in the fragments of the Prometheus Unbound of Æschylus, all goes forward as in a fall of Aërolites: Jupiter draws together a cloud, and "covers the land with a shower of rounded stones for rain." Posidonius allows himself to jest at the geological myth of the fragments and blocks. The Ligyan stone field for the rest is faithfully described by the ancients. The country is now called La Crau.—Guérin, Mesures barométriques dans les Alpes et Météorologie d'Avignon, 1829, ch. xii. p. 115.

32 (p. 39.)—The specific gravity of Aërolites varies between 1·9 (Alais) and 4·3 (Tabor). The more common density is about 3, water being assumed as 1. What is stated in the text in regard to the actual diameter of fire-balls, is based on the few satisfactory measurements we possess.

These for the fire-ball of Weston (Connecticut, Dec. 14, 1807) assign 500, for the one observed by Le Roi (10th July, 1771) about 1,000, and for that seen by Sir Charles Blagden (18th Jan. 1783) as many as 2,600 feet in diameter. Brandes (Unterhaltung. Bd. i. S. 42) assigns from 80 to 120 feet to shooting stars; their luminous tails being from 3 to 4 miles in length. But there are not wanting optical grounds for believing that the apparent diameter of fire-balls and shooting stars is greatly over-estimated. The volume of fire-balls is certainly not to be compared with the volume of Ceres (even supposing this planet to be no more than 70 English miles in diameter, as has been estimated): see the accurate and admirable treatise, On the Connexion of the Physical Sciences, 1835, p. 411. I here add in illustration of what has been said at page 39, of the great Aërolite of the bed of the river Narni, the passage from the Chronicon Benedicti, monachi Sancti Andreæ in Monte Soracte, which has been referred to by Pertz, a document of the 10th century, and that is preserved in the Biblioteca Chigi at Rome. The barbarous writing of the time is preserved unaltered: "Anno—921—temporibus domini Johannis Decimi pape, in anno pontificatus illius 7, visa sunt signa. Nam juxta urbem Roman lapides plurimi de cœlo cadere visi sunt. In civitate quæ vocatur Narnia tam diri ac tetri, ut nihil aliud credatur, quam de infernalibus locis deducti essent. Nam ita ex illis lapidibus unus omnium maximus est, ut decidens in flumen Narnus, ad mensuram unius cubiti super aquas fluminis usque hodie videretur. Nam et ignitæ faculæ de cœlo plurimæ omnibus in hac civitate Romani populi visæ sunt, ita ut pene terra contingeret. Aliæ cadentes," &c.—(Pertz, Monum. Germ. hist. Scriptores, t. iii. p. 715.) On the Aërolites of Aegos Potamos, whose fall the Pariscan Chronicle states to have happened in the 78·1 Olympiad (Böckh, Corp. Inscr. Græc. t. ii. p. 302, 320, and 340). Aristot. Meteor. i. 7 (Ideler, Comm. t. i. p. 404—407); Stob. Ecl. Phys. i. 25 p. 508, Heeren; Plut. Lys. c. 12; Diog. Laert. ii. 10. (And also under the Notes 39, 57, 58, and 59.) According to a Mongolian tradition, a black rocky mass, 40 feet in height, is said to have fallen from heaven in a plain near the sources of the Yellow River in Western China.—Abel-Rémusat, in Lamétherie, Journ. de Phys. 1819, Mai, p. 264.

33 (p. 39.)—Biot, Traité d'Astronomie physique, (3me éd.) 1841 t. i. p. 149, 177, 238, and 312. My immortal friend Poisson attempted the solution of the difficulty of spontaneous combustion occurring in an aërolite in a region where the density of the atmosphere is almost *nil*, in a very peculiar manner. He says, "A une distance de la terre où la densité de l'atmosphère est tout-à-fait insensible, il serait difficile d'attribuer, comme on le fait, l'incandescence des aërolithes à un frottement contre les molécules de l'air. Ne pourrait-on pas supposer que le fluide électrique à l'état neutre forme une sorte d'atmosphère, qui s'étend beaucoup au-delà de la masse d'air; qui est soumise à l'attraction de la terre, quoique physiquement impondérable; et qui suit, en conséquence, notre globe dans ses mouvements? Dans cette hypothèse, les corps dont il s'agit, en entrant dans cette atmosphère impondérable, décomposeraient le fluide neutre, par leur action inégale sur les deux électricités, et ce serait en s'électrisant qu'ils s'échaufferaient et deviendraient incandescents."—(Poisson, Rech. sur la Probabilité des Jugements, 1837, p. 6.)

34 (p. 39.)—Philos. Transact. vol. xxix. p. 161—163.

35 (p. 39.)—The first edition of Chladni's important treatise, On the origin of the masses of Iron discovered by Pallas and others, appeared two months before the shower of stones fell at Siena, and two years before Lichtenberg's proposition, "that stones come into our atmosphere from universal space," in the Göttingen Pocket-Book.—*Vide* Olbers' Letter to Benzenberg of Nov. 18th, 1837, and the latter's work on Falling Stars, p. 186.

36 (p. 39.)—Encke, in Poggend. Annalen, Bd. xxxiii. (1834) S. 213. Arago, in Ann. pour 1836, p. 291. Two letters of mine to Benzenberg, 19th May and Oct. 22d, 1837, on the supposed precession of the nodes in the orbit of the periodic streams of shooting stars (Benzenb. Sternsch. S. 207 and 209). Olbers, too, subsequently came into this opinion of the gradual retardation of the November phenomenon (Astron. Nachr. 1838, No. 372, S. 180). If I venture to connect two of the falls of shooting stars indicated by the Arabian writers with those discovered by Boguslawski, as having occurred in the 14th century, I obtain the following more or less accordant elements of the nodal movement:

In October 902, in the night of the death of King Ibrahim-ben-Ahmed, there was a great fall of stars, "like a fiery rain." This year was on this account called the year of the stars.—(Conde. Hist. de la domin. de los Arabes, p. 346.)

Oct. 19th, 1202.—Stars fell the whole night through; "they fell like locusts."—(Comptes-rendus, 1837, t. i. p. 294, and Fraehn, in Bull. de l'Acad. de St. Pétersbourg, t. iii. p. 308.)

Oct. 21, old style, 1366, die sequente post festum xi. millia Virginum ab hora matutina usque ad horam priman visæ sunt quasi stellæ de cælo cadere continuo, et in tanta multitudine, quod nemo narrare sufficit. This remarkable notice, of which more use will be made further on in the text, was discovered by M. von Boguslawski, Jun. in Benesse (de Horowic) de Weitmil or Weithmül's Chronicon Ecclesiæ Pragensis, p. 389. This chronicle is republished in the 2d part of the Scriptores rerum Bohemicarum von Pelzel und Dobrowsky, 1784 (Schum. Astr. Nachr. Dec. 1839).

Night of Nov. 9—10, 1787, many shooting stars observed by Hemmer in South Germany, particularly in Manheim.—(Kæmtz, Meteorol. iii. 237.)

Midnight, Nov. 12, 1799.—The extraordinary fall of stars, which Bonpland and I have described, and which was observed over the greater part of the globe.—(*Vide* Relat. Hist. t. i. p. 519—527.)

Nov. 12—13, 1822, shooting stars mingled with fire-balls, in great numbers, seen by Kloden, in Potsdam (Gilbert's Annalen, vol. 72, p. 291).

Nov. 13th, 1831, at four A.M., a great fall of stars seen by Captain Bérard on the coast of Spain, near Carthagena del Levante (Annuaire, 1836, p. 297).

Night of Nov. 12—13, 1833.—The remarkable North American phenomenon so admirably described by Denison Olmsted.

Night of Nov. 13—14, 1834.—The same phenomenon, but not so brilliant, observed in North America (Poggendorff, Ann. Bd. xxxiv. S. 129).

Nov. 13th, 1835, a stack was set on fire by a single fire-ball near Belley, Dép. de l'Ain (Annuaire, 1836).

In the year 1838, the stream of shooting stars showed itself most decidedly in the night from the 13th to the 14th November (Astronom. Nachrichten, 1838, No. 372).

37 (p. 39.)—It is not unknown to me that of the 62 shooting stars which were simultaneously observed in Silesia, at the instance of Prof. Brandes, some appeared to have had an elevation of $45\frac{7}{10}$, of 60 and even of 100 miles, vide Brandes, Unterhaltungen für Freunde der Astronomie und Physik, Heft i. S. 48. But Olbers, by reason of the smallness of the parallax, regards all determinations above 30 miles in height as doubtful.

38 (p. 39.)—The velocity of the planets in their orbits varies greatly; for Mercury it is 6·6, for Venus 4·8, and for the Earth 4·1 German miles per second.

39 (p. 40.)—Chladni discovered that an Italian natural philosopher, Paolo Maria Terzago, 1660, on the occasion of a fall of aërolites at Milan, in which a monk was killed, was the first who spoke of the possibility of aërolites being moon-stones: "Labant philosophorum mentes," says he, in his work, Musaeum Septalianum, Manfredi Septalae, Patricii Mediolanensis, industrioso labore constructum, Tortona, 1664, p. 44, "sub horum lapidum ponderibus; ni dicere velimus, lunam terram alteram, sive mundum esse, ex cujus montibus divisa frustra in inferiorem nostrum hunc orbem delabantur." Without having any knowledge of this conjecture, Olbers was led in the year 1795, after the great fall of stones that took place at Siena (16th June, 1794), to the inquiry of—how great the original projectile force must be to send masses from the moon to the earth? And a problem of this kind found occupation for such minds as Laplace, Biot, Brandes, and Poisson, for some ten or twelve years. The opinion once very generally entertained, but now abandoned, of the existence of active volcanoes in the moon without atmosphere and without water, favoured in the public mind the confusion of a mathematical possibility with a physical probability—an explanation of a physical fact preferable to other explanations. Olbers, Brandes, and Chladni, believed that they had discovered a refutation of the lunar origin of meteoric stones in the relative velocity of from 4 to 8 miles, with which fire-balls and shooting stars enter our atmosphere. To reach the earth, according to Olbers, without bringing the resistance of the air into the reckoning, an original velocity of 7780 feet per second were requisite; according to Laplace, the necessary velocity is 7377 feet; according to Biot, 7771 feet; and according to Poisson, 7123 feet. Laplace calls this primary velocity only from 5 to 6 times greater than that which a cannon-ball possesses as it leaves the gun; but Olbers has shown, "that with such a primary velocity of from 7500 to 8000 feet per second, meteoric stones would only reach the confines of our atmosphere with a velocity of 35,000 feet" (1·53 German geographical mile). But as the measured velocity of meteoric stones is in the mean 5 geographical miles, or more than 114,000 feet per second, they must originally have had a centrifugal force in the moon of 110,000 feet per second, fourteen times greater therefore than Laplace assumes. (Olbers in Schum. Jahrb. 1837, S. 52—58 und in Gehler's Nues physik. Wöterbuche, Bd. vi. Abth. 3, S. 2129—2136.) The absence of any resistance from the air would, however, give the projectile force of the lunar volcanoes an advantage beyond the projectile force of our volcanoes on the earth, supposing always that volcanic action is conceived as possible in the body of the moon; but upon the amount or measure of the power of these lunar volcanoes, we are still without any information. It is very

probable indeed that this amount has been greatly over-estimated. A very accurate observer and measurer of the power of Ætna, Dr. Peters, has found the greatest velocity of stones cast out from its crater to be but 1250 feet per second. Observations on the Peak of Teneriffe in 1798 gave 3000 feet. If Laplace, at the end of his work (Expos. du Syst. du Monde, 1824, p. 399), says very considerately, "que selon toutes les vraisemblances elles viennent des profondeurs de l'espace céleste;" we still find him in another place, probably unacquainted with the amazing, wholly planetary velocity of meteoric stones (Chap. vi. p. 233), reverting to the selenitic hypothesis with a kind of preference, but always premising that the stones cast out from the moon "deviennent des satellites de la terre, décrivant autour d'elle une orbite plus ou moins allongée, de sorte qu'ils n'atteignent l'atmosphère de la terre qu'après plusieurs et même un très-grand nombre de révolutions." In the same way as an Italian of Tortona conceived the fancy that aërolites came from the moon, Greek philosophers had a notion that they came from the sun. Diogenes Laertius (ii. 9) adverts to such an opinion when treating of the origin of the mass which fell at Ægos Potamoi (*vide* Note 32 above); and Pliny, who registers every thing, mentions the idea, and ridicules it the more willingly, because with earlier writers (Diog. Laert. ii. 3 and 5) he excuses Anaxagorus for having predicated a fall of stones from the sun: "Celebrant Græci Anaxagoram Clazomenium Olympiadis septuagesimæ octavæ secundo anno prædixisse cælestium litterarum scientia, quibus diebus saxum casurum esse e sole, idque factum interdiu in Thraciæ parte ad Aegos flumen. Quod si quis prædictum credat, simul fateatur necesse est, majoris miraculi divinitatem Anaxagoræ fuisse, solvique rerum naturæ intellectum, et confundi omnia, si aut ipse Sol lapis esse aut unquam lapidem in eo fuisse credatur; decidere tamen crebro non erit dubium." Anaxagoras is also said to have foretold the fall of the stone of smaller dimensions, which was preserved in the Gymnasium of Abydos. Falls of aërolites during sunshine, and when the disc of the moon was not visible, probably gave rise to the idea of the sun as their source. It was also one of the physical dogmas of Anaxagoras, and which, as in the case of the geologists of these our own times, exposed him to the persecution of the theologians, that the sun was "a molten fiery mass (μύδρος διάπυρος)." In the Phaeton of Euripides the sun, after the same views of the Clazomenæan, is called a "golden clod," *i. e.*, a fiery-coloured luminous mass of matter; from which, however, we are not to conclude that aërolites are "golden sun-stones." *Vide* Note 31, above; as also Valckenaer, Diatribe in Eurip. perd. dram. Reliquias, 1767, p. 30. Diog. Laert. ii. 40. We seem, then, to find four hypotheses among the Greek natural philosophers: a telluric origin of falling stars from ascending vapours; masses of stone raised by tempests, in Aristotle (Meteorol. lib. i. cap. IV. 2—13, and cap. VII. 9); an origin from the sun; an origin from celestial space, and as heavenly bodies that had long remained invisible. On the last view of Diogenes of Apollonia, which entirely agrees with our own, see the text (p. 43), and Note 58. It is remarkable that in Syria, as a learned Orientalist, my teacher of Persic, M. Andrea de Nericat, assured me, according to an old popular belief they are still solicitous about falls of stones from the sky in very clear moonlight nights. The ancients, on the contrary, were on the watch for the same event during eclipses of the moon (Plin. xxxvii. 10, p. 164; Solinus, c. 37; Salm. Exerc. p. 531); and the passages collected by Ukert, in his Geography of the Greeks and Romans (Th. ii. 1, S. 131, Note 14). On the improbability that aërolites arise from gases holding metallic matters dissolved, which, according to Fusinieri, exist in the upper strata of our atmosphere, and which previously dispersed in infinite space had suddenly coalesced, as well as on the penetration and miscibility of gases, see my Relation histor. t. i. p. 525.

40 (p. 40.)—Bessel in Schum. Astr. Nachr. 1839, Nr. 380 und 381, S. 222 und 346. At the close of the work there is a comparison of the sun's place in longitude with the epochs of the November phenomenon, since the first observations in Cumana, 1799.

41 (p. 40.)—Dr. Thomas Forster (the Pocket Encyclop. of Natural Phenomena, 1827, p. 17) informs us that in Christ Church College, Cambridge, there is preserved a MS. entitled "Ephemerides rerum naturalium," which is ascribed to a monk of the last century. In this MS. natural phenomena are noted as having occurred on every day of the year: the flowering of plants; the arrival of birds of passage, &c. The 10th of August is characterized by the word *meteorodes*. This indication and the tradition of the fiery tears of St. Lawrence, led Dr. Forster to pay particular attention to the August phenomenon.—Quetelet, Corresp. mathém., série III. tom. i. 1837, p. 433.

42 (p. 40.)—Humboldt, Rel. hist. t. i., p. 519—527. Ellicot, in the Transactions of the American Society, 1804, vol. vi. p. 29. Arago says of the November phenomenon, "Ainsi se confirme de plus en plus à nous l'existence d'une zone composée de millions de petits corps dont les orbites rencontrent le plan de l'écliptique vers le point que la terre va occuper tous les ans, du 11 au 13 Novembre. C'est un nouveau monde planétaire qui commence à se révéler à nous."—Annuaire, 1836, p. 296.

43 (p. 40.)—*Vide* Muschenbroek, Introd. ad Phil. Nat. 1762, t. ii. p. 1061. Howard, Climate of London, vol. ii. p. 23, Observations of the Year 1806, therefore, seven years after the earliest observations of Prof. Brandes (Benzenberg über Sternschnuppen, S. 240—244); August-Observations of Thomas Forster, *vide* Quetelet, loc. cit. 438—453; of Adolph Erman, Boguslawski und Kreil in Schum. Jahrb. 1838, S. 317—330. On the point in Perseus whence the stream proceeded on the 10th of August, 1839, see the accurate measurements of Bessel and Erman (Schum. Astr. Nachr. Nos. 385 and 428); on the 10th of August, 1837, however, the orbit did not appear to be retrograde; see Arago, in Comptes rendus, 1837, tom. ii. p. 183.

44 (p. 40.)—On the 25th of April, 1095, "innumerable eyes in France saw the stars fall as thick as hail from heaven" (ut grando, nisi lucerent, pro densitate putaretur; Baldr. p. 88); and this incident was regarded by the Council of Clermont as premonitory of a great movement in Christendom. (*Vide* Wilken, Gesch. der Kreuzzüge, Bd. i. S. 75.) On the 22d of April, 1800, a great fall of stars was observed in Virginia and Massachusetts; it was like a display of rockets that lasted for two hours. Arago first directed attention to this "trainée d'astéroides" as a recurring phenomenon (Annuaire, 1836, p. 297). The falls of aërolites in the beginning of December was also remarkable; their periodical recurrence is vouched for by the old observations of Brandes in the night from the 6th to the 7th of December, 1798, when he counted nearly 2000 falling stars, and perhaps by the extraordinary fall of aërolites of the 11th of December, 1836, at the village of Macao on the river Assu, Brazil (Brandes, Unterhalt. für Freunde der Physik, 1825, Heft i. S. 65, and Comptes rendus, tom. v. p. 211). Capocci, from 1809 to 1836, has found records of twelve actual falls of aërolites between the 27th and the 29th of November; and several others of the 13th of November, 10th of August, and 16th of July (Comptes rendus, tom. xi. p. 357). It is curious that in the part of the earth's orbit which corresponds to the months of January and February, and perhaps March, no periodical fall of shooting stars has yet been noticed; nevertheless I myself observed a remarkable number of shooting stars on the 15th of March, 1803, in the South Pacific Ocean; and a shower of the same was seen in the city of Quito shortly before the tremendous earthquake of Riobamba (4th Feb. 1797). The following epochs deserve the particular attention of observers:

22—25 April,
17 July (17—26 July?) (Quet. Corr. 1837, p. 435),
10 August,
12—14 November,
27—29 November,
6—12 December.

The frequency of these streams, however great the difference between isolated comets and rings filled with asteroids, ought not to excite astonishment when we think of the depths of universal space filled with myriads of comets.

45 (p. 41.)—Ferd. v. Wrangel, Reise längs der Nordküste von Sibirien in den Jahren 1820—1824, Th. ii. S. 259. On the return of the thicker shower of the November asteroids every thirty-four years, *vide* Olbers in Jahrb. 1837, S. 280. I was informed in Cumana, that shortly before the dreadful earthquake of 1766, just thirty-three years, therefore, before the great exhibition of shooting stars of November 11—12, 1799, the same display had been seen. But the earthquake of 1766 did not occur in November, but on the 21st of October. It were worth the while of travellers in Quito to investigate the particular day on which the volcano Cayambe appeared for an hour as if enveloped in a shower of falling stars, so that religious processions were set in motion to appease the heavens! (*vide* my Relat. histor. t. i. chap. iv. p. 307; chap. x. p. 520 and 527.)

46 (p. 41.)—From a letter to me of January 24, 1838. The extraordinary display of shooting stars of 1799 was observed almost exclusively in America, from New Herrnhut, in Greenland, to the Equator. The phenomena of 1831 and 1832 were only seen in Europe; those of 1833 and 1834 only in the United States of North America.

47 (p. 41.)—Lettre de Mr. Edouard Biot à Mr. Quetelet sur les anciennes apparitions d'etoiles filantes en Chine, in Bull. de l'Acad. de Bruxelles, 1843, t. x. No. 7, p. 8. On the notice from the Chronicon Ecclesiae Pragensis, *vide* Boguslawski, Jun., in Poggend. Annalen, Bd. xlviii. S. 612. To Note 12 should be added, that the orbits of four comets (568, 574, 1337, and 1385) have been reckoned exclusively from Chinese data. *Vide* John Russell Hind, in Schum. Astr. Nachr. 1844, Nr. 498.

48 (p. 41.)—"Il paraît qu'un nombre, qui semble inépuisable, de corps trop petits pour être observés, se meuvent dans le ciel, soit autour du soeil, soit autour des planètes, soit peut-être même autour des satellites. On suppose que

quand ces corps sont recontrés par notre atmosphère, la différence entre leur vitesse et celle de notre planète est assez grande pour que le frottement qu'ils éprouvent contre l'air, les échauffe au point de les rendre incandescents, et quelquefois de les faire éclater. Si le groupe des étoiles filantes forme un anneau continu autour du soleil, sa vitesse de circulation pourra être très-différente de celle de la terre; et ses déplacements dans le ciel, par suite des actions planétaires, pourrons encore rendre possible ou impossible, à différentes époques, le phénomène de la rencontre dans le plan de l'écliptique."—Poisson, Recherches sur la probabilité des jugements, p. 306, 307.

49 (p. 41.)—Humboldt, Essai politique sur la Nouv. Espagne, 2e édit.) t. iii. p. 310.

50 (p. 41.)—Pliny shows himself to have been attentive to the colour of the crust: *colore adusto*. The words, *lateribus pluisse*, also refer to the burned external appearance of Aërolites (ii. 56 and 58).

51 (p. 42.)—Humboldt, Rel. hist. t. ii. chap. xx. p. 299—302.

52 (p. 42.)—Gustav Rose, Reise nach dem Ural, Bd. ii S. 202.

53 (p. 42.)—*Vide* Poggend. Ann. 1825, Bd. iv. S. 173—192. Rammelsberg, Erstes Suppl. zum chem. Handwörterbuche der Mineralogie, 1843, s. 102. "It is," says the acute Olbers, "a remarkable though unnoticed fact, that *fossil* meteoric stones have been found, like fossil shells, in secondary and tertiary formations. Shall we thence feel at liberty to conclude, that before the last and present arrangement of the surface of our earth, meteoric stones had fallen upon it? Schreibers calculates that at this time there are about 700 falls of meteoric stones in each year." (Olbers, in Schum. Jahrb. 1838, s. 329.) Problematic nickeliferous masses of native iron have been lately found in North Asia, (Goldseiferwerk von Petropawlowsk, 20 miles south-east of Kusnezk,) at a distance of 31 feet deep, and in the Western Carpathians (Magura, near Szlanicz). Both of these masses are extremely like Aërolites—Vide Erman, Archiv für wissenschaftliche Kunde von Russland, Bd. i. S. 315, and Haidinger's Bericht über die Szlaniczer Schürfe in Ungarn.

54 (p. 42.)—Berzelius, Jahresber. Bd. xv. S. 217 and 231; Rammelsberg, Handwörterb. Abth. ii. S. 25—28.

55 (p. 42.)—"Sir Isaac said, he took all the planets to be composed of the same matter with this earth, viz., earth, water, and stones, but variously concocted."—Turner, Collections for the Hist. of Grantham, cont. authentic Memoirs of Sir Isaac Newton, p. 172.

56 (p. 43.)—Adolph Erman, in Poggend. Ann. 1839, Bd. xlviii. S. 582—601. Biot at a previous period, (Comptes rendus, 1836, t. ii. p. 670) raised doubts of the probability of the November phenomenon appearing again in the beginning of May. Mädler has taken the mean temperature of the three days of May that have been decried for the last 86 years, according to Berlin observations, (Verhandl. des Vereins zur Beförd. des Gartenbaues 1834, s. 377,) and finds the temperature of the 11th, 12th and 13th of May to recede 1°·22 C. precisely at a season when the advance in the temperature is the most rapid. It would be very desirable that this phenomena of a fall of temperature, which there has been an obvious disposition to ascribe to the fusion of masses of ice in the north-east of Europe, were investigated at very different points of the continent of America, or in the southern hemisphere. *Vide* Bull. de l'Acad. Imp. de St.-Pétersbourg 1843, t. i. No. 4.

57 (p. 43.)—Plut. Vitae par. in Lysandro, cap. 22. The account of Damachos (Daïmachos), according to which a fiery cloud was seen for 70 days in succession, and which emitted sparks like falling stars, and finally sinking down, deposited the stone of Ægos Potamos, "which was but an insignificant portion of the cloud," is extremely improbable, because the course and direction of the fire-ball must then have continued for many days like that of the earth. The fire-ball of the 19th of July, 1686, described by Halley, performed its visible course in minutes (Philos. Trans. vol. xxix. p. 163). Whether Daimachos, the writer, περὶ εὐσεβείας, is the same with the Daimachos of Platæa, who was sent by Seleucus to India to the son of Androkottos, and whom Strabo (p. 70, Casaub.) characterizes as a "vender of lies," remains uncertain. From another passage of Plutarch (Compar. Solonis c. Pop. cap. 4), we should almost be disposed to believe that he was. *Vide* Note 32.

58 (p. 43.)—Stob. ed. Heeren, i. 25, p. 508, Plut. de plac. Philos. ii. 13.

59 (p. 43.)—The remarkable passage of Plutarch (De plac. Philos. ii. 13) is the following: "Anaxagoras teaches that the surrounding ether is fiery in respect of its substance; and through the force of its circumvolution tears away masses of rock from the earth, sets them on fire, and turns them into stars." The Clazomenæan employs the same kind of force (centrifugal force) for bringing the Nemæan lion from the moon to the Peloponnesus. (Aelian xii. 7; Plut. de facie in orbe lunae, c. 24; Schol. ex. Cod. Paris. in Apoll. Argon. lib. i. p. 498, ed. Schaef. t. ii. p. 40; Meineke, Annal. Alex. 1843, p. 85.) We have therefore in this instance moon animals instead of moon stones. According to Böckh's acute remark, the old myth of the Nemæan lion of the moon has an astronomical origin, and is connected symbolically in chronology with the intercalary cycles of the lunar year, the worship of the moon at Nemæa, and the games there celebrated.

60 (p. 43.)—The following important passage, one of the many inspirations of Kepler on the radiation of heat by the fixed stars, slow combustion and the vital processes, occurs in the Paralipom. in Vitell. Astron. pars optica, 1604, Propos. xxxii. p. 25: "Lucis proprium est calor, sydera omnia calefaciunt. De syderum luce claritatis ratio testatur, calorem universorum in minori esse proportione ad calorem unius solis, quam ut ab homine, cujus est certa caloris mensura, uterque simul percipi et judicari possit. De cincindularum lucula tenuissima negare non potes, quin cum calore sit. Vivunt enim et moventur, hoc autem non sine calefactione perficitur. Sed neque putrescentium lignorum lux suo calore destituitur; nam ipsa puetredo quidam lentus ignis est. Inest et stirpibus suus calor." *Vide* Kepler, Epit. Astron. Copernicanae, 1618, t. i. lib. i. p. 35.

61 (p. 44.)—"There is another thing, which I recommend to the observation of mathematical men: which is, that in February, and for a little before, and a little after that month (as I have observed several years together), about six in the evening, when the Twilight had almost deserted the horizon, you shall see a plainly discernible way of the Twilight striking up towards the Pleiades, and seeming almost to touch them. It is so observed any clear night, but it is best *illic nocte*. There is no such way to be observed at any other time of the year (that I can perceive), nor any other way at that time to be perceived darting up elsewhere. And I believe it hath been, and will be constantly visible at that time of the year. But what the cause of it in nature should be, I cannot yet imagine, but leave it to further inquiry."—Childrey, Britannia Baconica, 1661, p. 183. This is the first and simple account of the phenomenon, Cassini, Découverte de la lumière céleste qui paroît dans le zodiaque, in the Mém. de l'Acad. t. viii. 1730, p. 276. Mairan, Traité phys. de l'Aurore boreale, 1754, p. 16. In the curious book of Childrey, quoted above, there are very rational views of the epochs of the occurrence of the maxima and minima in the distribution of the annual heat, as well as on the course of the daily temperature; and on the retardation of the extreme effects in meteorological processes. Unfortunately the Baconian philosophising Chaplain to Lord Henry Somerset, like Bernardin de St. Pierre, teaches that the earth is pointed at the poles. Originally he says it was globular, but the ceaseless accumulation of ice at the poles altered the figure of the body of the earth; and as ice is formed from water, so does the quantity of water go on decreasing everywhere else.

62 (p. 44.)—Dominic Cassini (Mém. de l'Acad. t. viii. 1730, p. 188), and Mairan (Aurore bor. p. 16) even maintained that the phenomenon seen in Persia, in 1668, was the zodiacal light. Delambre (Hist. de l'Astronomie moderne, t. ii. p. 742) ascribes the discovery of this light definitively to the traveller Chardin; but both in his Couronnement de Soliman and in many passages of his travels (ed. de Langlès, t. iv. p. 326; t. x. p. 97) Chardin refers the Persian niazouk (nyzek), ou petite lance, only to "la grande et fameuse comète qui parut presque par toute la terre en 1668, et dont la tête étoit cachée dans l'occident, de sorte qu'on ne pouvoit en rien apercevoir sur l'horizon d'Ispahan." (Atlas du Voyage de Chardin, tab. iv.), from observations at Schiraz. But the head or nucleus of this comet was seen in Brazil and in India (Pingré, Cométographie, t. ii. p. 22). On the conjectured identity of the last great comet of 1843 with that which Cassini mistook for the zodiacal light, see Schum. Astron. Nachr. 1843, No. 476 and 480. In the Persian, the words nizehi âteschîn (fiery spears or lances) are also used for the beams of the rising or setting sun; nayâzik, in Freytag's Arabic Lexicon, is interpreted stellæ cadentes. The comparison of comets with lances and spears was, however, extremely common in all languages in the middle ages. Even the great comet which was seen from April to June, 1500, is always spoken of by Italian writers of the time under the title of "il Signor Astone (*vide* my Exam. crit. de l'Hist. de la Géographie, t. v. p. 80). The statement variously made that Descartes (Cassini, p. 230, Mairan, p. 16) and Kepler (Delambre, t. i. p. 601) were acquainted with the zodiacal light appears to me altogether untenable. Descartes (Principes, iii. art. 136, 137) speaks in a very obscure manner of the production of comets' tails: "Par des rayons obliques qui, tombant sur diverses parties des orbes planétaires, viennent des parties latérales à notre oeil par une réfraction extraordinaire;" also how comets' tails can be seen morning and evening, "comme une longue poutre," if the sun be placed between the comet and the earth. This passage refers to the zodiacal light as little as the one in Kepler, in which he speaks of the existence of an atmosphere about the sun (limbus circa solem, coma lucida), which in total eclipses of the sun

hinders "that it become entirely night." Still more uncertain, or rather more erroneous, is the assertion that the "trabes quas δοκοὺς vocant" (Plin. ii. 22 and 27) was applied to the ascending tongue-shaped zodiacal light as Cassini (p. 231) and Mairan (p. 15) will have it. Everywhere with the ancients the word "trabes" is taken as synonymous with fire-balls and fiery-meteors generally, and even occasionally with streaming comets. On δοκὸς, δοκίας, δοκίτης, *vide* Schäfer, Schol. Par. ad Apoll. Rhod. 1813, t. ii. p. 206; Pseudo-Aristot. de Mundo, 2, 9; Comment. Alex., Joh. Philop. et Olymp. in Aristot. Meteor. lib. i. cap. vii. 3, p. 195, Ideler; Seneca, Nat. Quaest. i. 1.

63 (p. 44.)—Humboldt, Monumens des peuples indigènes de l'Amérique, t. ii. p. 301. The rare MS. which belonged to the Archbishop of Rheims, Le Tellier, contains very various extracts from an Aztekian ritual book, from an astrological calendar, and from historical annals from 1197 to 1549. The last include natural phenomena—dates of earthquakes, comets—as of 1490, 1592, and, for Mexican chronology, important eclipses of the sun. In the MS. Historia de Tlascala of Camargo, the light which appeared in the east and rose almost to the zenith is spoken of as "sparkling, and as if thickly sown over with stars." The account of the 40 days' phenomenon (Prescott, Hist. of the Conquest of Mexico, vol. i. p. 284) will by no means apply to an eruption of Popocatepetl which rises close by in the south-east. Later commentators have confounded this phenomenon, which Montezuma regarded as one foreboding him misfortune, to the "estrella que humeava" (properly which sparkled; Mexican choloa, to leap, to sparkle). On the connection of this vapour with the star Citlal Choloha (Venus) and the starry mountain (Citaltepetl, the volcano of Orizaba), see my Monumens, t. ii. p. 303.

64 (p. 44.)—Laplace, Expos. du Syst. du Monde, p. 270; Mécanique céleste, t. ii. p. 169 and 171. Schubert, Astr. Bd. iii. § 206.

65 (p. 44.) — Arago, in Annuaire, 1842, p. 408. See Sir John Herschel's Considerations on the Volume and Light of the Planetary Nebulæ, in Mary Somerville's Connexion of the Physical Sciences, 1835, p. 108. The opinion that the sun is a nebulous star, whose atmosphere has the appearance of the zodiacal light, was not advanced by Cassini, but by Mairan, 1731.—(*Vide* Traité de l'Aurore bor. p. 47 and 263. Arago, in Annuaire, 1842, p. 412.) It was a revival of the views of Kepler.

66 (p. 45.) — Cassini, as well as Laplace, Schubert and Poisson after him, adopted the hypothesis of a detached ring as an explanation of the figure of the zodiacal light. He says expressly: "Si les orbites de Mercure et de Vénus étoient visibles (matériellement dans toute l'étendue de leur surface), nous les verrions habituellement de la même figure et dans la même disposition à l'égard du Soleil et aux mêmes tems de l'année que la lumière zodiacale."—(Mém. de l'Acad. t. viii. 1730, p. 218, and Biot, in the Comptes rendus, 1836, t. iii. p. 666.) Cassini believed that the vaporiform ring of the zodiacal light was composed of an innumerable host of small planetary bodies, which revolve about the sun. He was himself not indisposed to believe that the fall of fire-balls might be connected with the passage of the earth through the zodiacal nebulous ring. Olmsted, and especially Biot (l. c. p. 673), have endeavoured to demonstrate this connection with the November phenomenon; any such connection, however, is doubted by Olbers (Schum. Jahrb. 1837, p. 281). On the question whether the plane of the zodiacal light perfectly agrees with the plane of the sun's equator, *vide* Houzeau, in Schum. Astr. Nachr. 1843, No. 492, p. 190.

67 (p. 45.)—Sir John Herschel, Astron. § 487.

68 (p. 45.) — Arago im Annuaire, 1832, p. 246. — Many physical facts appear to indicate that, with a mechanical division of matter into its minutest particles, when the mass becomes extremely small in comparison with the surface, the electrical tension may arise to the point of producing luminous and calorific rays. Experiments with a large concave mirror have not yet given any decisive indications of the existence of radiating heat in the zodiacal light.—(Lettre de Mr. Matthiessen à Mr. Arago, in the Comptes rendus, t. xvi. 1843, Avril, p. 687.)

69 (p. 45.)—"What you tell me of the variations in the light of the zodiacal light, and their causes within the tropics, has interested me by so much the more, as I have for a long time every spring given particular attention to the phenomenon in our northern latitudes. I have myself always believed that the zodiacal light rotates, but I concluded that it extended with constantly increasing intensity of lustre quite to the sun (in opposition to Poisson's view, which you communicate to me). The luminous ring, which shows itself about the sun under a total eclipse, I have regarded as constituted by this most brilliant portion of the zodiacal light. I have persuaded myself that this light is very different in different years; that for several years in succession it is extremely bright and extensive; in others, again, that it is not even to be seen. I fancy I can see the first indications of a recognition of the zodiacal light in a letter from Rothmann to Tycho, when he says, that in the spring, he had found the sun 24° below the horizon at the end of the evening twilight. Rothmann must certainly have confounded the disappearance of the sinking zodiacal light, in the vapours of the evening horizon, with the actual end of the evening twilight. I have not myself observed any risings and fallings, probably by reason of the weakness with which the zodiacal light appears in our latitudes. But you are assuredly correct when you ascribe such sudden alterations in the light of the heavenly bodies, which you observed within the tropics, to our atmosphere, especially to changes in its higher regions. This is especially shown in the tails of great comets. One frequently sees, especially in clear weather, pulsations in these tails, which begin from the head of the comet as the lowest point, and tremble through the entire length of the tail in 1 or 2 seconds, when the tail appears to be lengthened and immediately afterwards to be shortened by several degrees. That those pulsations, to which Hooke and Schroeter and Chladni paid particular attention, do not take place in the comets' tails themselves, but are produced by our atmosphere, becomes obvious when we reflect that the several portions of the tail (several millions of miles in length) lie at very different distances from us, and that its light can only reach us at intervals of time, several minutes apart from one another. Whether what you observed on the Orinoco, not at intervals of seconds, but of minutes, were true corruscations of the zodiacal light, or belonged wholly and solely to the upper strata of our light-circle, I will not pretend to determine. Neither do I know how the remarkable luminousness of entire nights, and the anomalous increase and protraction of the twilight in the year 1831, are to be explained, especially as it was observed that the brightest parts in these extraordinary twilights did not correspond with the sun's place below the horizon."—From a letter of Dr. Olbers to me, dated Bremen, 26th March, 1833.

70 (p. 45.) — Biot, Traité d'Astron. physique, 3me éd., 1841, t. i. p. 171, 238, and 312.

71 (p. 45.) — Bessel, in Schum. Jahrb. für 1839, S. 51; probably one million of miles daily; in relative velocity, 834,000 miles; and therefore more than twice the velocity of the earth in its orbit round the sun.

72 (p. 46.) — On the Motion of the Solar System, after Bradley, Tobias, Mayer, Lambert, Lalande, and William Herschel, see Arago, in Annuaire, 1842, p. 388—399; Argelander, in Schum. Astron. Nachr. Nr. 363, 364, 398; and in the treatise: Von der eigenen Bewegung des Sonnensystems, 1837, S. 43, on Perseus as the central constellation of the entire stratum of stars. See also Otho Struve, in Bull. de l'Acad. de St. Pétersb. 1842, t. x. No. 9, p. 137—139; according to whom, from a subsequent combination, the direction of the sun's motion was found to be 261° 23′ R. A. + 37° 36′ Decl.; and, as a mean from Argelander's and his own labours, from a combination of 797 stars, 259° 9′ R. A. + 34° 36′ Declination.

73 (p. 46.)—Aristot. de Cœlo, iii. 2, p. 301; Bekkèr, Phys. viii. 5, p. 256.

74 (p. 46.)—Savary, in the Connaissance des Tems, 1830, p. 56 and 163; Encke, Berl. Jahrb. 1832, S. 253; Arago, in Annuaire, 1834, p. 260, 295; John Herschel, in Mem. of the Astronom. Soc. vol. v. p. 171.

75 (p. 46.)—Bessel, Untersuchung des Theils der planetarischen Störungen, welche aus der Bewegung der Sonne entstehen, in Abh. der Berl. Akad. der Wissensch. 1824 (Mathem. Classe), S. 2—6. The question was opened up by Johann Tobias Mayer, in Comment. Soc. Reg. Götting. 1804—1808, vol. xvi. p. 31—68.

76 (p. 47.)—Philos. Transact. for 1803, p. 225; Arago, in Annuaire, 1842, p. 375. If the reader would form a more *tangible* idea of the distance of the fixed stars referred to some short way before in the text, let him suppose the earth to be at the distance of *one foot* from the sun, Uranus would then be 19 *feet*, and Vega in Lyra 34¼ German geographical (158·6 English) *miles* from that luminary.

77 (p. 47.)—Bessel, in Schum. Jahrbuche, 1839, S. 53.

78 (p. 47.) — Mädler, Astr. S. 476; Derselbe, in Schum. Jahrb. 1839, S. 95.

79 (p. 47.)—Sir William Herschel, in the Philos. Transact. for 1817, Pt. ii. p. 328.

80 (p. 47.)—Arago, in Annuaire, 1842, p. 459.

81 (p. 47.)—Sir John Herschel, in a letter from Feldhuysen of the 13th Jan. 1836; Nicholl, Archit. of the Heavens, 1838, p. 22. See also some observations of Sir William Herschel on the great starless space which, at a vast distance, separates us from the milky way, in the Philos. Trans. for 1817, Part ii. p. 328.

82 (p. 47.)—Sir John Herschel, Astron. § 624; and farther in Observations on Nebulæ and Clusters of Stars (Phil. Trans. 1833, Pt. ii. p. 479, fig. 25): "we have here a brother system, bearing a real physical resemblance and strong analogy of structure to our own."

83 (p. 48.)—Sir William Herschel, in the Transact. for 1785, Pt. i. p. 257; Sir John Herschel, Astr. § 616 ("the nebulous region of the heavens forms a nebulous milky way

composed of distinct nebulæ as the other of stars"), and farther in his letter to me of March 1829.

84 (p. 48.)—Sir John Herschel, Astron. § 585.

85 (p. 48.)—Arago, in Annuaire, 1842, p. 282—285, 409—411, and 439—442.

86 (p. 48.)—Olbers on the Transparency of Universal Space, in Bode's Jahrbuch, 1826, s. 110—121.

87 (p. 48.)—"An opening in the heavens." Sir William Herschel, in the Transact. for 1785, vol. lxxv. Pt. i. p. 256; Le Français Lalande, in the Connaissance des tems pour l'an viii. p. 383; Arago, in Annuaire, 1842, p. 425.

88 (p. 48.)—Aristot. Meteor. ii. 5, 1; Seneca, Natur. Quæst. i. 14, 2; "cœlum discessisse," in Cic. de Divin. i. 43.

89 (p. 48.)—Arago, in Annuaire, 1842, p. 429.

90 (p. 48.)—In December 1837 Sir John Herschel saw the star η Argo, which had hitherto appeared of the second magnitude, and quite unchanging, increase rapidly to the first magnitude. In January 1838 the intensity of its light was still equal to that of α Centauri. According to the latest intelligence, Maclear, in March 1843, found the star as brilliant as Canopus: α Crucis appeared quite misty beside η Argo.

91 (p. 48.)—"Hence it follows that the rays of light of the remotest nebulæ must have been almost two millions of years on their way, and that, consequently, so many years ago this object must have had an existence in the sidereal heaven, in order to send out those rays by which we now perceive it." William Herschel, in the Transact. for 1802, p. 498; John Herschel, Astr. § 590; Arago, in Annuaire, 1842, p. 334, 359, and 382—385.

92 (p. 49.)—From a beautiful sonnet of my brother, Freiheit und Gesetz (Wilhelm von Humboldt, Gesammelte Werke, Bd. iv. S. 358, No. 25).

93 (p. 49.)—Otfried Müller, Prolegomena, S. 373.

94 (p. 50.)—It is proper to distinguish between the *absolute* depth to which man has penetrated in his mining operations, or the depth from the surface of the earth at the place where the operations are carried on, and the *relative* depth, *i. e.* the depth below the level of the sea. The greatest relative depth that has been reached is, perhaps, the bore at New-Salzwerk, Minden, in Prussia. In June 1841 it was exactly 1844½ Parisian feet; the absolute depth was, however, 2094½ Par. feet. The temperature of the water in the deepest bore was 32·7° C. (90 8° F.) which, assuming 9·6° C. as the mean temperature of the air, gives a rise of 1·6° for 29·6 metres (upwards of 97·6 feet English). The Artesian well of Grenelle, at Paris, is only 1683 feet in absolute depth. From the accounts of the missionary Imbert from China, the depth of our Artesian wells is far surpassed by that of the fire-spring Ho-tsing, which yields inflammable gas employed in salt boiling. In the Chinese province Szütschuan, these fire-springs are said very commonly to reach a depth of from 1800 to 2000 feet; and at Tseu-lieu-tsing (place of perpetual flax), a Ho-tsing, bored with the rod in the year 1812, is reported to extend to the depth of 3000 feet (Humboldt, Asie centrale, t. ii. p. 521 and 525; Annales de l'Association de la Propagation de la Foi, 1829, No. 16, p. 369). The relative depth attained at Monte Massi, in Tuscany, south from Volterra, according to Matteucci, is but 1175 feet. The bore at New-Salzwerk approaches very nearly in relative depth the coal pit at Apendale, New-castle-under-Lyme (Staffordshire). There the works are carried on 725 yards, or 2045 French feet, under the surface (Th. Smith, The Miner's Guide, 1836, p. 160). Unfortunately, the height of the ground above the level of the sea is not accurately ascertained. The relative depth of the Monkwearmouth pit, near Newcastle-on-Tyne, is only 1404 feet (Phillips, Philos. Mag. vol. v. 1834, p. 446); that of the Esperance pit, at Liege, 1271; and that of the lately-worked pit Marihaye, at Val-St.-Lambert, is 1157 feet. The greatest absolute depths to which man has penetrated are in mines, that are either among lofty mountains or in mountain-valleys so much raised above the sea-level that this has either not been reached at all or has only been surpassed by a very small quantity.

The Eselschacht at Kuttenberg, Bohemia, before it was abondoned, had reached the enormous depth of 3545 feet (Schmidt, Berggesetze, Bd. i. S. 32). At St.-Daniel, and at Geist, on the Röhrerbühel, the works, in the 16th century, were 2916 feet deep. A drawing of these workings of the year 1539 is still preserved. Joseph von Sperges, Tyroler Bergwerksgeschichte, S. 121. See also Humboldt, Gutachten über Herantreibung des Meissner Stollens in die Freiberger Erzrevier, published in Herder über den jetzt begonnenen Erbstollen, 1838, S. 124.) It may be imagined that information of the extraordinary depth of the workings at Röhrerbühel had reached England at an early period, for in Gilbert's work, De Magnete, I find the statement that man had penetrated from 2400 to 3000 feet into the bowels of the earth: "Exigua videtur terrae portio, quae unquam hominibus spectanda emerget aut eruitur: cum profundius in ejus viscera, ultra eflorescentis extremitatis corruptelam, aut propter aquas in magnis fodinis, tanquam per venas scaturientes, aut propter aëris salubrioris ad vitam operariorum sustinendam necessarii defectum, aut propter ingentes sumptus ad tantos labores exantlandos, multasque difficultates, ad profundiores terrae partes penetrare non possumus; adeo ut quadringentas aut [quod rarissime] quingentas orgyas in quibusdam metallis, descendisse, stupendus omnibus videatur conatus" (Gulielmi Gilberti, Colcestrensis, de Magnete Physiologia nova, Lond. 1600, p. 40).

The absolute depth of the mines in the Saxon Erzgebirge are 1824 and 1714 feet; the relative depths of these respectively are only 626 and 260 feet. The absolute depth of the rich workings in Joachimsthal, Bohemia, is 1919 feet; but taking the height of the surface upon Dechen's estimate at 2250 feet above the level of the sea, it is obvious that there the sea-level has not even been attained. In the Harz, the workings in the Samson pit, at Andreasberg, are carried on at the absolute depth of 2062 feet. In Old Spanish America I know of no deeper mines than those of Valenciana, near Guanaxuato, Mexico: I found the Planes de San Bernard 1582 feet deep; but this mine does not reach the level of the sea by 5592 feet. If we compare the depth of the old Kuttenberg works (a depth which exceeds the height of the Brocken, and only falls short of that of Etna by 200 feet) with the heights of the loftiest buildings that have been reared by man (the Pyramid of Cheops and the Minster at Strasburg), we find that the mines are to these in the proportion of 8 to 1.

I have thought it important thus to bring together these data in relation to the absolute and relative depths that have been reached by man, a subject in connection with which many errors have been constantly committed, principally, as it seems, through faulty reductions of the measurements from one standard to another. On proceeding eastward from Jerusalem towards the Dead Sea, a prospect is gained which, according to our present hypsometrical knowledge, is unparalleled on the face of the earth: there, on approaching the chasm in which the Jordan flows, we advance, in open day, along beds of rock which, according to Berton's and Russegger's barometrical levellings, lie 1300 feet in perpendicular depth below the level of the Mediterranean Sea (*vide* Humboldt, Asie Centrale, t. ii. p. 323).

95 (p. 50.)—Bason-shaped curved strata, which dip down on one hand and rise again at a measurable distance, although not penetrated by mines or shafts, still suffice to give us accurate information of the constitution of the crust of the earth at great depths from the surface. I have to thank the excellent geologist M. von Dechen for the following. He writes to me: "The depth of the coal measures at Mont-St.-Gilles, Liege, which, with our friend M. von Oeyenhausen, I have estimated at 3650 feet below the surface, must lie at the depth of 3250 feet below the sea-level, inasmuch as Mont-St.-Gilles is certainly not 400 feet high; and the coal-bason at Mons lies fully 1750 feet deeper. These depressions, however, are trifling when compared with that of the coal strata of the Saar River (Saarbrück). After repeated trials, I have found that the lowest coal strata known in the country of Duttweiler, near Bettingen, northeastward from Saarlouis, dip 19,406 and 20,656 feet under the level of the sea." This conclusion exceeds by 8000 feet the estimate which I have given in the text of Cosmos for the bason of Devonian strata. These Belgian coal measures, therefore, lie as far below the level of the sea as Chimborazo rises above it, at a depth where the temperature of the earth must be 224° C. (435° F.). From the highest summit of the Himalaya to the bottom of this bason, containing the vegetable remains of the primeval world, we have a perpendicular depth of 45,000 feet, *i. e.* $\frac{1}{435}$ of the semi-diameter of the earth.

96 (p. 51.)—Plato, Phaedo, p. 97 (Aristot. Metaph. p. 985). See Hegel, Philosophie der Geschichte, 1840, S. 16.

97 (p. 51.)—Bessel, Allgemeine Betrachtungen über Gradmessungen nach astronomisch-geodätischen Arbeiten, at the end of Bessel und Baeyer's: Gradmessung in Ostpreussen, S. 427. On the accumulation of matter on the side of the moon which is turned to us, see farther, Laplace, Expos. du Syst. du Monde, p. 308.

98 (p. 51.)—Plin. ii. 68; Seneca, Nat. Quaest. Praef. c. ii. "El Mundo es poco" (the earth is small) writes Columbus from Jamaica to Queen Isabella on the 7th of June, 1503; not in the philosophical sense of the two Romans, but because it seemed politic to him to represent the passage from Spain as no great matter, in the same way that he spoke of "seeking the east from the west." *Vide* my Examen crit. de l'hist. de la Géogr. du 15me siècle, t. i. p. 83, and t. ii. p. 327; where I have, at the same time, shown that the opinion maintained by Delisle, Fréret, and Gosselin, according to which the extraordinary diversity in the estimates of the earth's perimeter among the Greeks is merely apparent, and depends on differences of the stadia, was already advanced by Jaime Ferrer, in the year 1495, in a proposal for the determination of the Papal line of demarcation.

99 (p. 52.)—Brewster, Life of Sir Isaac Newton, 1831, p 162: "The discovery of the spheroidal form of Jupiter by

Cassini had probably directed the attention of Newton to the determination of its cause, and, consequently, to the investigation of the true figure of the earth." Cassini stated the oblateness of Jupiter at $\frac{1}{15}$ in 1691 (Anc. Mém. de l'Acad. des Sciences, t. ii. p. 108); but we know, through Lalande (Astronom. 3me éd. t. iii. p. 335), that Maraldi possessed several printed sheets of a Latin work which Cassini began, "On the Spots of the Planets," from which it is obvious that Cassini was aware of the oblateness of Jupiter previous to 1666; 21 years, therefore, before the appearance of Newton's Principia.

100 (p. 52.)—According to Bessel's investigation of ten measurements of degrees, in which the error in the French measurement discovered by Puissant was taken into account (*vide* Schumacher, Astron. Nachr. 1841, No. 438, S. 116), the semi-axis major of the elliptical spheroid of rotation which comes nearest to the irregular figure of the earth is 3272077·14 toises; the semi-axis minor is 3261139·33 toises; the oblateness, $\frac{1}{299 \cdot 152}$; the length of the mean degree of the meridian, 57013·109 toises, with an error of + 2·8403 toises; whence the length of a geographical mile comes out 3807·23 toises. Earlier estimates of measurements of degrees vary between $\frac{1}{302}$ and $\frac{1}{297}$: thus, Walbeck, de forma et magnitudine telluris in demensis arcubus meridiani definiendis, makes it $\frac{1}{302 \cdot 78}$ in 1819; Ed. Schmidt (Lehrbuch der mathem. und phys. Geographie, S. 5), $\frac{1}{297 \cdot 48}$ in 1829 from seven measurements of degrees. On the influence of great differences of latitude upon the polar flattening, *vide* Bibliothèque universelle, t. xxxiii. p. 181, and t. xxxv. p. 56; also, Connaissance des tems, 1829, p. 290. From the moon's equation alone, Laplace found, first (Expos. du Syst. p. 229), from the older tables of Bürg $\frac{1}{304 \cdot 5}$; subsequently, from the lunar observations of Burckhardt and Bouvard $\frac{1}{299 \cdot 1}$ (Mécan. celeste, t. v. p. 13 and 43).

101 (p. 52.)—Pendulum experiments, as general results, have given, after the great expedition of Sabine (1822 to 1823, from the equator to 80° N. lat.), $\frac{1}{288 \cdot 7}$; from Freycinet (excluding the observations of Ile de France, Guam, and Mowi), $\frac{1}{286 \cdot 2}$; after Foster, $\frac{1}{289 \cdot 5}$; after Duperrey, $\frac{1}{266 \cdot 4}$; after Lütke, $\frac{1}{269 \cdot 0}$. Against these we have the observations between Formentera and Dunkirk (Connais. des tems, 1816), according to Mathieu, $\frac{1}{298 \cdot 2}$; and between Formentera and Unst Island, according to Biot, $\frac{1}{304 \cdot 7}$. *Vide* Baily, Report on Pendulum Experiments, in the Memoirs of the Royal Astronom. Society, vol. vii. p. 96; also Borenius, in Bulletin de l'Acad. de St.-Pétersbourg, 1843, t. i. p. 25.

The first proposal to apply the length of the pendulum to the determination of mass, and to take the third part of the seconds pendulum as an universal pes horarius, or standard measure for all nations, occurs in Huygens' Horologium Oscillatorium, 1673, prop. 25. The same wish was reiterated anew in a public monument raised under the equator by La Condamine, Bouguer, and Godin. On the beautiful marble tablet which I found uninjured in the quondam Jesuits' College at Quito are these words: "Penduli simplicis aequinoctialis unius minuti secundi archetypus, mensurae naturalis exemplar, utinam universalis!" From what La Condamine says, in his Journal du Voyage à l'Equateur, 1751, p. 163, of passages unfilled up in the inscription, and a slight difference with Bouguer concerning the numbers, I expected to have found notable differences between the inscription of the marble tablet and the statement published at Paris. On carefully comparing them, however, I only found two of any importance—"ex arcu graduum 3½" instead of "ex arcu graduum plus quam trium," and for 1742 the year 1745. This last statement is singular, inasmuch as La Condamine returned to Europe in Nov. 1744, and Bouguer had preceded him in June, and Godin in July. The most necessary and useful correction in the figures of the inscription would be that of the astronomical longitude of the town of Quito (*vide* my Recueil d'Obs. Astron. t. ii. p. 319—354). Nouet's latitudes, cut into the Egyptian monuments, afford a more recent instance of the danger of all solemn attempts to perpetuate erroneous or ill-calculated results.

102 (p. 52.)—On the increased intensity of attraction in the volcanic islands, St. Helena, Ualan, Fernando de Noronha, Isle of France, Guaham, Mowi, and Galapagos, with the exception of the island of Rawak, perhaps in consequence of their vicinity to the high land of New Guinea, *vide* Mathieu in Delambre, Hist. de l'Astron, au 18me siècle, p. 701.

103 (p. 52.)—Many observations also show great irregularities in the length of the pendulum, which are ascribed to local attractions (*vide* Delambre, Mesure de la Méridienne, t. iii. p. 548; Biot in the Mém. de. l'Académie des Sciences, t. viii. 1829, p. 18, 23). When we proceed from west to east in the south of France and Lombardy, we find the least intensity in the force of gravitation at Bordeaux; the intensity increases rapidly in places situated to the east, Figeac, Clermont-Ferrand, Milan, and Padua, in which last city the maximum force is observed. The influence of the southern flanks of the Alps is not merely to be ascribed to the general magnitude of their volume, but as M. Elie de Beaumont (Recher. sur les Révol. de la surf. du globe, 1830, p. 729.) believes, in principal part to the melaphyre and serpentine which have raised the chain. On the flanks of Mount Ararat, which with Caucasus lies as it were in the centre of gravity of the old world, consisting of Europe, Asia, and Africa, Fedorow's careful pendulum experiments proclaim not hollows, but dense volcanic masses (Parrot, Reise zum Ararat, Bd. ii. S. 143). In the geodetic operations of Carlini and Plana in Lombardy, differences of from 20″ to 47″·8 were found between the immediate observations of latitude and the results of these operations. *Vide* the examples of Andrate and Mondovi, Milan and Padua, in the Opérations géodés. et astron. pour la mesure d'un arc du parallèle moyen, t. ii. p. 347; Effemeridi astron. di Milano, 1842, p. 57. Milan estimated by Berne, as it stands in the French trigonometrical survey, is in latitude 45° 27′ 52″; whilst immediate astronomical observations make it 45° 27′ 35″. As the perturbations extend far to the south of the Po towards Parma (Plana, Opérat. géodés. t. ii. p. 847), we may conjecture that even in the constitution of the soil of the plain, there are causes producing deviations. Struve has met with the same thing in the flattest parts of the east of Europe (Schum. Astron. Nachr. No. 164). On the influence of dense masses which are conceived to lie at a moderate depth, corresponding with the point of mean elevation of the Alps, see the analytical expressions (after Hossard and Rozet) in Comptes rendus, t. xviii. 1844, p. 292, which may be compared with Poisson (Traité de Mécanique, t. i. p. 282, 2me éd.) The earliest indications of the influence of rocks of different kinds on the vibrations of the pendulum, are those of Dr. Thomas Young (Phil. Trans. 1819, p. 70—96). In the conclusions, from the length of the pendulum in regard to the curve of the earth, the possibility is not to be overlooked of the crust of the earth having become consolidated before metallic and dense basaltic masses, forced from the interior, had approached the surface.

104 (p. 52.)—Laplace, Expos. du Syst. du Monde, p. 231.

105 (p. 52.)—La Caille's pendulum experiments at the Cape of Good Hope, which were calculated with great care by Mathieu (Delambre, Hist. de l'Astr. 18me siè. p. 479), indicate an oblateness of $\frac{1}{284 \cdot 4}$; but from numerous comparisons of observations under similar parallels of latitude in both hemispheres (New Holland and the Maldives compared with Barcelona, New York, and Dunkirk), there are no grounds for estimating the mean oblateness of the south pole as greater than that of the north pole (Biot, in Mém. de l'Acad. des Sciences, t. viii. 1829, p. 39—41).

106 (p. 53.)—The three methods of conducting the observations, give the following results: 1st. From deflection of the plumb-line in the neighbourhood of Shehallien in Perthshire, 4·713 by Maskelyne, Hutton, and Playfair (1774—1776 and 1810) according to a method already proposed by Newton; 2d. From vibrations of the pendulum on mountains, 4·837 (Carlini's observations on Mont Cenis compared with Biot's observation at Bordeaux, Effemer. astr. di Milano, 1824, p. 184): 3rd. From the torsion balance of Cavendish, after an apparatus originally imagined by Mitchell, 5·48 (from Hutton's revision of the calculation 5·32, and from Ed. Schmidt's revision 5·52: Lehrb. des mathem. Geographaphie, Bd. i. S. 487); from the torsion balance of Reich, 5·44. In the calculation of this experiment carried through in a most masterly manner by Prof. Reich, the original mean result was 5·43 (with a probable error of but 0·0233); a result which, increased by the quantity by which the centrifugal force of the earth diminishes the force of gravitation, for the latitude of Freiburg (50° 55′), must be changed unto 5·44. The employment of masses of cast-iron instead of lead gave no difference of result that might not safely be ascribed to error of observation; there was no evidence of magnetic attraction (Reich. Versuche über die mittlere Dichtigkeit der Erde, 1838, S. 60, 62, and 66). By the assumption of too small a degree of oblateness of the earth, and the uncertain estimate of the density of the rocks composing its surface, a mean density of the earth was come to, as in the experiments on mountains, which was by ⅙ too small, viz., 4·761 (Laplace, Mécan. cél. t. v. p. 46) or 4,785 (Eduard Schmidt, Lehrb. der. math. Geogr. Bd. i. § 387 and 418). On the hypothesis of Halley, on the earth as a hollow sphere—the germ of Franklin's idea of earthquakes, *vide* Phil. Transact. for the year 1693, vol. xvii. p. 563. (On the structure of the internal parts of the earth and the concave habited arch of the shell.) Halley held it more worthy of the Creator "that the earth, like a house of several stories, should be inhabited both within and without. For light in the hollow sphere (p. 576) provision could also be made in a certain way."

107 (p. 53.)—Here belong the admirable analytical labours of Fourier, Biot, Laplace, Poisson, Duhamel, and Lamé. In his work, Théorie mathématique de la Chaleur, 1835, p. 3, 428—430, 436, and 521—524 (see also the abstract of La Rive, in the Bibliothèque universelle de Genève, t. lx. p. 415), Poisson has developed an hypothesis totally different from the view advocated by Fourier (Théorie. analyt. de la Chaleur). He denies the present fluid state of the centre of the earth; he believes "that in cooling by radiation to the medium surrounding the earth, the parts first consolidated on the surface sank downwards, and that by a double upward and downward current, the great inequality was lessened which would have taken place in a solid body cooling from the surface." The great geometrician thinks it more probable that the consolidation commenced in the parts lying nearer to the centre; "the phenomenon of the increase of heat with the depth does not extend to the whole mass of the earth, and is a mere consequence of the motion of our planet in universal space, the several parts of which, by reason of their stellar heat (chaleur stellaire) have very different temperatures." The heat of the water of our Artesian wells, according to Poisson, is therefore heat which has penetrated the body of the earth from without; the earth may be viewed as we should a mass of rock transported from the equator to the pole in so short a time, that it could not cool completely. The increase of temperature in this block would not extend completely to its centre. The physical doubts which may reasonably be raised against this extraordinary cosmical hypothesis, (an hypothesis which ascribes to heavenly space what must much rather belong to matter in its first transition from the gaseous to the solid state) may be found collected in Poggendorff's Annalen, Bd. xxxix. s. 93—100.

108 (p. 54.)—See above, pages 9, 15, and 16. The increase in temperature is found in the Puits de Greneille from 98·4 feet; in the bore of New-Salzwerk, Minden, almost 91 feet; at Prégny, Geneva, also 91 feet, although there the outlet is 1510 feet above the level of the sea. This agreement of results, from bores that are severally 1683, 2094, and 680 feet in absolute depth, by a method first suggested in 1821 by Arago, (Annuaire, 1835, p. 234) is very striking. The two points of the earth at a short perpendicular distance from one another, whose annual temperature is ascertained with the greatest precision, are probably the external atmosphere of the Observatory of Paris and of the cellar under the Observatory. The former is 10°·822, the latter 11°·834 C.; difference 1°·012 C. for 86 feet of depth (Poisson, Théorie, &c., p. 415 and 462). In the course of the last 17 years, from causes which have not been ascertained, the thermometer of the *Caves* has risen 0°·220 C. If the penetration of waters from lateral channels into the main bore of Artesian wells produces some disturbance, it must be admitted that in reference to mines there are many more perturbing causes at work, and that interfere with the accuracy of conclusions in reference to their temperature at different depths. The general result of Reich's great work on the temperature of the mines of the Saxon Erzgebirge is the somewhat slow increase of 1° C. for 128½ feet of descent. (Reich, Beob. über die Temperature des Gesteins in verschiedenen Tiefen, 1834, S. 134.) Yet Phillips (Poggend. Ann. B. 34, S. 191), in a shaft of the Monkwearmouth coal-pit, found an increase of 1° C. for $99\frac{6}{10}$ feet of descent, exactly what Arago found in the Puits de Grenelle.

109 (p. 54.)—Boussingault sur la Profondeur à laquelle se trouve la Couche de Température invariable entre les tropiques, in the Annales de Chimie et de Physique, t. liii. 1833, p. 225—247.

110 (p. 55.)—Laplace, Exp. du Syst. du Monde, p. 229 and 263; Mécanique cél. t. v. p. 18 and 72. It is to be observed that the fraction $\frac{1}{170}$ of a centigrade degree of a mercurial thermometer, which is given in the text as the limit of stability of the heat of the earth since Hipparchus's time, rests on the assumption that the dilatation of the materials of which the body of the earth consists is the same as that of glass $= \frac{1}{100000}$ for 1° C. of heat. *Vide* on this point Arago, in Annuaire pour 1834, p. 177—190.

111 (p. 55.)—William Gilbert of Colchester, whom Galileo calls "great to a degree that might excite envy," says, "Magnus magnes ipse est globus terrestris." He ridicules the magnetic mountain of Fracastoro, the great contemporary of Christopher Columbus, as the magnetic pole: "Rejicienda est vulgaris opinio de montibus magneticis, aut rupe aliqua magnetica, aut polo phantastico a polo mundi distante." He assumes the variation of the magnetic needle over the surface of the earth as unchanging: "Variatio uniuscujusque loci constans est;" and explains the isogonic lines from the configuration of continents and the relative position of the sea basin, which has a weaker magnetic attractive force than the solid masses that rise above the ocean (Gilbert de Maguete, ed 1633, p. 42, &c.)

112 (p. 55.)—Gauss, Allgemeine Theorie des Erdmagnetismus, in den Resultaten aus den Beob. des magnet. Vereins im Jahr. 1838, § 41, S. 56.

113 (p. 55.)—There are also perturbations which do not extend to any distance, which are more local, and perhaps have their seat less deeply. A rare example of such extraordinary perturbations, which are felt in the Freiburg mines and not in Berlin, was published by me now many years ago (Lettre de M. de Humboldt à S. A. R. le Duc de Sussex sur les moyens propres à perfectionner la connaissance du Magnétisme terrestre, in Becquerel's Traité expérimental de l'Electricité, t. vii. p. 442). Magnetic storms that were experienced simultaneously from Sicily to Upsal, did not extend from Upsal to Altona (Gauss and Weber, Resultate des magnet. Vereins, 1839, s. 128; Lloyd, in the Comptes rendus de l'Académie des Sciences, t. xiii. 1843, Sém. ii. p. 725 and 827). Among the many perturbations which in recent times have been observed simultaneously over extensive districts of country, and which are collected in Sabine's important work (Observ. on days of unusual magnetic disturbance, 1843), one of the most remarkable is that of the 25th September, 1841, which was noticed at Toronto in Canada, at the Cape of Good Hope, at Prague, and partially in Van Diemen's Land. The English Sunday, on which it is sinful after Saturday night at 12 o'clock to read off a scale, and to follow the grand phenomena of nature in their course, intervening, broke off the observations in Van Diemen's Land, and so made our information on this remarkable storm incomplete!

114 (p. 55.)—The application of the magnetic inclination to the determination of the latitude along a coast running north and south, and which, like the shores of Chili and Peru, is enveloped in fog (garua) for a portion of the year, I published in Lamétherie's Journal de Physique, 1804, t. lix. p. 449. The application in the locality indicated is the more important, as, in consequence of the rapid current from south to north as far as Cape Parisia, it occasions a great loss of time to the shipping when the coast has to be first approached northward from the destined port. In the South Sea, from Callao de Lima harbour to Truxillo, with a difference of 3° 57′ of latitude, I have observed a variation of the needle of 9° C.; and from Callao to Guayaquil, with a difference of 9° 50′ of latitude, a variation of 23·05° (*vide* my Relat. Hist. t. iii. p. 622). From Guarmey (10° 4′ S. lat.), Huaura (11° 3′ S. lat.) to Chancay (11° 32′ S. lat.), the inclinations were 6·80°, 9·00°, and 10·35°. The determination of places by means of the magnetic inclination had this remarkable feature about it, that where the ship's course cuts the isoclinal lines almost perpendicularly, it is the only one that is independent of all determination of time, and so of the sight of the sun and other heavenly bodies. I very lately, and for the first time, discovered proposals to determine the latitude by the inclination of the magnetic needle in Gilbert's work, De Magnete (lib. v. cap. 8, p. 200). This was scarcely 20 years after the discovery of magnetic inclination by Robert Norman. Gilbert even points to the method as available "aëre caliginoso;" and Wright, in the preface which he has added to the great work of his teacher, speaks of such a proposal as "worth much gold." As he, with Gilbert, presumed erroneously that the isoclinal magnetic lines ran parallel with the geographical circles of latitude, as also that the magnetic equator coincided with the geographical equinoctial line, he did not perceive that the proposed method was only capable of a local and much more limited application than that he imagined.

115 (p. 55.)—Gauss and Weber, Resultate des magnetischen Vereins in J. 1838, § 31, s. 46.

116 (p. 55.)—According to Faraday (London and Edinburgh Philosophical Magazine, 1836, vol. viii. p. 178), pure cobalt is totally without magnetic power. Rose and Wöhler, again, do not admit this as absolutely ascertained. If one of two masses of cobalt (both of which are believed to be pure) shows itself totally indifferent to magnetism, it seems to me likely that the other which shows magnetic properties does so in virtue of some impurity.

117 (p. 55.)—Arago, in the Annales de Chimie, tom. xxxii. p. 214; Brewster, Treatise on Magnetism, 1837, p. 111; Baumgartner, in the Zeitschrift für Phys. und Mathem. Bd. ii. s. 419.

118 (p. 55.)—Humboldt, Examen critique de l'hist. de la Géographie, tom. iii. p. 36.

119 (p. 55.) — Asie centrale, tom. i. Introduction, p. xxxvii—xlii. The western nations, the Greeks and the Romans, knew that magnetism could be communicated for a great length of time to iron ("sola haec materia ferri vires a magnete lapide accipit *retinetque longo tempore*," Plin. xxxiv. 14). The great discovery of the terrestrial directive force therefore depended alone on this, that no one in the west happened to observe a longish piece of magnetic iron ore or a magnetized iron rod, floated at liberty upon water by means of a piece of wood, or balanced and suspended freely in the air by means of a thread.

120 (p. 56.) — A very slow secular progression or a local invariability of the magnetic declination may be of great

consequence in connection with the boundaries of property: "The whole mass of West India property," says Sir John Herschel, "has been saved from the bottomless pit of endless litigation by the invariability of the magnetic declination in Jamaica and the surrounding archipelago during the whole of the last century; all surveys of property there having been conducted solely by the compass." *Vide* Robertson, in the Phil. Trans. for 1806, pt. ii. p. 348, On the Permanency of the Compass in Jamaica since 1660. In the parent country (England) the magnetic declination has varied by 14° in the same period of time.

[121] (p. 56.)—I have elsewhere shown that from the documents which have come down to us in connection with the voyages of Columbus, we can with great certainty fix upon three places in the Atlantic line of no variation for the 13th September, 1492, the 21st May, 1496, and the 16th August, 1498. This line ran at these dates from North-East to South-West. It touched the American continent somewhat to the east of Cape Codera, whilst at present the conjunction is observed on the north coast of Brazil. (Humboldt, Examen critique de l'hist. de la Géogr. tom. iii. p. 44—48.) From Gilbert's Physiologia nova de Magnete, we see plainly (and this fact is very remarkable) that in the year 1600 the variation was still *nil* in the region of the Azores (lib. iv. cap. 1), precisely as in Columbus's time. I believe that, from documents in my Examen critique (tom. iii. p. 54), I have demonstrated that the celebrated line of demarcation, by means of which Pope Alexander VI. divided the western hemisphere between Spain and Portugal, was not drawn through the most western of the Azores, because Columbus wished to turn a physical division into a political one. He indeed laid great stress upon the zone (raya), "on which the compass showed no variation, where the air and the ocean, the latter covered with sea-weed, show themselves differently constituted, where cooling winds begin to blow, and (for so erroneous observations of the polar star made him imagine) where the figure (the sphericity) of the earth is no longer the same."

[122] (p. 56.)—It is a question of the highest interest in the problem of the physical cause of the terrestrial magnetism, whether the two oval systems of isogonal lines, so singularly included each within itself, will continue to advance for centuries in the same form, or will resolve themselves and expand. In the eastern Asiatic coil, the variation increases from without inwards: in the coil or oval of the South Sea, the opposite holds good; at present, indeed, no line without variation is known in the whole Southern Ocean; to the east of the meridian of Kamtschatka, no line has less variation than 2° (Erman, in Poggend. An. b. xxi. s. 129). Yet Cornelius Schouten appears, on Easter-day of the year 1616, somewhat to the south of Mukahiva, in 15° S. Lat., 132° W. Long., in the middle of the present closed isogonal system, consequently, to have found the variation *nil* (Hansteen, Magnetism. der Erde, 1819, S. 28). It must not be forgotten, that in all these considerations we can only follow the direction of the magnetic lines in their advances as they are projected upon the surface of the earth.

[123] (p. 56.)—Arago, in Annuaire, 1836, p. 284; and 1840, p. 330—338.

[124] (p. 56.)—Gauss, Allg. Theorie des Erdmagnetismus, § 31.

[125] (p. 56.)—Duperrey, de la configuration de l'équateur magnétique, in the Annales de Chemie, tom. xlv. p. 371 and 379 (see also Morlet, in Mémoires présentés par divers savans à l'Acad. roy. des Sciences, tom. iii. p. 132).

[126] (p. 57.)—See the remarkable mass of isoclinal lines in the Atlantic Ocean for the years 1825 and 1837, in Sabine's Contributions to Terrestrial Magnetism, 1840, p. 139.

[127] (p. 57.)—Humboldt, über die seculäre Veränderung der magnetischen Inclination, in Poggend. Annalen, Bd. xv. S. 322.

[128] (p. 57.)—Gauss. Resultate der Beob. des magn. Vereins im Jahr. 1838, § 21; Sabine, Report on the Variations of the Magnetic Intensity, p. 63.

[129] (p. 57.)—The following is the history of the discovery of the law of the (general) increase of intensity in the magnetic force with magnetic latitude. When in 1798 I was anxious to attach myself to the expedition of Captain Baudin, fitting out for a voyage round the world, I was requested by Borda, who took a warm interest in my project, in different latitudes of both hemispheres, to observe the swing of the vertical needle in the magnetic meridian, with a view to determine whether the intensity of the force was the same or different in different places. This investigation I made one of the principal points in the course of my voyage to the tropical countries of America. I observed that the same needle which in Paris performed 245, in Havannah 246, in Mexico 242 oscillations, in the course of ten minutes; at San Carlos, Rio Negro (10° 53′ N. lat., 80° 40′ W. long.), in the same interval of time, performed 216 oscillations; on the magnetic equator, *i. e.* the line on which the inclination is = 0, in Peru (7° 1′ S. lat., 80° 40′ W. long.), it performed only 211 oscillations; in Lima (12° 2′ S. lat.) it again performed 219 oscillations. I found further, from 1799 to 1803, that the whole force taken at 1,0000 on the magnetic meridian in the Peruvian Andes, between Micuipampa and Caxamarca, at Paris will be represented by 1,3482; in Mexico by 1,3155; in San Carlos by 1,0480; in Lima by 1,0773. When I made known this law of the variable intensity of the terrestrial magnetic force, and adduced the numerical value of observations made in 104 different places, in illustration of the conclusions, in a paper which was read before the Parisian Institute at its sitting of the 26th Frimaire, An. xiii., and of which the mathematical portion belongs to M. Biot, the subject was regarded as entirely new. It was only after the reading of this paper, as Biot himself says expressly, (Lamétherie, Journ. de Physique, t. lix. p. 446, note 2,) and as I repeat the statement in my Relation Historique (t. i. p. 262, note 1), that M. de Rossel communicated to M. Biot his observations on oscillation made six years previously in Van Dieman's Land, Java, and Amboyna; from these observations was deduced the same law of declining intensity in the Indian Archipelago. It is almost to be supposed that this excellent man, in his own work, was not aware of the regularity of the increase and decrease of the intensity, as before the reading of my paper he never mentioned this certainly not unimportant physical law to our common friends, La Place, Delambre, Prony, and Biot. It was only in 1808, four years after my return from America, that the observations made by M. de Rossel were published in the Voyage de l'Entrecasteaux, t. ii. p. 287, 291, 321, 480, 644. Up to the present time it has still been usual in all the tables of magnetic intensity that have been published in Germany by Hansteen, Magnet. der Erde 1819, s. 71; Gauss, Beob. des magnet. Vereins 1838, S. 36—39; Erman, Physikal. Beob. 1841, S. 529—579; in England (Sabine, Report on Magnet. Intensity, 1838, p. 43—62; Contributions to Terrestrial Magnetism, 1843,) and in France (Becquerel, Traité d'électr. et de magnét. t. vii. p. 354—367), to reduce the oscillations observed in any part of the earth to the measure of the force which I found on the magnetic equator in North Peru; so that from the unity thus arbitrarily assumed, the intensity of the magnetic force at Paris is always set down at 1,348. Still older than the observations of Admiral Rossel, however, are those that were made in the unfortunate expedition of La Pérouse by Lamanon, during the stay at Teneriffe (1785) and to the arrival at Macao (1787), and which were sent to the Academy of Sciences. It is known for certain that these papers were in the hands of Condorcet in the July of 1787 (Becquerel, t. vii. p. 320). In spite of searching, however, they have not again been found; but from the copy of a letter of Lamanon, now in the possession of Ad. Duperrey, addressed to the then perpetual secretary of the Academy of Sciences, which has been omitted in the account of the Voyage of La Pérouse, it is stated expressly, "Que la force attractive de l'aimant est moindre dans les tropiques qu'en avançant vers les poles, et que l'intensité magnétique déduite du nombre des oscillations de l'aiguille de la boussole d'inclinaison change et augmente avec la latitude." Had the Academy of Sciences, still anticipating the return of La Pérouse, felt itself at liberty, in the course of 1787, to publish an account of observations made by three different individuals unknown to one another, the theory of terrestrial magnetism would have been extended by a new class of observations eighteen years sooner than it was. This simple statement of facts will perhaps justify the assertion which the third volume of my Relation historique (p. 615) contains: "Les observations sur les variations du magnétisme terrestre auxquelles je me suis livré pendant 32 ans, au moyen d'instrumens comparables entre eux en Amérique, en Europe et en Asie, embrassent, dans les deux hémisphères, depuis les frontières de la Dzoungarie chinoise jusque vers l'ouest à la Mer du Sud qui baigne les côtes du Mexique et du Pérou, un espace de 188° de longitude, depuis les 60° de latitude nord jusqu'aux 12° de latitude sud. J'ai regardé la loi du décroissement des forces magnétiques, du pôle à l'équateur, comme le résultat le plus important de mon voyage Américain." It is not certain, but extremely probable, that Condorcet read the letter of Lamanon of July, 1787, at a meeting of the Academy of Sciences of Paris; and such a simple reading I myself regard as a sufficient act of publication (Annuaire du Bureau des Longit. 1842, p. 463). The first recognition of the law, therefore, belongs indisputably to the companion of La Pérouse; but, long unheeded or forgotten, I believe that the knowledge of the law of the variation in the intensity of the magnetic force with the latitude, first acquired a scientific existence with the publication of my observations from 1798 to 1804. The subject, and the length of this note, will not appear indifferent to him who is familiar with the recent history of magnetism, and the doubts that have been started in connection with it, and who from personal experience knows that we are apt to attach some value to that which has been the object of our uninterrupted attention for five long years, under the pressure of tropical climates, and engaged in hazardous mountain expeditions.

[130] (p. 57.)—The maximum intensity for the whole sur-

face of the earth, according to the observations hitherto collected, appears to be 2;052, the minimum 0,706. Both phenomena belong to the Southern hemisphere; the first to 73° 47′ S. lat., 169° 30′ E. long., near Mount Crozier, West North-West of the South magnetic pole, at a place where Sir James Ross found the inclination of the needle 87° 11′ (Sabine, Contributions to Terrestrial Magnetism, 1843, No. 5, p. 231); the second, observed by Erman under 90° 59′ S. lat, 37° 24′ W. long., 80 miles eastward from the coast of the province of Espiritu Santo, Brazil (Erman Phys. Beob. 1841, S. 570), at a point where the inclination is only 7° 55′. The accurate relations of the intensity to one another are therefore as 1 to 2·906. It was long believed that the greatest intensity of the magnetic force was only two and a half times as great as the weakest which the surface of our earth manifests (Sabine, Report on Intensity, p. 82).

131 (p. 57.)—On Amber (succinum, glessum) Pliny says, xxxvii. 3, "Genera ejus plura. Attritu digitorum accepta caloris anima trahunt in se paleas ac folia arida quæ levia sunt, ac ut magnes lapis ferri ramenta quoque." (Plato, in Timaeo, p. 80; Martin, Etudes sur le Timée, t. ii. p. 343—346; Strabo, xv. p. 703, Casaub.; Clemens Alex. Strom. ii. p. 370, where, strangely enough, τὸ σούχιον and τὸ ἤλεκρον are distinguished.) When Thales, in Aristot. de anima 1, 2, and Hippias in Diog. Laertio 1, 24, attribute a soul to the magnet and to amber, this animation only refers to a moving principle.

132 (p. 57.)—"The magnet attracts iron in the same way as amber attracts the smallest grains of mustard. It is like a breath of wind which penetrates through both, and is communicated with the rapidity of an arrow." These words are Kuopho's, a Chinese orator on the magnet, and writer of the beginning of the fourth century. (Klaproth, Lettre à M.A. de Humboldt, sur l'invention de la boussole, 1834, p. 125.)

133 (p. 58.)—"The phenomena of periodical variations depend manifestly on the action of solar heat, operating probably through the medium of thermoelectric currents induced on the earth's surface. Beyond this rude guess, however, nothing is as yet known of the physical cause. It is still a matter of speculation, whether the solar influence be a principal or only a subordinate cause in the phenomena of terrestrial magnetism." (Observ. to be made in the Antarctic Exped. 1840, p. 35.)

134 (p. 58.)—Barlow, in the Philos. Transact. for 1822, P. i., p. 117; Sir David Brewster, Treatise on Magnetism, p. 129. Long before Gilbert and Hooke, it was taught in the Chinese work, Ou-thsa-tsou, that heat lessened the directive property of the magnet. (Klaproth, Lettre à M. A. de Humboldt, sur l'invention de la boussole, p. 96.)

135 (p. 58.)—*Vide* the paper on Terrestrial Magnetism in the Quart. Review, 1840, vol. lxvi. p. 271—312.

136 (p. 58.)—As the first demand for the establishment of these observatories (a net-work of stations provided with similar instruments) took its rise with me, I dare not cherish the hope that I shall live long enough to see both hemispheres covered in equal and due measure with magnetical stations, under the control of able naturalists and astronomers, and especially under the liberal and continued support of the British and Russian governments. In the years 1806 and 1807 at Berlin, with my friend and fellow labourer, Oltmanns, particularly at the times of the solstices and equinoxes, I frequently observed the movements of the needle from hour to hour, and even from half hour to half hour, during five or six days and nights in succession. I had persuaded myself that continuous, uninterrupted observations of several days and nights were preferable to the single observations of many months. The apparatus, a magnetic telescope by Prony, suspended in a glass case from a thread without torsion, enabled angles of 7 and 8 seconds to be read off upon a finely-divided scale, fixed at a proper distance and illuminated at night with lamps. Magnetic perturbations (storms) which occasionally returned on several successive nights at the same hours, led me even at that time to desire most anxiously that similar apparatuses should be used to the east and west of Berlin, for the sake of distinguishing general telluric phenomena from those of a local nature, and that may depend on perturbations in the unequally-heated body of the earth, or in the cloud-forming atmosphere. My removal to Paris, and the lengthened political disturbances which spread over the whole of the west of Europe, prevented my wish from being accomplished at this time. The light diffused by the great discovery of Orsted (1820), of the intimate connection between electricity and magnetism, finally aroused the general interest after its long sleep, in the periodical change of the electro-magnetic charge of the earth. Arago, who many years before had begun the longest unbroken series of hourly observations which we possess in Europe in the observatory of Paris, with an admirable declination instrument by Gambey, showed, by means of simultaneous observations of perturbation made at Kasan, what advantages resulted from corresponding measurements of variation. When I returned to Berlin, after a residence of eighteen years in France, I had a small magnetic house erected in the autumn of 1828, not only with a view to carrying out the work begun in 1806, but especially that simultaneous observations, at hours previously agreed upon, might be made at Berlin, Paris, and Freiburg (at a depth of 35 fathoms under the surface). The simultaneousness of the perturbations, and the parallelism of the movements for October and December, 1829, were there graphically represented (Poggend. Annal. Bd. xix. S. 357, Tab. I.—III.). An expedition into the North of Russia, undertaken in 1829 by command of the Emperor, gave me an opportunity of extending my plan upon a great scale. This plan was unfolded to a committee especially named in one of the imperial academies of science; and under the protection of the chief of the mining corps, Count von Cancrin, and the excellent superintendence of Prof. Kupffer, magnetic stations were fixed over the whole of the north of Asia, from Nicolajeff by Catherinenburg, Barnaul, and Vertschinsk, to Peking.

The year 1832 (*vide* Götting. gelehr. Anzeig. St. 206) marks the great epoch in which the profound author of a new theory of terrestrial magnetism, Frederick Gauss, erected apparatus, constructed upon new principles, in the Gottingen Observatory. In 1834 the magnetic observatory was finished, and in the same year Gauss spread his instruments and his method of conducting observations, in which the distinguished natural philosopher, William Weber, took great interest, over a large portion of Germany, Sweden, and Italy (Resultate der Beob. des magnetischen Vereins im Jahr. 1838, S. 135, and Poggend. Annalen, Bd. xxxiii. S. 426). In the magnetical association that was now formed, with Gottingen for its centre, at four periods of the year, ever since 1836, hourly observations for an entire day were regularly instituted, but which were not those of the equinoxes and solstices which I had proposed and followed in 1830. Up to this time, Great Britain, in possession of the largest commerce in the world, and with her wide-spread navy, had taken no part in the movement, which, since 1828, had begun to afford important results towards the determination of terrestrial magnetism. I was so fortunate, in a public appeal from Berlin to the Duke of Sussex, then President of the Royal Society, by my letter of April 1836, headed, "Lettre de M. de Humboldt à S. A. R. le Duc de Sussex sur les moyens propres à perfectionner la connaissance du magnétisme terrestre par l'établissement de stations magnétiques et d'observations correspondantes," to excite a lively interest in the undertaking which had so long been the object of my warmest wishes. In my letter to the Duke of Sussex I urged the erection of permanent stations in Canada, St. Helena, the Cape of Good Hope, the Isle of France, Ceylon, and New Holland, all of which I had, however, pointed out as advantageous positions five years previously. There was a joint physical and meteorological committee appointed in the Royal Society, which, besides fixed magnetic observatories in both hemispheres, proposed to the government to fit out a naval expedition for magnetic observations in the Antarctic Seas. I need not here proclaim all that science owes in this conjuncture to the zeal and activity of Sir John Herschel, Col. Sabine, Professor Airy, and Mr. Lloyd, as well as the powerful support that was given by the British Association for the Advancement of Science assembled at Newcastle in 1838. In June, 1839, the Antarctic expedition, under the command of Captain James Clarke Ross, was resolved on; and now, since its fortunate return, we enjoy the double fruits of important geographical discoveries in the neighbourhood of the South Pole, and a series of simultaneous observations in eight or ten new magnetic stations.

137 (p. 58.)—Instead of ascribing the internal heat of the earth to the transition of matter from a state of gaseous fluidity to the solid condition on the formation of the planets, Ampère has broached what to me appears a very improbable opinion, viz., that it might be a consequence of an incessant chemical action of a central mass of earth and alkali-metals upon the external crust undergoing oxydation. "On ne peut douter," he says, in his masterly Théorie des phénomènes électro-dynamiques (1826, p. 199), "qu'il existe dans l'intérieur du globe des courants électro-magnétiques, et que ces courants sont la cause de la chaleur qui lui est propre. Ils naissent d'un noyau métallique central composé des métaux que Sir Humphrey Davy nous a fait connaître, agissant sur la couche oxidée qui entoure le noyau."

138 (p. 58.)—The remarkable connection between the curvature of magnetic lines and that of my isothermal lines was first observed by Sir David Brewster (Transactions of the Royal Society of Edinburgh, vol. ix. 1821, p. 318, and Treatise on Magnetism, 1837, p. 42, 44, 47, and 268). This distinguished natural philosopher admits two "poles of maximum cold" in the northern hemisphere; one American (73° N. Lat., 102° W. Long., near Cape Walter); another Asiatic (73° N. lat., 78° E. Long.); whence, according to him, arise two hot and two cold meridians, *i. e.* meridians of greatest heat and greatest cold. In the 16th century, however, Acosta (Hist. nat. de las Indias, 1589, lib. i. cap. 17), resting what he says on the observations of a highly

experienced Portuguese pilot, taught that there were four lines without variation. This view appears, if we may judge from the controversy of Henry Bond (author of the work—The Longitude Found, 1676) with Beckborrow, to have had some influence upon Halley's Theory of magnetic poles. *Vide* my Examen critique de l'hist. de la Géographie, t. iii. p. 60.

[139] (p. 58.)—Halley, in the Philosophical Transactions, vol. xxix. (for 1714—1716, No. 341).

[140] (p. 58.)—Dove, in Poggendorff's Annalen, Bd. xx. S. 341, Bd. xix. S. 388: "The dipping needle comports itself very nearly as an atmospherical thermometer, whose difference in like manner shows the increased tension of the electricity before this has risen to such a height that a spark is elicited. *Vide* also the excellent observations of Prof. Kaemtz, in his Lehrbuch der Meteorologie, Bd. iii. S. 511—519; Sir David Brewster, Treatise on Magnetism, p. 280. On the magnetic properties of the galvanic flame or luminous bow from a Bunsen's charcoal and zinc battery, *vide* Casselmann's Beob. (Marburg, 1844,) S. 56—62.

[141] (p. 59.)—Argelander's important observations on the Northern Lights, embodied in his Vorträge, gehalten in der physikalisch-ökonomischen Gesellschaft zu Königsberg, Bd. i. 1834, S. 257—264.

[142] (p. 59.)—On the results of the observations of Lottin, Bravais, and Siljerström, who passed a winter at Bosekop, on the coast of Lapland (70° N. Lat.), and in 210 nights saw 160 Auroræ boreales, *vide* Comptes rendus de l'Acad. des Sciences, tom. x. p. 289, and Martin's Météorologie, 1843, p. 453. See also Argelander, in his Vorträge, geh. in der Königsberg. Gesellschaft, Bd. i. S. 259.

[143] (p. 59.)—John Franklin (Narrative of a Journey to the Shores of the Polar Sea in the years 1819—1822, p. 552 and 597; Thienemann, in Edinburgh Philosophical Journal, vol. xx. p. 366; Farquharson, ib. vol. vi. p. 392; Wrangel, Phys. Beob. S. 59; Parry, Journal of a Second Voyage, performed in 1821—1823, p. 156) saw a great Aurora continue through the day. Something of the same kind was seen in England, 9th Sept. 1827. At mid-day, a luminous arch, 20° high, and rays shooting from it, were perceived after rain, in a part of the heavens that had become clear. Journal of the Royal Institution of Great Britain, 1828, Jan., p. 429.

[144] (p. 59.)—After my return from my American travels, I described the cirro-cumulus cloud—when it appears very regularly divided into rounded masses as if by the agency of repulsive forces—under the name of polar streaks (bandes polaires), because their perspective point of convergence is mostly in the magnetic meridian in the first instance, so that the parallel rows of cumuli follow the magnetic meridian. One peculiarity of this enigmatical phenomenon is, the swaying hither and thither of the point of convergence. Usually the streaks are only completely developed in one region of the sky, and in their motion they are seen directed first from south to north, and then gradually veering round from east to west. I cannot ascribe the advance of the zones to any change in the quarter of the wind in the superior strata of the atmosphere. They arise when the air is extremely calm and the heaven is particularly serene, and under the tropics are far more common than in the temperate and frigid zones. I have observed the phenomenon among the Andes, when I was at the height of 14,000 feet above the level of the sea, as well as in Northern Asia, in the plains of Krasnojarski, southward from Buchtarminsk, and in both instances so much alike, that the natural process in virtue of which it takes place must be regarded as one of very extensive prevalence. See the important observations of Kaemtz (Vorlesungen über Meteorologie, 1840, S. 146); also those of later date, or Martins and Bravais' Météorologie, 1843, p. 117. In an exhibition of south polar streaks of very delicate clouds, which Arago observed by day on the 23d of June, 1844, at Paris, dark rays shot upwards from an arch which was directed from east to west. We have above (p. 59) referred to darker polar lights—to rays bearing some resemblance to dusky smoke.

[145] (p. 60.)—The northern lights are called "the merry dancers" by the inhabitants of the Shetland Islands. Kendal, in Quarterly Journ. of Science, new series, vol. iv. p. 395.

[146] (p. 60.)—See the admirable work of Muncke, in the new edition of Gehler's Physik. Wörterbuch, Bd. vii. 1, S. 113—268, particularly S. 156.

[147] (p. 60.)—Farquharson, in Edinb. Philos. Journal, vol. xvi. p. 304; Philos. Transact. for 1829, p. 113.

[148] (p. 60.)—Kämtz, Lehrb. der Meteorologie, Bd. iii. S. 498, 501.

[149] (p. 61.)—Arago on the dry fog of 1783 and 1831, which illuminated the night, in Annuaire for 1842; and on extraordinary luminous phenomena in clouds without storms, *vide* Annuaire for 1838, p. 279.

[150] (p. 62.)—Herodotus, iv. 28. The old prejudice (Pliny, ii. 80), that Egypt never suffers from earthquakes, is answered by the colossal statue of Memnon, which has been again restored (Letronne, La Statue vocale de Memnon, 1833); but the valley of the Nile does lie without the circle of concussion of Byzantium, the Archipelago, and Syria (Ideler ad Aristot. Meteor. p. 584).

[151] (p. 62.)—Saint-Martin, in the learned notes to Lebeau, Hist. du Bas Empire, t. ix. p. 401.

[152] (p. 62.)—Humboldt, Asie centrale, t. ii. p. 110—118. On the difference between concussion of the surface and the strata lying under it, *vide* Gay-Lussac, in the Annales de Chimie et de Physique, t. xxii. p. 429.

[153] (p. 62.)—Tutissimum est cum vibrat crispante ædificiorum crepitu; et cum intumescit assurgens alternoque motu residet, innoxium et cum concurrentia tecta contrario ictu arietant; quoniam alter motus alteri renititur. Undantis inclinatio et fluctus more quaedam volutatio infesta est, aut cum in unam partem totus se motus impellit (Plin. ii. 82).

[154] (p. 62.)—Even in Italy they have begun to acknowledge the independence of earthquakes of the state of the weather, *i. e.* the appearance of the heavens immediately before the concussion. F. Hoffmann's numerical results accord in all respects with the experience of the Abbé Scina, of Palermo (Posthum. Works, vol. ii. p. 386—395). I have myself several times observed reddish clouds on the day of shocks, and shortly before they happened; on the 4th Nov. 1799, indeed, I experienced two smart shocks at the moment of a loud clap of thunder (Relat. Hist. liv. iv. chap. 10). Vasalli Eandi, of Turin, observed Volta's electrometer much agitated during the protracted earthquake of Pignerol, April 2 to May 17, 1808 (Journ. de Physique, t. lxvii. p. 291). But these signs from clouds, from altered aërial electricity, and from calms, cannot be regarded as universally significant, as necessarily connected with earthquakes. In Quito, Peru, and Chili, as well as in Canada and Italy, many earthquakes are observed along with the clearest skies, with the freshest land and sea-breezes. But if no meteorological indication present itself on the day of the shock, or shortly before this occurs, it seems impossible to overlook the influence of particular seasons (the vernal and the autumnal equinoxes), *i. e.* the commencement of the rainy season after long drought within the tropics, and the change of the monsoons according to popular belief, although we cannot perceive the genetical connection of meteorological processes with what takes place in the interior of the earth. Numerical inquiries on the distribution of earthquakes throughout the course of the year, such as have been instituted with great industry by Von Hoff, Merian, and Fried. Hoffmann, vouch for their frequency at the epochs of the equinoxes. It is very remarkable that Pliny designates an earthquake a subterraneous thunder-storm, not so much by reason of the rolling noise as because he holds that the elastic concussive forces acting through their tension accumulate in the interior of the earth when they are absent in the atmosphere: Ventos in causa esse non dubium reor. Neque enim unquam intremiscunt terrae, nisi sopito mari caeloque adeo tranquillo, ut volatus avium non pendeant, subtracto omni spiritu qui vehit; nec unquam nisi post ventos conditos, scilicet in venas et cavernas ejus occulto afflatu. Neque aliud est in terra tremor, quam in nube tonitruum; nec hiatus aliud quam cum fulmen erumpit, incluso spiritu luctante et ad libertatem exire nitente (Plin. ii. 79; in Seneca, Nat. Quaest. vi. 4—31). In these words we see the germ of all that has since been said soberly, or dreamed on the causes of earthquakes.

[Mr. Edmonds—Cornwall Journal (?)—has endeavoured to connect the occurrence of earthquakes with the period of the moon. He shows that a great number of the most disastrous have occurred the day after the first quarter. —Tr.]

[155] (p. 62.)—I have given data which show that the hourly variation of the barometer is not affected before or after earthquakes, in my Relat. Hist. t. i. p. 311 and 513.

[156] (p. 62.)—Humboldt, Rel. Hist. t. i. p. 515—517.

[157] (p. 63.)—On the *Bramidos* of Guanaxuato, *vide* my Essai polit. sur la Nouv. Espagne, t. i. p. 303. The subterraneous noises without any appreciable movement of the earth in the deep mines or on the surface (6420 feet above the level of the sea) were not heard in the lofty table-lands in the neighbourhood, but only in the hilly parts of the Sierra, from the Cuesta de los Aguilares, not far from Marsili northward, to Santa Rosa. And the waves of sound did not reach to particular parts of the Sierra 6 or 7 miles north-west of Guanaxuato to the other side of Chichimequillo, near the boiling spring of San José de Comangillas Very severe measures were taken by the magistracy of th mountain town, when the alarm at the sounds was at its height. "14th Jan. 1784.—The flight of a family of wealthy persons shall be punished with a fine of 100 piastres; that of poor persons with two months' imprisonment. The militia are empowered to bring back fugitives." Not the least remarkable point is the opinion which the gentry (el Cabildo) are to form from their better knowledge: "The gentry, in their wisdom (en su Sabidura), will know when there is any danger, and then they may recommend flight; for the present, processions are all that are requisite." A famine was the consequence of the alarm for the truenos;

no one would venture down into the Sierra from the plateaus where corn abounded.

The ancients were also acquainted with noises without earthquakes (Arist. Meteor. ii. ; Plin. ii. 80). The strange noise which was heard from March 1822 to September 1824, in the Dalmatian island Meleda (4 miles from Ragusa), and on which Partsch has thrown so much light, was accompanied by shocks from time to time.

158 (p. 64.)—Drake, Nat. and Stat. View of Cincinnati, p. 232—238 ; Mitchell, in the Transactions of the Lit. and Philos. Soc. of New York, vol. i. p. 281—308. In the Piedmontese county of Pignerol, glasses of water which were filled to the brim continued for hours in incessant motion.

159 (p. 64.)—In Spanish they say : "rocas que hacen puente." With this phenomenon of non-transmission through superior strata, is connected the remarkable fact that, in the beginning of the present century, shocks of an earthquake were felt in the deep silver mines of Marienberg, in the Saxon Erzgebirge, which were not perceived at all on the surface. The miners rushed up in alarm. Contrariwise, the people at work in the mines of Falun and Persberg felt nothing of the smart shocks (Nov. 1823) which threw all the inhabitants above ground into a state of great alarm.

160 (p. 64.)—Sir Alex. Burnes, Travels into Bokhara, vol. i. p. 18 ; and Wathen, Mem. on the Usbek State, Journal of the Asiatic Soc. of Bengal, vol. iii. p. 337.

161 (p. 64.)—Philos. Transact. vol. xlix. p. 414.

162 (p. 65.)—On the frequency of earthquakes in Cashmir, *vide* Troyer's Uebersetzung des alten Radjatarangini, vol. ii. p. 279 ; and the Reise von Carl v. Hügel, Bd. ii. S. 184.

163 (p. 65.) — Strabo, lib. i. p. 100, Casaub. That the phrase πηλοῦ διαπύρου ποταμόν does not mean mud, but lava, appears plainly from Strabo, lib. vi. p. 412. *Vide* Walter über Abnahme der vulkanischen Thätigkeit in historischen Zeiten, 1844, S. 25.

164 (p. 66).—Bischoff's comprehensive work, Wärmelehre des inneren Erdkörpers.

165 (p. 66.)—On the Artesian fire-springs (Ho-tsing) in China, and the ancient use of *portable* gas, in bamboo tubes, in the city of Khiung-tscheu, *vide* Klaproth, in my Asie centrale, t. ii. p. 519—530.

166 (p. 66.)—Boussingault (Annales de Chimie, t. lii. p. 181) observed no escape of hydrochloric acid in the volcanoes of New Granada, whilst Monticelli found this acid in enormous quantities during the eruption of Vesuvius of 1813.

167 (p. 66.)—Humboldt, Recueil d'Observ. astronomiques, t. i. p. 311 (Nivellement barométrique de la Cordillère des Andes, No. 206).

168 (p. 66.)—Adolph Brongniart, in the Annales des Sciences naturelles, t. xv. p. 225.

169 (p. 66.)—Bischoff, op. cit. 324, Anm. 2.

170 (p. 66.)—Humboldt, Asie centr. t. i. p. 43.

171 (p. 66.)—On the Theory of the Isothermal lines, see the clever papers of Kupffer in Poggend. Ann. Bd. xv. S. 184, and Bd. xxxii. S. 270 ; in the Voyage dans l'Oural, p. 382—398 ; and in the Edinb. Journ. of Science, new series, vol. iv. p. 355. See also Kämtz, Lehrb. der Meteor. Bd. ii. S. 217 ; and on the ascent of the Chthonisothermal lines in mountainous countries, Bischoff, S. 174—198.

172 (p. 66.)—Leop. v. Buch in Poggend. Ann. Bd. xii. S. 405.

173 (p. 66.)—On the temperature of the drops of rain in Cumana, which falls to 22·3° C. (72·1° F.) when the temperature of the air shortly before had been 30°—31° C. (86°—87·8° F.), and sinks during the rain to 23·4° C. (75·1° F.), *vide* my Relat. Hist. t. ii. p. 22. The rain-drops as they fall change the temperature they had on their production, which depends on the height of the clouds whence they come, and the heating of these on their upper surface by the sun's rays. After the rain-drops, on their first formation, by reason of the latent caloric of the vapour becoming sensible, have acquired a higher temperature than the surrounding medium, they still rise somewhat in temperature, whilst, as they fall through lower, warmer, and moister strata of air, vapour continues to be precipitated upon them, and they increase in size (Bischoff, Wärmelehre, S. 73) ; but this rise is compensated by evaporation. Cooling of the air by rain is effected (setting aside what probably belongs to the electrical processes attending thunder-storms) by the drops, which are themselves of lower temperature, in consequence of the place of their formation, and farther bring down a portion of the higher colder air ; and then by moistening the ground and giving occasion to evaporation. Such are the usual relations of the phenomenon. When, in rare cases, the rain-drops are warmer than the lower strata of the atmosphere (Humboldt, Relat. Hist. t. iii. p. 513), the reason may perhaps be sought for in superior warmer currents, or in a higher temperature acquired by extended and not very dense clouds exposed to the action of the rays of the sun. How, for the rest, the phenomena of supplementary rainbows (explained by the interferences of light) are connected with the size of the falling drops and their increase, and how an optical phenomenon, when rightly observed, may enlighten us in regard to a meteorological process, according to diversity of zone, has been shown with great acuteness by Arago, in the Annuaire for 1836, p. 300.

174 (p. 66.)—Boussingault's careful experiments satisfy me that in the tropics the temperature of the ground a very short way below the surface corresponds exactly with the mean temperature of the air. I have pleasure in quoting the following table :

Stations.	One foot under the surface.	Mean temperature of the air.	Height above the level of the sea, in Parisian feet.
Guayaquil . .	26·0° C.	25·6° C.	0
Anserma nuevo	23·7	23·8	3231
Zupia	21·5	21·5	3770
Popayan . . .	18·2	18·7	5564
Quito	15·5	15·5	8969

The doubt about the temperature of the earth within the tropics, which I have perhaps myself contributed to raise by my observations in the Cave of Caripe (Cueva del Guacharo), are resolved by the consideration that I compared the presumed mean temperature of the air of the convent of Caripe (18 5°), not with the temperature of the air of the cavern (18·7°), but with the temperature of the subterranean stream (16·8°) ; I have, however, said, that it was very possible that mountain water from a great height might be mixed with the water of the cavern (Relat. hist. t. iii. 146—194).

175 (p. 67.)—Boussingault, in Annales de Chimie, t. lii. p. 181. The spring of Chaudes Aigues in Auvergne, is only 80° C. It is also to be observed that, whilst the aguas calientes de las Trincheras burst out from a granite rock, split into regular blocks, and far from all volcanoes, and have fully a temperature of 97° C., the whole of the springs that rise on the flanks of still active volcanoes, Pasto, Cotopaxi, and Tunguragua, only show a temperature of from 36° to 54°.

176 (p. 67.)—The Cassotis, or spring of St. Nicholas, and the Castalia, foot of the Phædriadæ (Pausanias, x. 24, 25, and x. 8, 9) ; the Pirene, Acrocorinth (in Strabo. p. 379) ; the Erasinos-spring, Mount Chaon, South from Argos (in Herodotus, vii 67, and Pausanias, ii. 24, 7) ; the spring of Aedepsos, Cubœa, some of which have a temperature of 31°, others one of from 62° to 75° (in Strabo, p. 60 and 447, Athenæus, ii. 3, 73) ; the hot springs of Thermopylæ, foot of Oeta, 65° (in Pausan. x. 21, 2) ; all from MS. notices by Professor Curtius, the learned companion of Otfried Müller.

177 (p. 67.) — Plin. ii. 106 : Seneca, Epist. 79, § 3, ed. Ruhkopf. (Beaufort, Survey of the Coast of Karamania, 1820, Art. Yanar, next Deliktasch, the ancient Phaselis, p. 24). See also Ctesias, Fragm. cap. x. p. 250, ed. Bähr. Strabo, lib. xiv. p. 665, Casaub.

178 (p. 67.)—Arago, in Annuaire for 1845, p. 234.

179 (p. 67.)—Acta S. Patricii, p. 555, ed. Ruinart, t. ii. p 385, Mazochi. Dureau de la Malle first directed attention to this remarkable passage, in his Recherches sur la Topographie de Carthage, 1835, p. 276. (*Vide* Seneca, Nat. Quaest. iii. 24.)

180 (p. 68.)—Humboldt, Rel. hist. t. iii. p. 562—567 ; Asie centrale, t. i. p. 43, t. ii. p. 505—515 ; Vues des Cordillères, pl. xli. On the Macalubi (the Arabic Makhlub, cast down), and how the earth ejected liquid earth, *vide* Solinus, cap. v. ; idem ager Agrigentinus eructat limosas scaturigines, et ut venae fontium sufficiunt rivis subministrandis, ita in hac Siciliae parte solo nunquam deficiente, aeterna rejectatione terram terra evomit.

181 (p. 68.)—See the interesting little map of the island Nisyros, in Rose, Reise auf den griechischen Inseln, Bd. ii. 1843, S. 69.

182 (p. 68.)—Leopold von Buch, Phys. Beschreibung der Canarischen Inseln, S. 326 ; and on Erhebungscratere und Vulcane, in Poggend. Ann. Bd. 37, S. 189. Strabo distinguishes very finely between the two modes in which islands are produced, when he speaks of the separation of Sicily from Calabria. "Some islands," he says (lib. vi. p. 258, ed. Casaub.), "are fragments of the continent ; others have arisen from the sea—an event that still happens at the present day : for the islands of the great ocean have probably been lifted from its bosom, those that lie off promontories have probably been detached from the main land."

183 (p. 68.)—Ocre Fisove (Mons Vesuvius) in the Umbrian language, (Lassen, Deutung der Eugubinischen Tafeln, im Rhein. Museum, 1832, S. 387) ; the word ocre is probably genuine Umbrian, and means, as Festus informs us, Mountain. Ætna, if Αἴτνη be, as Voss says, an Hellenic sound, and be connected with αἴθω and αἴθινος, may signify a burning and shining mountain. But this etymological derivation seems doubtful. The word Ætna would probably be found a Sicilian word, had we but any remains of the Sicilian language. The oldest eruption of Etna spoken of is that referred to in Pindar and Æschylus under Hiero (Olymp. 75, 2). But it is probable that Hesiod was aware of eruptions of the mountain before the settlement of the

Greek Colony. The word Αἴτνη in the text of Hesiod, is of doubtful origin, as I have shown elsewhere. (Humboldt, Examen. crit. de la Géogr. t. i. p. 168.)

184 (p. 68.)—Seneca, Epist. 79.

185 (p. 68.)—Aelian. Var. hist. viii. 11.

186 (p. 69.)—Petri Bembi Opuscula (Aetna Dialogus), Basil. 1556, p. 63; "Quicquid in Aetnae matris utero coalescit, nunquam exit ex cratere superiore, quod vel eo incedere gravis materia non queat, vel, quia inferius alia spiramenta sunt, non fit opus. Despumant flammis urgentibus ignei rivi pigro fluxu totas delambentes plagas, et in lapidem indurescunt."

187 (p. 69.)—See my drawing of the volcano of Jorullo, of its *Hornitos* and of the uplifted *Malpays*, in my Vues de Cordillères, Pl. xliii. p. 239.

188 (p. 69.)—Humboldt, Essai sur la Géogr. des plantes et Tableau phys. des Régions équinoxiales, 1807, p. 130, und Essai géogn. sur le gisement des Roches, p. 321. But that the total absence of streams of lava, along with incessant activity of volcanoes, is not connected solely with the configuration, position, and absolute height of the mountains, we are assured by the phenomenon of the greater number of the volcanoes of Java. (*Vide* Leop. von Buch, Descr. phys. des Iles Canaries, p. 419; Reinwardt and Hoffmann in Poggend. Ann. Bd. xii. S. 607.)

189 (p. 70.)—See the bases of my measurements compared with those of Saussure and Lord Minto, in the Abhandlungen der Académie der Wiss. zu Berlin aus den J. 1822 and 1823, S. 30.

190 (p. 70.)—Pimelodes Cyclopum s. Humboldt, Recueil d'Observations de Zoologie et d'Anatomie comparée, t. i. p. 21—25.

191 (p. 71.)—Leop. von Buch, in Poggend. Ann. Bd. xxxvii. S. 179.

192 (. 71.)—On the chemical origin of iron glance in volcanic masses, *vide* Mitscherlich in Poggend. Ann. Bd. xv. S. 630; and on the extrication of hydrochloric acid gas, Gay-Lussac in the Annales de Chimie et de Phys. t. xxii. p. 423.

193 (p. 71.)—See the beautiful experiments on the refrigeration of rocky masses in Bischoff's Wärmelehre, S. 384, 443, 500—512.

194 (p. 71.)—Berzelius and Wöhler in Poggend. Annalen, Bd. i. S. 221, and Bd. xi. S. 146; Gay-Lussac, in the Annales de Chimie, t. xxii. p. 422; Bischoff, Reasons against the Chemical Theory of Volcanoes, in the English edition of his Wärmelehre, p. 297—309.

195 (p. 72.)—According to Plato's geognostic notions, as they are exposed in the Phædo, Periphlegethon, in respect of the activity of volcanoes, plays nearly the same part which we now ascribe to the increased heat of the earth with the greater depth, and the melted state of the internal strata of the earth. (Phaedo, ed. Ast. p. 603 and 607, Annot. p. 808 and 817.) "Within the earth, all around, there are greater and smaller caverns. There water flows in abundance; and also much fire, great fire-streams, and streams of wet mud (here purer, there more filthy) as in Sicily the streams of mud that are poured out before and along with the fire-stream itself; all places are filled with these, according as each of the streams takes its several way. Periphlegethon flows out into an extensive district burning with fierce fire, where it forms a lake larger than our sea, boiling with water and mud. From hence it moves in circles round the earth turbid and muddy." This stream of melted earth and mud is so much the general cause of volcanic phenomena, that Plato adds: "Thus is Periphlegethon constituted, from which also the fire-streams (οἱ ῥύακες) inflate small or detached portions wherever these are met with on the earth (ὅπῃ ἂν τύχωσι τῆς γῆς). Volcanic scoriæ and lava streams are therefore portions of periphlegethon itself, portions of the subterranean melted and ever-moving mass. That οἱ ῥύακες are lava streams, and not, as Schneider, Passow, and Schleiermacher, will have it, "fire-vomiting mountains," appears from many passages that have been already collected by Ukert (Geogr. der Griechen und Römer, Th. ii. I. S. 200); ῥύαξ is the volcanic phenomenon seized from its most remarkable point of view, the lava stream. Whence the expression the ῥύακες of Ætna. Aristot. Mirab. Ausc. t. ii. p. 833, § 38, Bekker; Thucyd. iii. 116; Theophr. de Lap. 22, p. 427; Schneider, Diod. v. 6, and xiv. 59, where the remarkable words: "many places near the sea, not far from Ætna, were destroyed," ὑπὸ τοῦ καλουμένου ῥύακος; Strabo, vi. p. 269, xiii. p, 628, and of the celebrated glowing mud of the Lelantine plain in Cubæa (Strabo, i. p. 58, Casamb.); lastly Appian. de bello civili, vi. 114. The blame which Aristotle throws on the geological fancies of the Phædo (Meteor. ii. 2, 19) attaches only to the rivers which flow over the surface of the earth. The expression, so distinct in reference to the "eruptions of wet mud in Sicily preceding the glowing (lava) streams" is very remarkable. Observations on Ætna could not have led to such language, unless torrents of ashes or pumice mixed with the melted snow and water of the cone during an eruption, were taken for ejected mud. It seems more probable that the ὑγροῦ πηλοῦ ποταμοί of Plato, the "moist mud streams," are an obscure recollection of the mud-volcanoes of Agrigentum, which I have already referred to (Note 89), which eject mud with loud noises. The loss of one among the many lost writings of Theophrastus: περὶ ῥύακος τοῦ ἐν Σικεγίᾳ, of which Diogenes Laertius (v. 39) makes mention, is much to be regretted in connection with this subject.

196 (p. 72.)—Leopold von Buch, Physical. Beschreib. der Canarischen Inseln, S. 326—407. I doubt whether we can, with the able Darwin (Geological Observations on the Volcanic Islands, 1844, p. 127), regard Central volcanoes in general as Rank-volcanoes of small compass developed on parallel fissures. Fried. Hoffmann believed that he perceived in the group of the Lipari islands, which he has so well described, and in which two eruption-fissures cross each other near Panaria, an intermediate member between the two principal modes in which volcanoes appear, the central, and the rank or row-volcanoes of Leopold von Buch (*vide* Poggend. Annal. 26, p. 81).

197 (p. 72.)—Humboldt, Geognost. Beob. über die Vulkane des Hochlandes von Quito, in Poggend. Annalen, Bd. xliv. S. 194.

198 (p. 72.)—Seneca, whilst he speaks very pointedly on the problematical lowering of Ætna, says, in his 79th letter: "Potest hoc accidere, non quia montis altitudo desedit, sed quia ignis evanuit et minus vehemens ac largus effertur: ob eandem causam, fumo quoque per diem segniore. Neutrum autem incredibile est, nec montem qui devoretur quotidie minui, nec ignem non manere eundem; quia non ipse ex se est, sed in aliqua inferna valle conceptus exaestuat et alibi pascitur: in ipso monte non alimentum habet sed viam." (Ed. Ruhkopfiana, t. iii. p. 32.) The subterraneous communications, "by means of galleries," between the volcanoes of Sicily, Lipari, Pithecuse (Ischia), and Vesuvius, which may be conjectured to have been formerly on fire, are fully recognized by Strabo, who calls the whole country "subigneous." (Lib. i. p. 247, 248.)

199 (p. 72.)—Humboldt, Essai polit. sur la Nouv. Espagne, t. ii. p. 173—175.

200 (p. 73.)—On the Eruption of Methone, *vide* Ovid. Metamorphos. xv. 296—306):

"Est prope Pittheam tumulus Troezena sine ullis
Arduus arboribus, quondam planissima campi
Area, nunc tumulus; nam—res horrenda relatu—
Vis fera ventorum, caecis inclusa cavernis,
Exspirare aliqua cupiens, luctataque frustra
Liberiore frui coelo, cum carcere rima
Nulla foret toto nec pervia flatibus esset,
Extentam tumefecit humum; ceu spiritus oris
Tendere vesicam solet, aut direpta bicorni
Terga capro. Tumor ille loci permansit, et alti
Collis habet speciem, longoque induruit aevo."

This description of a dome-shaped elevation of the land, so important in a geological point of view, accords remarkably with what Aristotle says, (Meteor. ii. 8, 17—19) on the upliftment of an Eruption island. "The quaking of the earth does not cease until the wind (ἄνεμος) which occasions the shocks has made its escape into the crust of the earth. So did it happen lately at Heraclea in Pontus, and formerly too in Hiera, one of the Æolian islands. In this a portion of the earth swelled up and rose into the shape of a hill with loud noises, until the powerful lifting breath (πνεῦμα) found a vent, and threw out sparks and ashes, which covered the neighbouring town of the Liperians, and even extended to several towns of Italy." In this description, the vesicular-like distension of the crust of the earth (a state in which many trachytic mountains have remained) is very well distinguished from the eruption itself. Strabo (lib. i. p. 59, ed. Cas.) likewise describes the phenomenon of Methone: "Near the town in the Hermionian bay, a flaming eruption took place; a fiery mountain was thrown up, seven (?) stadia high, inaccessible during the day from heat and sulphureous odours, but sweet-smelling (?) in the night, and so hot that the sea boiled five stadia off, and was turbid full twenty stadia out, and was also filled full of detached masses of rock." On the present mineralogical constitution of the peninsula of Methone, *vide* Fiedler, Reise durch Griechenland, Th. i. S. 257—263.

201 (p. 73.)—Leop. von Buch, Physik. Beschr. der Canar. Inseln, S. 356—358, particularly the French translation of this excellent work, p. 402; also in Poggendorff's Annalen, Bd. xxxvii. S. 183. A submarine island was again in the most recent times formed in the crater of Santorin. In 1810 this island was still 15 fathoms under the surface of the sea; but in 1830 only 3 or 4 fathoms. It rises steeply like a great cone from the bottom of the sea; and the persistence of the submarine activity is proclaimed by the admixture of sulphuric acid vapours with the sea-water, so that ships which are coppered, lying at anchor in the bay of Neo-Kammeni, as well as at Wromolimni near Methana, have their bottoms cleansed and made bright without further trouble. (*Vide* Virlet in Bulletin de la Société géolo-

gique de France, t. iii. p. 109, and Fiedler, Reise durch Griechenland. Th. ii. S. 469 and 584.)

202 (p. 73.)—Appearances of new islands near San Miguel, one of the Azores: 11th June, 1638, 31st December, 1719, 13th June, 1811.

203 (p. 73.)—Prévost, in Bulletin de la Société géologique, t. ii. p. 34; Friedrich Hoffmann, Hinterlassene Werke, Bd. ii. S. 451—456.

204 (p. 73.)—"Accedunt vicini et perpetui Aetnae montis ignes et insularum Aeolidum, veluti ipsis undis alatur incendium; neque enim aliter durare tot seculis tantus ignis potuisset, nisi humoris nutrimentis aleretur." (Justin, Hist. Philipp. iv. i.) The volcanic theory with which the physical description of Sicily here begins is extremely intricate. Deep-lying beds of sulphur and rosin, an extremely thin crust, full of cavities and readily divided; violent motion of the waves of the sea, which, as they strike together, draw down air (the wind) for the maintenance of the fire: such are the elements of the theory of Trogus. As he presents himself as a physiognomist in Pliny (xi. 52), we may presume that he did not limit himself to history alone; but many of his works are lost to us. The view according to which air was forced into the interior of the earth, there to influence the volcanic force, is moreover connected with the notions of the ancients on the influence exerted by the direction of the wind upon the intensity of the fire which burns in Ætna, in Hiera and Stromboli (see the remarkable passage in Strabo, lib. vi. p. 275 and 276). The mountainous island of Stromboli (Strongyle) was therefore regarded as the seat of Æolus, "the controller of the winds," as the sailors foretold the weather from the violence of the volcanic eruptions of Stromboli. Such a connection between the eruptions of a small volcano and the state of the barometer and the quarter of the wind is still recognised (*vide* Leop. von Buch, Descr. phys. des Iles Canaries, p. 334; Hoffmann in Poggend. Ann. Bd. xxvi. S. 8); although it must be allowed that all our present knowledge of volcanic phenomena, and the slight alterations in the pressure of the air that accompany our winds, do not enable us to offer any satisfactory explanation of the fact. Bembo, brought up as a youth by Greek exiles in Sicily, gives a pleasant narrative of his wanderings, and in his "Ætna Dialogus" (middle of the 16th century) advances the theory of the penetration of sea water to the focus of the volcano, and of the necessity of the neighbourhood of the sea. On ascending Ætna the following question is thrown out: "Explana potius nobis quae petimus, ea incendia unde oriantur et orta quomodo perdurent? In omni tellure nuspiam majores fistulae aut meatus ampliores sunt quam in locis, quae vel mari vicina sunt, vel a mari protinus alluuntur: mare erodit illa facillime pergitque in viscera terrae. Itaque cum in aliena regna sibi viam faciat, ventis etiam facit; ex quo fit, ut loca quaeque maritima maxime terraemotibus subjecta sint, parum mediterranea. Habes quum in sulfuris venas venti furentes inciderint, unde incendia oriantur Aetnae tuae. Vides, quae mare in radicibus habeat, quae sulfurea sit, quae cavernosa, quae a mari aliquando perforata ventos admiserit aestuantes, per quos idonea flammae materies incenderetur.

205 (p. 73.)—See Gay-Lussac, sur les Volcans, in den Annales de Chimie, t. xxii. p. 427; and Bischoff, Wärmelehre, S. 372. Reactions of the volcanic hearth through tensive columns of water, viz., when the expansive force of the vapour surpasses the hydrostatic pressure, are proclaimed by the eruptions of smoke and aqueous vapour, which are observed at different times in Lancerote, Iceland, and the Kurile islands during eruptions of the neighbouring volcanoes.

206 (p. 73.)—Abel-Rémusat, Lettre à Mr. Cordier, in the Annales des Mines, t. v. p. 137.

207 (p. 73.)—Humboldt, Asie centrale, t. ii. p. 30—33, 38—52, 70—80, and 426—428. The existence of active volcanoes in Cordofan, 135 miles from the Red Sea, has lately been denied by Rüppell (Reise in Nubien, 1829).

208 (p. 74.)—Dufrénoy et Elie de Beaumont, Explication de la Carte géologique de la France, t. i. p. 89.

209 (p. 74.)—Sophocl. Philoctet. v. 971 and 972. On the conjectural epoch of the extinction of the Lemnian fire in the time of Alexander, *vide* Buttmann in Museum der Alterthumswissenschaft, Bd. i. 1807, S. 295; Dureau de la Malle in Malte-Brun, Annales des Voyages, t. ix. 1809, p. 5; Ukert in Bertuch, Geogr. Ephemeriden, Bd. xxxix. 1812, S. 361; Rhode, Res Lemnicae, 1829, p. 8, and Walter über Abnahme der vulkan. Thätigkeit in historischen Zeiten, 1844, S. 24. The hydrographical conception of Lemnos by Choiseul makes it extremely probable that the extinct foundations of Moschylos, together with the island Chryse, Philoctetes' desolate abode (Otfried Müller, Minyer, S. 300), have been long swallowed up by the sea. Reefs and shoals to the North-east of Lemnos still show the spot where the Ægæan Sea possessed an active volcano like Ætna, Vesuvius, Stromboli, and that of the Lipari isles.

210 (p. 74.)—*Vide* Reinwardt and Hoffmann in Poggendorff's Annalen, Bd. xii. S. 607; Leop. von Buch, Descr. des Iles Canaries, p. 424, 426. The argillaceous mud eruptions of Carguairazo, when the volcano crumbled together in 1698, the Lodazales of Igualata, and the Moya of Pelileo, are volcanic appearances of the same nature in the highlands of Quito.

211 (p. 74.)—In a profile of the environs of Tezcuco, Totonilco, and Moran, (Atlas géographique et Physique, Pl. vii.) which I originally (1803) designed for a Pasigrafia geognostica destinada al uso de los Jovenes del Colegio de Mineria de Mexico, but which was never published, I entitled (1832) the Plutonic and volcanic eruptive rocks *endogenous*, (that which is engendered in the interior,) the sedimentary and flœtz rocks *exogenous* (externally engendered). Pasigraphically the former were indicated by an arrow directed upwards, ↑, the latter by an arrow directed downwards, ↓, signs which had certain pictorial advantages, and permitted the nature of the rock to be shown without having recourse to those very unpicturesque and arbitrarily-shaped cones which are generally seen in such profile drawings. The titles endogenous and exogenous were borrowed from Decandolle, who uses the former in connection with monocotyledonous, the latter with dicotyledonous plants. But Mohl's more careful vegetable anatomy has shown, that in the strict sense of the words the growth of monocotyledonous vegetables does not proceed *from within*, nor that of decotyledonous plants *from without*. (*Vide* Link, Elementa philosophiae botanicae, t. i. 1837, p. 287; Endlicher und Unger, Grundzüge der Botanik, 1843, S. 89; and Jussieu, Traité de Botanique, t. i. p. 85.) What I call endogenous, Lyell, in his Principles of Geology, 1833, vol. iii. p. 374, characterises by the expression "netherformed" or "hypogene rocks."

212 (p. 74.) — *Vide* Leop. von Buch über Dolomit als Gebirgsart, 1823, S. 36; and farther, Ueber den Grad der Flüssigkeit, welchen man plutonischen Felsarten bei ihrem Heraustreten zuschreiben soll, wie über Entstehung des Gneuss aus Schiefern durch Einwirkung des Granits und der mit seiner Erhebung verbundenen Stoffe, as well as in the Abhandl. der Akad. der Wissench. zu Berlin aus dem Jahre 1842, S. 58 und 63, and in the Jahrb. für wissenschaftliche Kritik, 1840, S. 195.

213 (p. 75.)—Darwin, Volcanic Islands, 1844, p. 49 and 154.

214 (p. 75.)—Moreau de Jonnès, Hist. phys. des Antilles, t. i. p. 136, 138, and 543; Humboldt, Relation historique, t. iii. p. 367.

215 (p. 75.)—At Teguiza; Leop. von Buch, Canarische Inselna, S. 301.

216 (p. 75.)—*Vide* above, p. 4.

217 (p. 75.)—Bernhard Cotta, Geognosie, 1839, S. 273.

218 (p. 75.)—Leop. von Buch über Granit und Gneuss in den Abhandl. der Berl. Akad. aus dem J. 1842, S. 60.

219 (p. 75.)—In the granite of the Kolivan Lake, which rises like walls, and is divided into parallel narrow ledges, felspar and albite predominate, titanitic crystals are rare. Humboldt, Asie centrale, t. i. p. 295; Gustav Rose, Reise nach dem Ural, Bd. i. S. 524.

220 (p. 75.)—Humboldt, Relation historique, t. ii. p. 99.

221 (p. 75.)—See the drawing of Biri-tau, which I took from the south, with Kirghish tents pitched, in Rose, Reise, Bd. i. S. 584. On granite balls scaling off concentrically, *vide* Humboldt, Rel. hist. t. ii. p. 597; and Essai géogn sur le Gisement des Roches, p. 78.

222 (p. 75.)—Humboldt, Asie centrale, t. i. p. 299—311, and the drawings in Rose's Reise, Bd. i. S. 611, in which the curves of the granitic layers pointed out by Leop. von Buch as characteristic, are repeated.

223 (p. 75.)—This remarkable stratification was first described by Weiss, in Karsten's Archiv für Bergbau und Hüttenwesen, Bd. xvi. 1827, S. 5.

224 (p. 76.)—Dufrénoy et Elie de Beaumont, Géologie de la France, t. i. p. 130.

225 (p. 76.)—An important part is played by these substratified diorites near Steben, in the Nailaer Mountain district, a country where I was engaged in mining work in the last century, and with which some of the happiest associations of my youth are connected. *Vide* Friedr. Hoffmann in Poggendorff's Annalen, Bd. xvi. S. 558.

226 (p. 76.)—In the southern and Baschkir-Ural; *vide* Rose, Reise, Bd. ii. S. 171.

227 (p. 76.)—G. Rose, Reise nach dem Ural, Bd. ii. S. 47—52. On the identity of Elæolite and Nepheline (in the latter the quantity of lime is somewhat larger), *vide* Scheerer, in Poggend. Annalen, Bd. xlix. S. 359—381.

228 (p. 77.)—See the admirable papers of Mitscherlich, in the Abhandlungen der Berl. Akad. for the years 1822 and 1823, S. 25—41; in Poggendorff's Annalen, Bd. x. S. 137—152, Bd. xi. S. 323—332, Bd. xli. S. 213—216 (Gustav Rose über Bildung des Kalkspaths und Aragonits in Poggend. Ann. Bd. xlii. S. 353—366; Haidinger, in the Transactions of the Royal Society of Edinburgh, 1827, p. 148).

229 (p. 77.)—Lyell, Principles of Geology, vol. iii. p. 353 and 359.

230 (p. 78.)—The statements here made of the relations

of granite in reference to stratification, express the general or principal character of the whole formation. In some places (*vide* p. 75, and the description of the Narym chain, near the boundary of China, Rose's Reise, Bd. I. S. 599) granite indeed shows configurations which lead us to conjecture that at the period of its eruption it was not always without fluidity, just as happens in the case of Trachyte (Dufrénoy et Elie de Beaumoat, Description géologique de la France t. i. p. 70). As we have in the text mentioned the narrow fissures through which basalt has generally flowed, I take the opportunity in this place of referring to the wide chasms which have served the melaphyres (which must not be confounded with the basalts) as channels of efflux. See the interesting account by Murchison, in his Silurian System, p. 126, of a chasm 450 feet wide, in the coal-pit at Cornbrook, Hoar-Edge, through which the melaphyre has made its way.

231 (p. 78.)—Sir James Hall, in the Edinb. Transact. vol. v. p. 43, vol. vi. p. 71; Gregory Watt, in the Philos. Transactions of the Royal Society of London, for 1804, pt. ii. p. 279; Dartigues and Fleuriau de Bellevue, in the Journ. de Phys. t. lx. p. 456; Bischoff, Wärmelehre, S. 313 and 443.

232 (p. 78.)—Gustav Rose, in Poggendorff's Annalen der Physik, Bd. xlii. S. 364.

233 (p. 78.)—On the dimorphism of sulphur, *vide* Mitscherlich, Lehrbuch der Chimie, § 55—63.

234 (p. 78.)—On gypsum as monuaxal crystal, sulphate of magnesia, oxides of zinc and nickel, *vide* Mitscherlich, in Poggend. Ann. Bd. xi. S. 328.

235 (p. 78.)—Coste, Versuche, in Creusot über das brüchig werden des Stabeisens, in Elie de Beaumont, Mém. géol. t. ii. p. 411.

236 (p. 78.)—Mitscherlich über die Ausdehnung der krystallisirten Körper durch die Wärme. in Poggend. Ann. Bd. x. S. 151.

237 (p. 78.)—On double stratification cleavage, *vide* Elie de Beaumont, Géologie de la France, p. 41; Credner, Geognosie Thüringens und des Harzes, S. 40; Römer, das Rheinische Uebergangsgebirge, 1844, S. 5 und 9.

238 (p. 78.)—With addition of clay, lime, and potash, not silicic acid simply coloured with oxide of iron; Rose, Reise, Bd. ii. S. 169, 187, and 192; *vide* also Bd. i. S. 427, where the porphyry balls are represented between which the jasper occurs in the calcareous gray wacke mountains of Bogoslowsk, also as a consequence of the Plutonic effects of Augitic rock: Rose, Bd. ii. S. 545; also Humboldt, Asie centrale, t. i. p. 486.

239 (p. 78.)—Rose, Reise, nach dem Ural, Bd. i. S. 586—588.

240 (p. 78.)—For the volcanic origin of mica, it is important to remember that crystals of mica occur in the basalt of the Bohemian Middle Mountains; in the lava of Vesuvius of 1822 (Monticelli, Storia del Vesuvio negli anni 1821 e 1822, § 99); in clay-slate fragments of Hohenfels, not far from Gerolstein in the Eifel, enveloped in scoriaceous basalt, *vide* Mitscherlich, in Leonhard, Basalt-Gebilde, S. 244. On the production of felspar in clay slate, through the contact of porphyry between Urval and Poïet (Forez), *vide* Dufrénoy, in Géol. de la France, t. i. p. 137. A similar contact gives the slate at Paimpol, in Brittany, an amygdaloidal and cellular character, an appearance which amazed me very much on a geological journey which I made on foot, in company with Prof. Kunth, through that interesting country.

241 (p. 78.)—Leopold von Buch in the Abhandlungen der Akad. der Wissensch. zu Berlin aus dem J. 1842, S. 63; and in the Jahrbücher für wissenschaftliche Kritik, Jahrg. 1840, S. 196.

242 (p. 78.)—Elie de Beaumont, in the Annales des Sciences naturelles, t. xv. p. 362—372: "En se rapprochant des masses primitives du Mont Rose et des montagnes situées à l'ouest de Coni, on voit les couches secondaires perdre de plus les caractères inhérents à leur mode de dépôt. Souvent alors elles en prennent qui semblent provenir d'une toute autre cause, sans perdre pour cela leur stratification, rappelant par cette disposition la structure physique d'un tison à moitié charbonné dans lequel on peut suivre les traces des fibres ligneuses, bien au-delà des points qui présentent encore les caractères mutuels du bois." *Vide* also Annales des Sciences naturelles. t. xiv. p. 118—122; and H. von Dechen, Geognosie, S. 553. Among the most remarkable evidences of the transformation of rocks under the influence of Plutonic agency, are the belemnites in the schists of Nuffenen (Alpine valley of Egine and the Gries-glacier), as well as the belemnites in the so-called primitive limestone, which M. de Charpentier discovered on the western flank of the Col de Seigne, between Enclove de Monjovet and the Alpine-hut de la Lanchette (Ann. de Chimie, t. xxiii. p. 262), and which he showed me in Bex, in the autumn of 1822.

243 (p. 78.)—Hoffmann, in Poggend. Annalen, Bd. xvi. S. 552. "Strata of the transition clay slate of the Fichtelgebirge, which can be followed for four miles, and only at either extremity, where they come into contact with the granite converted into gneiss. There we can trace the gradual formation of gneiss, and the internal development of mica and of felspar amygaloids in clay slate, which indeed contains almost all the elements of those substances."

244 (p. 78.)—In the works of the ancient Greeks and Romans that have come down to us we observe the want of jasper columns and large vessels of jasper, a substance which the Ural mountains almost exclusively yield in masses of any magnitude. The stone that is worked as jasper in the Altai (Ravennaja Sopka, the Rhubarb mountains) is a magnificent striped porphyry. Theophrastus and Pliny reckon jasper among the number of non-transparent gems; and the latter thinks it incumbent on him to mention a piece of the mineral eleven inches long which he had seen: "Magnitudinem jaspidis undecim unciarum vidimus, formatamque inde effigiem Neronis thoracatam." The stone which Theophrastus calls smaragd or emerald, and from which the great obelisks were hewn, he regards as an unripe jasper.

245 (p. 78.)—Humboldt, Lettre à M. Brochant de Villiers, in the Annales de Chimie et de Physique, t. xxiii. p. 261; Leop. von Buch, Geogn. Briefe über das südliche Tyrol, S. 101, 105, and 273.

246 (p. 79.)—On the transformation of compact into granular limestone through contact with granite in the Pyrenees (Montagne de Rancie), *vide* Dufrénoy, in the Mémoires géologiques, t. ii. p. 440; and in the Montagnes de l'Oisans, *vide* Elie de Beaumont, Mém. géol. t. ii. p. 379—415; by Dioritic and Pyrorexic Porphyries (Ophite; Elie de Beaumont, Géol. de la France, t. i. p. 72), between Toulouse and St. Sebastian, *vide* Dufrénoy, in Mém. géol. t. ii. p. 130; through Syenite, in the island of Elba, in which petrefactions still continue visible in the limestone, in spite of the changes it has suffered, M. von Dechen, Geognosie, S. 573. In the metamorphosis of chalk, through contact with basalt, the dislocation of the minute particles through the production of crystals and the granulation is the more remarkable, since we have been made aware, by Ehrenberg's discoveries, of the fact, that these chalk particles previously consisted of articulated rings (*vide* Poggendorff's Annal. Bd. xxxix. S. 105; and on the rings of Aragonite precipitated from a state of solution, Gustav Rose, ib. Bd. xlii. S. 354).

247 (p. 79.)—Beds of granular limestone in granite at Port d'Or and Mont de Laboural, *vide* Charpentier, Constitution géologique des Pyrénées, p. 144, 146.

248 (p. 79.)—Leop. von Buch, Descr. des Canaries, p. 394; Fielder, Reise durch das Königreich Griechenland, Th. ii. S. 181, 190, and 516.

249 (p. 79.)—I have already referred to the remarkable passage in Origen's Philosophumena, cap. 14 (Opera ed. Delarue, t. i. p. 893). From the whole context it is not very unlikely that Xenophanes meant "an impression of laurel," (*τύποη δάφνης,*) not an "impression of a fish," (*τύπον ἀφύης*). Delarue blames Gronovius unfairly, who made the correction that "turned the laurel into an anchovy." The petrified fish is a far more likely object than the natural image of Silenus, which the quarry-men insisted they had dug out of the marble quarries of Paros (the mountain Marpessos, Servius ad Virgil, Æn. vi. 471), Plin. xxxvi. 5.

250 (p. 79.)—On the geological relations of the town of Carrara Luna, Selene civitas, *vide* Strabo, lib. v. p. 222; Savi, Osservazioni sui terreni antichi Toscani, in the Nuovo Giornale de' Letterati di Pisa, No. 63; and Hoffmann, in Karsten's Archiv für Mineralogie, Bd. vi. S. 258—263, as also his Geogn. Reise durch Italien, S. 244—265.

251 (p. 79.)—According to the view of an excellent and experienced observer, Karl von Leonhard; see his Jahrbuch für Mineralogie, 1834, S. 329, and Bernhard Cotta, Geognosie, S. 310.

252 (p. 79.)—Leop. von Buch, Geognostische Briefe an Alex. von Humboldt, 1824, S. 36 and 82; also in the Annales de Chimie, t. xxiii. p. 276, and the Abhandl. der Berliner Akad. aus den J. 1822 und 1823, S. 83—136; H. von Dechen, Geognosie, S. 574—576.

253 (p. 79.)—Hoffmann, Geogn. Reise bearbeitet von H. von Dechen, S. 113—119, 380—386; Poggend. Ann. der Physik, Bd. xxvi. S. 41.

254 (p. 80.)—Dufrénoy, in Mémoires géologiques, t. ii. p 145 and 179.

255 (p. 80.)—Humboldt, Essai géogn. sur le Gisement des Roches, p. 93; Asie centrale, t. iii. p. 532.

256 (p. 80.)—Elie de Beaumont, in Annales des Sciences naturelles, t. xv. p. 362; Murchison, Silurian System, p. 286.

257 (p. 80.)—Rose, Reise nach dem Ural, Bd. i. S. 364 and 367.

258 (p. 80.)—Leop. von Buch, Briefe, S. 109—129. *Vide* also Elie de Beaumont on the Contact of Granite with Jurastrata, in Mém. géol. t. ii. p. 408.

259 (p. 80.)—Hoffmann, Reise, S. 30 and 37.

260 (p. 80.)—On the chemical process in the formation of iron glance, *vide* Gay-Lussac in Annales de Chimie, t. xxii. p. 415; and Mitscherlich in Poggend. Ann. Bd. xv. S. 630. In the cavities of the Obsidian of the Cerro del Jacal, which I brought with me from Mexico, crystals of olivine have also been formed (apparently deposited from vapour, *vide* Gustav Rose, in Poggend. Ann. Bd. x. S. 323). Olivine

therefore occurs: in basalt, in lava, in obsidian, in artificial scoriæ, in meteoric stones, in the syenite of Elfdale, and (as hyalosiderite) in the Wacke of Kaiserstuhle.

261 (p. 80.)—Constantin von Beust über die Porphyrgebilde, 1835, S. 89—96; his Beleuchtung der Werner'schen Gangtheorie, 1840, S. 6; C. von Weissenbach, Abbildungen merkwürdiger Gangverhältnisse, 1836, fig. 12. The bandlike structure of the veins is however as little general, as is the sequence in respect of age of the several members of these masses. *Vide* Friesleben über die sächsischen Erzgänge, 1843, S. 10—12.

262 (p. 80.)—Mitscherlich über die künstliche Darstellung der Mineralien, in the Abhandlungen der Akademie der Wiss. zu Berlin aus den Jahren 1822 und 1823, S. 25—41.

263 (p. 80.)—In scoriæ: crystals of felspar discovered by Heine, after the extinction of a roasting copper ore furnace, not far from Sangerhausen, analysed by Kersten (Poggend. Annalen, Bd. xxxiii. S. 337); of augite in the scoriæ of Sahle (Mitscherlich in den Abhandl. der Akad. zu Berlin, 1822 and 1823, S. 40); of Olivine (Sefström in Leonhard, Basalt-Gebilde, Bd. ii. S. 495); of Mica in old scoriæ of Schloss Garpenberg (Mitscherlich in Leonhard, loc. cit. S. 506); of magnetic iron in scoriæ of Chatillon sur Seine (Leonhard, S. 441); of iron-glance arising in potter's clay (Mitscherlich in Leonhard, S. 234).

264 (p. 80.)—Produced on purpose: Idokras and garnet (Mitscherlich in Poggendorff's Annalen der Physik, Bd. xxxiii. S. 340); ruby (Gaudin in Comptes rendus de l'Académie des Sciences, t. iv. pt. iv. p. 999); olivine and augite (Mitscherlich and Berthier, in Annales de Chimie et de Physique, t. 24, p. 376). Although, according to Gust. Rose, augite and hornblende show the greatest similarity in the form of their crystals, and their chemical composition is almost identical, still hornblende has never been found by the side of augite in scoriæ: even as little have chemists succeeded in their attempts at producing hornblende or felspar (Mitscherlich in Poggend. Annalen, Bd. xxxiii. S. 340, and Rose, Reise nach dem Ural, Bd. ii. S. 358 and 363). See also Beudant, in Mém. de l'Acad. des Sciences, t. viii. p. 221, and Becquerel's able inquiries, in his Traité ed l'Electricité, t. i. p. 334; t. iii. p. 218; t. v. 1, p. 148 and 185.

265 (p. 80.)—D'Aubuisson, in Journal de Physique, t. lxviii. p. 128.

266 (p. 81.)—Leop. von Buch, Geognost. Briefe, S. 75—82; where it is at the same time shown, that the red sandstone (the dead layer of the Thuringian flœtz formations) and the coal formation must be viewed as products of eruptive porphyritic rocks.

267 (p. 81.)—On Hooke's "hope to raise a chronology" out of the study of fossil shells, and to state the intervals of the time wherein such or such catastrophes or mutations have happened, *vide* Posth. Works, Lecture, Feb. 29, 1688.

267* (p. 81.)—A discovery of Miss Mary Anning, who also first discovered the coprolites of fishes. These, and the excrements of the Ichthyosaurus, have been found in such quantities at Lyme Regis, that they seem to lie, according to Buckland's expression, "heaped like potatoes upon the ground." *Vide* his Geology with reference to Natural Theology, vol. i. p. 188—202, and 305.

268 (p. 81.)—Leop. von Buch, in Abhandlungen der Akad. der Wiss. zu Berlin aus dem J. 1837, S. 64.

269 (p. 82.)—The same, Gebirgsformationen von Russland, 1840, S. 24—40.

270 (p. 82.)—Agassiz, Monographie des Poissons fossiles du Vieux Grès Rouge, p. vi. and 4.

271 (p. 82.)—Leop. von Buch in Abhandl. der Berl. Akad. 1838, S. 149—168; Beyrich, Beitr. zur Kenntniss des Rheinischen Uebergangsgebirges. 1837, S. 45.

272 (p. 82.)—Agassiz, Recherches sur les Poissons fossiles, t. i. Introd. p. xviii. (Davy, Consolations in Travel, Dial. iii.)

273 (p. 82.)—According to Hermann von Meyer, a protosaurus (Palæologica, S. 229). The rib of a saurian, said to be from the mountain limestone of Northumberland, is, according to Lyell, extremely doubtful (Geology, vol. i. p. 148). The discoverer himself ascribes it to alluvial strata which cover the limestone.

274 (p. 82.)—F. von Alberti, Monographie des Bunten Sandsteins, Muschelkalks und Keupers, 1834, S. 119 und 314.

275 (p. 82.)—See the acute considerations of H. von Meyer (Palæologica, S. 228—252) on the organization of the flying reptiles. In the petrified specimen of Pterodactylus crassirostris, which, as well as the longer known Pterod. longirostris, was found in the lithographic limestone of Solenhofen, Professor Goldfuss has found "traces of the membrane which served for flight," as well as "impressions of the curled, flocky, in some places inch-long hair, which covered the skin."

276 (p. 82.)—Cuvier, Recherches sur les Ossemens fossiles, t. i. p. lii.—lvii. See also the geological scale of epochs in Phillips's Geology, 1837, p. 166—185.

277 (p. 82.)—Agassiz, Poissons fossiles, tom. i. pt. xxx. and tom. iii. p. 1—52; Buckland, Geology, vol. i. p. 273—277.

278 (p. 83.)—Ehrenberg, über noch jetzt lebende Thierarten der Kreidebildung in den Abhandl. der Berliner Akad. aus dem J. 1839, S. 164.

279 (p. 83.)—Valenciennes, in Comptes rendus de l'Acad. des Sciences, tom. vii. 1838, pt. ii. p. 580.

280 (p. 83.)—The Weald-Clay; Beudant, Géologie, p. 173. The ornitholites increase in number in the gypsum of the tertiary formation (Cuvier, Ossemens fossiles, tom. iii. p. 302—328).

281 (p. 83.)—Leop. von Buch, in Abhandl. der Berl. Akad. aus dem J. 1830, S. 135—187.

282 (p. 83.)—Quenstedt, Flözgebirge Würtembergs, 1843, S. 135.

283 (p. 83.)—Ibid. S. 13.

284 (p. 83.)—Murchison divides the variegated sandstone into two divisions, the upper of which remains the Trias of Alberti, whilst out of the lower, to which the Voges-sandstone of Elie de Beaumont belongs, the Zechstein and the Todtliegendes, he forms his Permian System. With the upper trias, *i. e.*, with the upper division of our variegated sandstone, he begins the secondary formations: the Permian system, the mountain or carboniferous limestone, the Devonian and Silurian strata, are with him palæozoic formations. According to these views, chalk and jura are called the upper, keuper, muschelkalk, and variegated sandstone, the inferior secondary formations; the Permian system and the carboniferous lime are entitled the upper, the devonian and silurian strata together the inferior palæozoic formations. The basis of this general classification is developed in the great work in which the unwearied British geologist gives an account of a great portion of the east of Europe.

285 (p. 83.)—Cuvier, Ossemens fossiles, 1821. tom. i. p. 157, 262, and 264. *Vide* Humboldt, über die Hochebene von Bogota in der Deutchen Bierteljahrs-Schrift, 1839, Bd. i. S. 117.

286 (p. 83.)—Journal of the Asiatic Society, No. xv. p. 109.

287 (p. 84.)—Beyrich, in Karsten's Archiv für Mineralogie, 1844, Bd. xviii. S. 218.

288 (p. 84.)—Through the admirable labours of Count Sternberg, Adolph Brongniart, Göppert, and Lindley.

289 (p. 84.)—*Vide* Robert Brown's Botany of Congo, p. 42, and the unfortunate d'Urville, in the Memoir: De la distribution des Fougères sur la surface du globe terrestre.

290 (p. 84.)—To this belong the Cycadeæ of the old coal formation of Radnitz, Bohemia, discovered by Count Sternberg, and described by Corda. Two species, Cycadites et Zamites Cordai, *vide* Goppert, fossile Cycadeen in den Arbeiten der Schles. Gesellschaft, für vaterl. Cultur im J 1843, S. 33, 40, and 50. In the coal formation of Königshütte, Upper Silesia, a Cycadea (Pterophyllum gonorrhachis, Goep.) has also been found.

291 (p. 84.)—Lindley, Fossil Flora, No. xv. p. 163.

292 (p. 84.)—Fossil Coniferae, in Buckland, Geology, p. 483—490. Mr. Witham has the merit of having first detected the existence of coniferæ in the earlier vegetation of the old coal formations. All the stems of trees discovered in these formations had previously been regarded as palms. The species of the genus Araucarites, however, is not peculiar to the coal fields of Great Britain; they are also met with in Upper Silesia.

293 (p. 84.)—Adolph Brongniart, Prodrome d'une Hist. des Végétaux fossiles, p. 176; Buckland, Geology, p. 479; Endlicher and Unger, Grundzüge der Botanik, 1843, S. 455.

294 (p. 84.)—"By means of Lepidodendron a better passage is established from Flowering to Flowerless Plants than by either Equisetum or Cycas, or any other known genus."—Lindley and Hutton, Fossil Flora, vol. ii. p. 53.

295 (p. 84.)—Kunth, Anordung der Pflanzenfamilien, in his Handb. der Botanik, S. 307 and 314.

296 (p. 84.)—That fossil coal consists of vegetable fibres carbonized not through fire, but in the moist way, and under the co-agency of sulphuric acid, is vouched for particularly by Göppert's able observations, of a piece of Ambertree wood converted into coal (*vide* Karsten, Archiv für Mineralogie, Bd. xviii. S. 530). The coal lies close to the wholly unaltered amber. On the part which the lower vegetables may have had in the production of coal, *vide* Link in the Abhandl. der Berliner Akademie der Wissenschaften, 1838, S. 38.

297 (p. 84.)—See the excellent paper of Chevandier, in the Comptes rendus de l'Acad. des Sciences, 1844, tom. xviii. pt. i. p. 285. In order to compare the half-inch thick layer of carbonaceous matter with the coal strata, regard must also be had to the enormous pressure which these strata have suffered from the superincumbent beds, and which is even attested by the generally flattened form of the fossil stems of trees that are dug up. "The wood-hills, as they are called, of the southern shore of the island of New Siberia, discovered in 1806 by Sirowatskoi, consist, according to Hedenström, of elevations of about 30 fathoms, made up of horizontal layers of sandstone interchangingly with bituminous trunks of trees. On the tops of the hillocks the stems stand erect. The stratum of drift wood is

visible for five wersts." *Vide* Wrangel, Reise längs der Nordküste von Siberien in den Jahren 1220—1824, Th. i. S. 202.

298 (p. 84.)—This corypha is the sopato (zoyatl, Aztekian) or palma dulce of the natives; *vide* Humboldt and Bonpland, Synopsis Plant. Æquinoct. Orbis Novi, tom. i. p. 302. One deeply versed in the American languages, Professor Buschmann, observes that the palma soyate is also named in Yepe's Vocabulario de la Lengua Othomi, and that the Aztekian word zoyatl (Molina, Vocabulario) also occurs in the local names zoyatitlan and zoyapanco near Chiapa.

299 (p. 85.)—Near Baracoa and Cayos de Moa; *vide* Tagebuch des Admirals vom 25 and 27 November, 1492, and Humboldt, Examen critique de l'Hist. de la Géogr. du Nouveau Continent, tom. ii. p. 252, and tom. iii. p. 23. Columbus was so observant of all natural objects, that he distinguished—and was, indeed, the first to do so—Podocarpus from Pinus. "I find," he says, "en la tierra aspera del Cibao pinos que no llevan piñas [fir-tops or cones], pero por tal orden compuestos por naturaleza, que (los frutos) parecen azeytunas del Axarafe de Sevilla." The great botanist, Richard, when he produced his excellent work on the Cycadeæ and Coniferæ, was not aware that long before L'Héritier, at the close of the 15th century, Podocarpus had already been distinguished from the pines—by a seafaring man, too.

300 (p. 85.)—Charles Darwin, Journal of the Voyages of the Adventure and Beagle, 1839, p. 271.

301 (p. 85.) — Göppert describes other three Cycadeæ (species of Cicaditeæ and Pterophyllum) from the lignitic clay-shists of Altsattel and Commotau in Bohemia, perhaps from the Eocene period (Göppert, in the work quoted in Note 90).

302 (p. 85.)—Buckland, Geology, p. 509.

303 (p. 85.)—Leopold von Buch, in Abhandl. der Akad. der Wiss. zu Berlin aus den J. 1814—1815, S. 161, and in Poggendorff's Annalen, Bd. ix. S. 575; Elie de Beaumont, in Annales des Sciences nat. t. xix. p. 60.

304 (p. 86.)—*Vide* Elie de Beaumont, Descr. géol. de la France, t. i. p. 65; Beudant, Géologie, 1844, p. 209.

305 (p. 87.)—Transactions of the Cambridge Philosophical Society, vol. vi. pt. 2, 1837, p. 297. According to others, as 100 : 284.

306 (p. 87.)—In the middle ages the prevalent opinion was that the sea covered but one seventh of the surface of the globe, an opinion which Cardinal d'Ailly (Imago Mundi, cap. 8) founded on the Apocryphal 4th Book of Ezra. Columbus, who always derived much of his cosmological knowledge from the Cardinal's work, was much interested in upholding this idea of the smallness of the sea, to which the misunderstood expression of "the ocean stream" contributed not a little. *Vide* Humboldt, Examen critique de l'Hist. de la Géographie, t. i. p. 186.

307 (p. 87.) — Agathemeros, in Hudson, Geographi minores, t. ii. p. 4. *Vide* Humboldt, Asie centr. t. i. p. 120, 125.

308 (p. 87.)—Strabo, lib. i. p. 65, Casaub. *Vide* Humboldt, Examen crit. t. i. p. 152.

309 (p. 87.)—On the mean latitude of the Northern Asiatic shores, and the true name of Cape Taimura (Cape Siewero—Wostotschnoi), and Cape North-East (Schalagskoi Mys), *vide* Humboldt, Asie centrale, t. iii. p. 35 and 37.

310 (p. 88.)—Ib. t. i. p. 198—200. The southern point of America and the Archipelago, which we call Terra del Fuego, lies in the meridian of the north-western part of Baffin's Bay, and of the great uncircumscribed polar land, which perhaps belongs to West Greenland.

311 (p. 88.)—Strabo, lib. ii. p. 92 and 108, Casaub.

312 (p. 88.)—Humboldt, Asie centrale, t. iii. p. 25. I had already, at an early period of my work, De distributione geographica plantarum secundum coeli temperiem et altitudinem montium, directed attention to the important influence of compact or divided continents on climate and human civilization: "Regiones vel per sinus lunatos in longa cornua porrectae, angulosis littorum recessibus quasi membratim discerptae, vel spatia patentia in immensum, quorum littora nullis incisa angulis ambit sine anfractu Oceanus" (p. 81 and 182). On the relations of the extent of coast to the area of a continent (at the same time as a measure of the accessibility of the interior), *vide* the Inquiries in Berghaus, Annalen der Erdkunde, Bd. xii. 1835, S. 490, and Physikal. Atlas, 1839, No. III. S. 69.

313 (p. 88.)—Strabo, lib. ii. p. 92 and 198, Casaub.

314 (p. 88.)—Of Africa, Pliny says (v. 1.)—"Nec alia pars terrarum pauciores recipit sinus." The small Indian peninsula this side the Ganges, in its triangular outline, presents another analogous form. In Ancient Greece there prevailed an opinion of the regular configuration of the dry land. There were four gulphs or bays, among which the Persian was placed in opposition to the Hyrcanian (*i. e.* the Caspian Sea) (Arrian, vii. 16; Plut. in vita Alexandri, cap. 44; Dionys. Perieg. v. 48 und 630, pag. 11 und 38, Bernh.) These four bays and the isthmuses of the land, according to the optical fancies of Agesianax, were reflected in the moon (Plut. de Facie in Orbe Lunae, p. 921, 19). On the terra quadrifida, or four divisions of the dry land, of which two lay north, two south of the equator, *vide* Macrobius, Comm. in Somnium Scipionis, ii. 9. I have submitted this portion of the geography of the ancients, on which great confusion prevails, to a new and careful examination, in my Examen crit. de l'Hist. de la Géogr. t. i. p. 119, 145, 180—185, as also in Asie centr. t. ii. p. 172—178.

315 (p. 88.)—Fleurieu, in Voyage de Merchand autour du Monde, t. iv. p. 38—42.

316 (p. 88.)—Humboldt, in the Journal de Physique, t. liii. 1799, p. 33, and Rel. hist. t. ii. p. 19, t. iii. p. 189 and 198.

317 (p. 88.)—Humboldt, in Poggendorff's Annalen der Physik, Bd. xl. S. 171. On the remarkable Fiord formation of the south-east end of America, *vide* Darwin's Journal (Narrative of the Voyages of the Adventure and Beagle, vol. iii.) 1839, p. 266. The parallelism of the two mountain chains is maintained from 5° North to 5° South latitude. The change in the direction of the coast at Arica appears to be a consequence of the altered course of the chasm upon or through which the Andes have arisen.

318 (p. 89.)—De la Beche, Sections and Views illustrative of Geological Phenomena, 1830, Tab. 40; Charles Babbage, Observations on the Temple of Serapis at Pozzuoli, near Naples, and on certain causes which may produce Geological Cycles of great extent, 1834. A bed of sandstone, five English miles thick, heated to 100° Fahr., would rise on its surface about 25 feet. Clay strata heated, on the contrary, would occasion a contraction or sinking of the ground. See the calculation for the secular rise of Sweden, on the presumption of a rise by so small a quantity as 3° Reaum., in a stratum 140,000 feet thick, heated to the melting point, in Bischoff, Wärmelehre des Innern unseres Erdkörpers, S. 303.

319 (p. 89.)—The presumption of the stability—which has hitherto been so implicit—of the point of gravity, has at all events been shaken to a certain extent by the gradual rise of large portions of the earth's surface. *Vide* Bessel über Maass und Gewicht, in Schumacher's Jahrbuch für 1840, S. 134.

320 (p. 89.) — Th. ii. (1810), S. 389. *Vide* Hallström, in Kongl. Vetenskaps-Academiens Handlingar (Stockh.), 1823, p. 39; Lyell, in the Philos. Trans. for 1835, p. 1; Blom (Amtmann in Budskerud), Stat. Beschr. von Norwegen, 1843, S. 89—116. If not before Von Buch's travels through Scandinavia, still before the publication of the account of them, Playfair, in his Illustrations of the Huttonian Theory, § 393, as well as Keilhau (Om Landjordens Stigning in Norge in dem Nyt Magazin for Naturvidenskaberene), and even before Playfair, the Dane Jessen, had expressed an opinion that it was not the sea which fell in level, but the firm land of Sweden which rose: these ideas remained wholly unknown to our great geologist, and exerted no influence on the progress of physical geography. Jessen, in his work, Kongeriget Norge fremstillet efter nets naturlige og borgerlige Tilstand, Kjöbenh. 1763, sought to explain the changes in the relative levels of the land and sea, upon the old notions of Celsius, Kalm, and Dalin. He broaches some confused notions about the possibility of an internal growth of rocks, but finally declares himself in favour of an upliftment of the land by earthquakes. "All along," he observes, "no such rising was apparent immediately after the earthquake of Egersund; still, other causes producing such an effect may have been brought into operation by it."

321 (p. 89.)—Berzelius, Jahresbericht über die Fortschritte der physischen Wiss. No. 18, S. 686. The island Saltholm, over against Copenhagen, and Bornholm, however, rise but very little—Bornholm scarcely 1 foot in a century; *vide* Forchhammer, in Philos. Magazine, 3d Series, vol. ii. p. 309.

322 (p. 89.)—Keilhau, in Nyt Mag. for Naturvid. 1832, Bd. i. p. 105—254, Bd. ii. p. 57; Bravais, sur les lignes d'ancien niveau de la Mer, 1843, p. 15—40. See also Darwin on the Parallel Roads of Glen-Roy and Lochaber, in the Philos. Transactions for 1839, p. 60.

323 (p. 89.)—Humboldt, Asie centrale, t. ii. p. 319—324, t. iii. p. 549—554. The depression of the Dead Sea has been again and again determined by the barometrical measurements of Count Bertou, the more careful ones of Russegger, and the trigonometrical survey of Lieut. Symond, of the Royal Navy, who specifies 1506 feet as the difference of level between the surface of the Dead Sea and the highest houses in Jaffa. Mr. Alderson, who communicated this result to the Geographical Society of London, in a letter, of the contents of which I was informed by my friend Captain Washington, Mr. Alderson then imagined (Nov. 28. 1841) that the Dead Sea lay about 1314 feet under the level of the Mediterranean. In another and later communication from Lieut. Symond (Jameson's Edinburgh New Philosophical Journal, vol. xxxiv. 1843, p. 178), as a final result, two trigonometrical operations are detailed. which agree remarkably with each other, and assign 1231 feet (Paris measure) as the depression of the level of the Dead Sea below that of the Mediterranean.